Innovative Lightweight High-performance Steel-concrete Sandwich Composite for Marine Structures

Zhenyu Huang • Yingwu Zhou

Innovative Lightweight High-performance Steel-concrete Sandwich Composite for Marine Structures

Springer

Zhenyu Huang
School of Mechanics
and Construction Engineering
Jinan University
Guangzhou, China

Yingwu Zhou
College of Civil and Transportation Engineering
Shenzhen University
Shenzhen, China

ISBN 978-981-95-4680-0 ISBN 978-981-95-4681-7 (eBook)
https://doi.org/10.1007/978-981-95-4681-7

This Springer imprint is published by the registered company Springer Nature Singapore Pte Ltd.
The registered company address is: 152 Beach Road, #21-01/04 Gateway East, Singapore 189721, Singapore

To the relentless pioneers of materials science and marine engineering who illuminate the path to resilience, and to future generations, may these innovations forge safer marine infrastructure beneath every horizon.

Preface

The relentless pursuit of resilient marine infrastructure demands materials that harmonize structural efficiency with uncompromising durability. This book addresses this critical need by presenting groundbreaking research on lightweight high-performance cement composite–steel sandwich systems engineered specifically for marine environments. These innovative composites offer transformative solutions to challenges posed by static sustained loads, ice impact, vessel collisions, and corrosive seawater exposure—conditions where conventional materials often fall short. Through rigorous integration of experimental testing, computational modeling, and analytical derivation, we establish a holistic framework for understanding and deploying these advanced structural systems.

Purpose and Scope

This volume synthesizes a decade of pioneering research into four interconnected domains:

1. Material innovation: Development of ultra-lightweight cement composites (ULCCs) with enhanced ductility, rubberized modifications for energy absorption, and sustainable LC^3-based formulations (Chaps. 1 and 2)
2. Structural mechanics: Fundamental behavior under flexural, shear, punching, and impact loading (Chaps. 3, 4, 5, 6, and 7), including failure mechanisms unique to marine dynamic conditions
3. Performance validation: Robust numerical frameworks (H-J-C models, FE simulations, AI-aided prediction) and analytical methods (SDOF/TDOF models) validated through destructive testing
4. Design integration: Ready-to-implement design methodologies for beams, walls, and curved shells (Chaps. 4, 5, 6, and 7), culminating in connector optimization strategies

Intended Audience

This work serves as an indispensable resource for:

- Researchers exploring composite materials science and marine structural dynamics
- Civil/marine engineers designing offshore platforms, Arctic vessels, and coastal protections
- Graduate students specializing in advanced composite structures and impact mechanics
- Industry practitioners seeking validated solutions for prefabricated marine components

Organizational Framework

- The book's architecture progresses systematically from material fundamentals to composite component applications:
- Foundation (Chaps. 1 and 2): Microstructural design and dynamic behavior of ULCC
- Component behavior (Chaps. 3 and 4): Axial response, compression and bending interaction of sandwich walls, flexural/shear behavior of sandwich beams, and punching resistance of single/double-layer panels
- Dynamic loading (Chaps. 5 and 6): Impact/repeated-impact performance and stainless-steel composite resilience
- Connector innovation (Chaps. 7): Slotted perforated plate connectors for enhanced shear transfer
- Synthesis (Chap. 8): Design guidelines and future research trajectories for offshore implementation

Guangzhou, China
Shenzhen, China

Zhenyu Huang
Yingwu Zhou

Acknowledgments

This work would not have been possible without the support and contributions of numerous individuals and organizations.

First and foremost, we extend our deepest gratitude to the National University of Singapore, Shenzhen University, and Jinan University for providing the academic environment and resources essential to this research. Special thanks to the Guangdong Provincial Key Laboratory of Durability for Marine Civil Engineering, Shenzhen University, for their intellectual guidance and infrastructural support.

This book was generously funded by Guangdong Outstanding Youth Fund (No. 2022B1515020037), Shenzhen Natural Science Foundation Basic Research Program-General Program (No. JCYJ20240813143019025), Fundamental and Interdisciplinary Disciplines Breakthrough Plan of the Ministry of Education of China (JYB2025XDXM802), Fundamental Research Funds for the Central Universities and National Natural Science Foundation of China (Nos. 51708360 and 51978407). Their financial support enabled the extensive experimental test, data collection, and analysis that were critical to this book's completion.

We would like to express our sincere gratitude to Academician Xing Feng (now President of Jinan University), Professor Zhou Yingwu, Professor Sui Lili, Professor Li Dawang, and Professor Han Ningxu for their meticulous guidance and tremendous support over the past 8 years since Professor Huang Zhenyu was introduced to Shenzhen University, which greatly shaped his academic career. It is a great honor to continue to work closely with Academician Xing Feng to embark on a new academic journey. We will continue to delve deeper into the field of marine engineering materials and structures, contributing our efforts to building China into a maritime power.

We gratefully acknowledge Dr. Iftikhar Azim for his substantial and leading contributions to this book. His meticulous review and refinement of both theoretical frameworks and experimental validation were instrumental to the quality and completion of this book. We also thank him for his careful proofreading of the proofs to ensure clarity and technical accuracy. Our sincere gratitude is also extended to our dedicated students—Fu Zhanxia, Wang Fang, Li Ren, Yan Xu, Liang Tingting,

Zhao Xiaolong, Lin Gan, Deng Weixiong, Wang Hong, Zhou Yu, Zhao Xu and Zhang Shiyan, among others—for their significant contributions to the research presented in this work.

We also wish to acknowledge our esteemed collaborators—Professor J.Y. Richard Liew, Professor Ye Jianqiao, Associate Professor Qian Xudong, Dr. Chen Lijie, Dr. Zhang Wei, and others—for their invaluable insights, spirit of cooperation, and constructive feedback throughout the development of this project. Their expertise in materials science and structural engineering has profoundly enriched the discussions and methodological approaches documented herein. My heartfelt thanks go to my parents and my wife for their unwavering support.

Competing Interests

The authors have no competing interests to declare that are relevant to the content of this manuscript.

Contents

Authors and Contributors

About the Authors

Zhenyu Huang is a Professor in Key Laboratory of Marine Civil Engineering Materials and Structures (Ministry of Education), Research Institute of Near-Zero Carbon Coastal Infrastructure and Intelligent Operation and Maintenance Technology, and School of Mechanics and Construction Engineering, Jinan University, Guangzhou. He has been consecutively listed in the "World's Top 2% Scientists" ranking jointly published by Stanford University and Elsevier and is a recipient of the Guangdong Provincial Natural Science Foundation for Outstanding Youth. He received his PhD from the National University of Singapore in 2015, working as a Postdoctoral Research Fellow in the SembCorp-NUS Corporate Lab in NUS. He serves as Deputy Director of Guangdong Provincial Key Laboratory of Durability for Marine Civil Engineering, Shenzhen University. He is currently a member of the board of steel–concrete composite structure branch of China Steel Structure Association and a member of the board of China Steel Structure Association.

Dr. Huang's research focuses on marine engineering materials and structures, including lightweight high-performance concrete, marine floating structures, modular prefabricated structural systems (MiC, PPVC), dynamic protection and resilience enhancement of composite structures, and AI data-driven prediction. He has published 82 academic papers, including 63 SCI/core journal papers (55 as first/corresponding author in journals such as ASCE-JSE, Engineering Structures, and Composite Structures). He holds 1 authorized US patent and 22 authorized Chinese invention patents. He has led or participated in over 10 national, provincial, and municipal research projects, including serving as PI for projects such as the NSFC General Fund and Youth Fund, the Guangdong Provincial Natural Science Foundation Outstanding Youth Fund, and the General Program. His research achievements have been recognized with the first and second prizes of the Guangdong Provincial Science and Technology Progress Award, the Gold and Silver Awards in the ACI (Singapore Chapter) High-Performance Concrete Project Competition, the Best Composite Structure Paper Award at the 8th International Conference on Steel and

Aluminum Structures, and the Best Paper Award at the inaugural International Conference on Engineering Research and Practice for Steel Construction. He serves as a Youth Editorial Board Member for journals such as Ocean, Sustainable Structures, Progress in Steel Building Structures, Bulletin of the Chinese Ceramic Society, and Materials Reports and as a reviewer for over 40 renowned journals in the field of civil and marine engineering.

Yingwu Zhou, a distinguished professor and doctoral supervisor, is a recipient of the National Excellent Youth Fund and National Outstanding Youth Fund. He is recognized as a highly cited researcher by Elsevier and has been consistently listed among the "Top 2% of Scientists Globally" (2020–2023). He serves as the Executive Deputy Director of the Key Laboratory of Marine Civil Engineering in Guangdong Province and holds editorial positions in prestigious journals such as Construction and Building Materials and Advances in Structural Engineering. Additionally, he serves as a board member of the Guangdong Society of Mechanics; a member of the International Union of Laboratories and Experts in Construction Materials, Systems, and Structures (RILEM); and a member of the International Federation for Structural Concrete (fib). He obtained his bachelor's, master's, and doctoral degrees in structural engineering from Dalian University of Technology in 2009. Prior to his current position, he worked as a researcher at City University of Hong Kong (2007–2010) and Hong Kong Polytechnic University (2011–2013). He has long been engaged in research on novel high-performance low-carbon cement-based composite materials, new composite structures, and structural performance enhancement and assurance. He has led eight national-level projects and nine provincial- and municipal-level projects in the past 5 years, with a total funding of 3.46 million RMB. He has published over 170 academic papers, including 130 SCI papers. His papers have been highly cited, with 3035 citations in the Web of Science SCI and an h-index of 36. He has been granted 25 invention patents (including 2 US patents and 1 Australian patent). He has also been awarded the first and second prizes of the Guangdong Provincial Science and Technology Progress Award and the second prize of the Shenzhen Natural Science Award.

Contributors

Iftikhar Azim Shenzhen University, Shenzhen, China

Zhenyu Huang Jinan University, Guangzhou, China
Shenzhen University, Shenzhen, China

Shiyan Zhang Shenzhen University, Shenzhen, China

Yingwu Zhou Shenzhen University, Shenzhen, China

Chapter 1
Ultra-Lightweight High Performance Cement Composite for Marine Structures

1.1 Introduction

Lightweight concrete (LWC) has been a subject of sustained research interest due to its unique advantages in modern construction. Historical records indicate its use dates back to ancient civilizations, with early applications emerging around 3000 BC [1]. Over the past three decades, advancements in concrete technology have enabled significant progress in LWC development, particularly for applications where reduced density and enhanced thermal properties are critical. The material's low self-weight offers distinct benefits, including improved transportation efficiency, reduced foundation loads, and lower construction costs through decreased reinforcement requirements. Furthermore, the inherent thermal insulation properties of LWC, attributable to its porous microstructure, make it particularly suitable for energy-efficient building envelopes, such as floor and wall panels. These characteristics have driven its adoption in high-rise buildings, long-span bridges, and offshore structures, where dead load reduction and energy conservation are paramount [2, 3].

Traditional LWC formulations employ lightweight aggregates such as expanded shale, perlite, polystyrene, clay, glass, vermiculite, foamed slag, or fly ash. While these materials effectively reduce density, their low compressive strength and energy-intensive production methods—often involving high-temperature sintering—pose limitations in terms of structural performance and environmental sustainability [4–10].

Recent research has identified fly ash cenospheres (FAC) as a promising alternative due to their exceptional mechanical and thermal properties. FACs are hollow, spherical particles derived as byproducts of coal combustion in thermal power plants, exhibiting high crushing strength, low water absorption, and superior thermal resistance [11]. Pioneering work by Chia et al. [12] demonstrated the potential of FAC-based ultra-lightweight cement composites (ULCC), achieving densities as low as 1430 kg/m^3 alongside 28-day compressive strengths of approximately

Z. Huang, Y. Zhou, *Innovative Lightweight High-performance Steel-concrete Sandwich Composite for Marine Structures*,
https://doi.org/10.1007/978-981-95-4681-7_1

60 MPa. These ULCC formulations, produced using conventional mixing techniques with Portland cement, silica fume, and chemical admixtures, exhibit specific strengths up to 42.8 kPa m^3 kg^{-1}, making them ideal for demanding applications such as Arctic offshore structures [13].

Despite these advancements, the existing literature lacks a comprehensive synthesis of ULCC's raw materials and microstructural properties, mix design, mechanical properties, durability properties and functional properties, as well as its structural behavior. This part addresses this gap by systematically evaluating ULCC's performance relative to conventional LWCs, with a focus on the role of FACs in enhancing material properties. By consolidating current knowledge, this part aims to facilitate broader adoption of ULCC and inspire further innovations in sustainable construction materials. Figure 1.1 presents the numbers of papers published on (a) fly ash cenosphere and (b) ultra-lightweight cement composite since 2000.

1.2 Raw Materials and Microstructural Properties of ULCC

ULCC is formulated using a specialized blend of constituents including ordinary Portland cement (OPC) as the primary binder; silica fume (SF) as a supplementary cementitious material; fly ash cenospheres as lightweight micro-aggregates; water, superplasticizer for workability enhancement; shrinkage-reducing admixture (SRA) to mitigate cracking; and fibers for reinforcement. The distinctive low-density characteristic of ULCC is principally achieved through the incorporation of cenospheres, which serve as partial replacement for conventional mineral aggregates.

Cenospheres represent a sustainable material solution, being industrial byproducts obtained from coal combustion processes in thermal power plants [14–17]. These unique particles feature a hollow spherical morphology with diameters typically ranging from 10 μm to 300 μm [11]. Structural analysis reveals that each cenosphere comprises a thin outer shell, with wall thickness representing approximately 2.5–10.5% of the particle diameter [18, 19]. The hollow internal structure, filled with inert gas, contributes to their remarkably low particle density of 600–900 kg/m^3.

From a chemical perspective, cenospheres consist primarily of amorphous aluminosilicate phases, with combined SiO_2 and Al_2O_3 content reaching up to 90% of their composition [20]. Table 1.1 presents the detailed chemical analysis of typical cenospheres. The significant amorphous silica content enables cenospheres to participate in pozzolanic reactions during cement hydration, similar to conventional fly ash [21]. This reactivity contributes to the development of the composite's long-term strength.

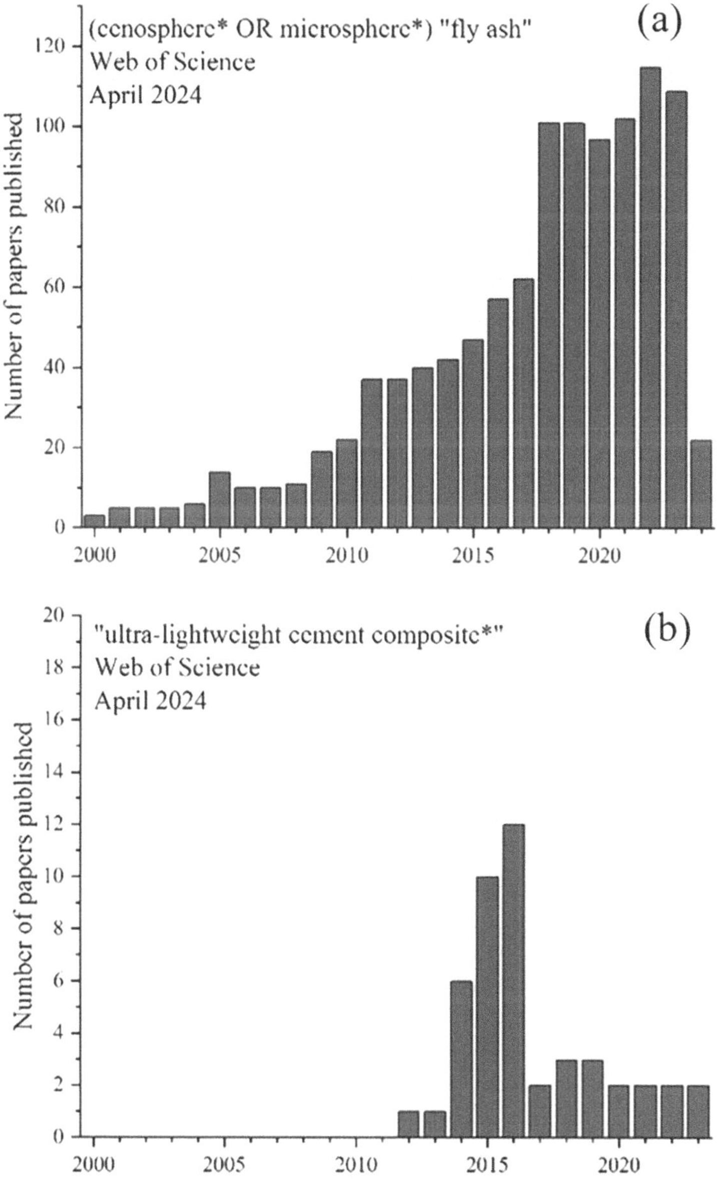

Fig. 1.1 Numbers of papers published on (**a**) fly ash cenosphere and (**b**) ultra-lightweight cement composite

Table 1.1 Chemical composition of cenospheres

SiO_2 (%)	Al_2O_3 (%)	CaO (%)	K_2O (%)	Na_2O (%)	Fe_2O_3 (%)	MgO (%)
58.1	32.0	0.9	3.1	0.8	3.7	1.4

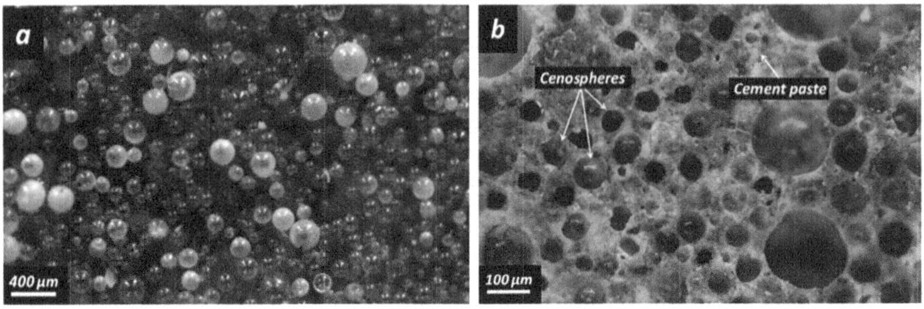

Fig. 1.2 Microstructural characterization of ultra-lightweight cement composite (**a**) image of cenosphere (**b**) image of ultra-lightweight cement composite incorporating cenospheres. (Reprinted from Huang et al. (2017), copyright 2025, with permission from Elsevier)

The microstructural characteristics of ULCC are illustrated in Fig. 1.2a, b, which respectively show the morphology of individual cenospheres and their distribution within the cementitious matrix. Unlike conventional concrete, ULCC represents a unique class of cement-based composite that completely eliminates the need for both coarse and fine sand aggregates, relying instead on the optimized packing of cenospheres within the binder system.

The macroscopic mechanical properties of ULCC are fundamentally governed by its microstructural architecture. A comprehensive investigation of pore structure evolution at different curing ages provides critical insights into the material's growth patterns, failure mechanisms, and long-term performance. These microstructural features serve as key indicators for evaluating ULCC's density, strength development, durability, and permeability characteristics. Cenospheres, the primary lightweight aggregate in ULCC, possess unique structural attributes including a hollow interior and a robust outer shell. While their pozzolanic reactivity is relatively limited at ambient temperatures, the reaction between cenospheres and cementitious components produces gelatinous hydration products that significantly enhance the interfacial transition zone (ITZ) between the cement matrix and aggregates [20, 22]. This reaction-induced densification process effectively mitigates microcrack formation at the ITZ, contributing to improved mechanical performance. Advanced characterization techniques have revealed critical microstructural features of ULCC, which are discussed below in detail:

- Material Homogeneity: Synchrotron-based high-resolution microtomography demonstrates uniform dispersion of cenospheres within the cement matrix (Fig. 1.3). This technique enables clear differentiation between reacted and unreacted cenosphere portions through pore distribution analysis and reaction product identification [23].
- Porosity Characteristics: Comparative μ-CT studies show that cenosphere-modified mortars exhibit 18–25% lower porosity and 30–40% reduced capillary water absorption compared to expanded perlite systems [24]. The discontinuous pore structure created by cenospheres significantly decreases water permeability, as confirmed through moisture absorption/desorption testing [25].

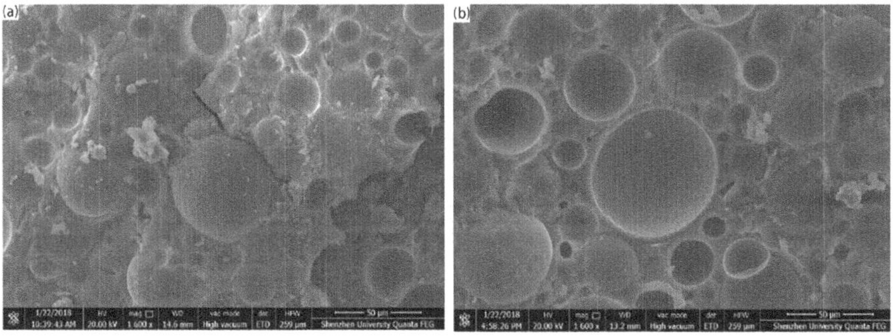

Fig. 1.3 Microtomography images in a cenosphere-modified cement matrix (**a**) crack propagation (**b**) interfacial transition zone (ITZ). (Reprinted from Huang et al. (2019), copyright 2025, with permission from J. Xi'an Univ. of Arch. & Tech. (Natural Science Edition) (In Chinese))

- Fracture Behavior: Microstructural observations reveal distinctive crack propagation patterns (Fig. 1.3a). Cracks preferentially propagate through weaker matrix regions rather than penetrating cenospheres, demonstrating their effective stress-redistribution capability. The ITZ between cenospheres and cement paste shows strong adhesion (Fig. 1.3b), with pozzolanic reaction products creating a robust interfacial bond that resists crack advancement.

Overall, these observations highlight the importance of engineering the cenosphere-paste interface and pore structure to achieve high-performance ULCC [26]. However, material performance shows complex dependencies: (1) at elevated cement matrix strengths, crack penetration through cenospheres becomes more probable; (2) achieving an optimal mix design requires careful balancing of matrix-aggregate strength compatibility, ITZ quality control, water-to-binder ratio optimization, and admixture fineness selection.

These findings underscore the importance of microstructural engineering in developing high-performance ULCC formulations. The relationship between processing parameters, microstructural evolution, and macroscopic properties presents critical opportunities for material optimization.

1.3 Mix Design of ULCC

Table 1.2 lists a summary of the mix proportions of ULCC in existing literature. The content of OPC in ULCC ranges from 297 kg/m^3 to 972 kg/m^3. FA, GGBS, and LC2 have been tried to partially replace OPC in ULCC with replacement levels ranging from 15% to 58%. Various water/binder ratios ranging from 0.2 to 0.7 have been explored with FAC/binder ratios ranging from 0.17 to 1.13. Most of the studies utilized shrinkage reducing admixture (SRA), superplasticizer (SP), and fibers to improve the performances of ULCC. SRA was used at dosages from 0.90% to 2.57% by binder weight. Polycarbonxylated-based SP was mostly used with lower dosages ranging from 0.33% to 2.79%, while only one study adopted naphthalene-based

Table 1.2 Summary of mix design of ULCC

Water	OPC	SF	RL of FA	RL of GGBS	RL of LC2	*w/b*	FAC/binder	SRA/binder	SP	SP/binder	Type of fiber	Fiber	Density	28-day compressive strength	Curing regime	Ref.
kg/m^3	kg/m^3	kg/m^3	%	%	%			%		%		% in volume	kg/m^3	MPa		
258	742	64.5	–	–	–	0.3	0.45	2.49	P	0.78	PVA	0.2–0.9	1329–1368	52.0–63.3	–	[27]
268–298	710–972	80.2–84.6	–	–	–	0.3	0.35–0.57	1.10–1.79	–	0.33–0.94	PVA	0.5	1216–1424	46.5–55.5	–	[28]
250–302	514–720	62.6–76.9	–	–	–	0.3–0.5	0.46–0.76	2.5–3.9	–	–	PVA	0.5	1283–1361	37.8–60.9	Air curing in laboratory	[29]
282–305	499–846	43.4–73.6	–	–	–	0.3–0.5	0.42–0.73	0–1.68	P, N	0.43–1.22	PE	0–0.07	1154–1471	33.0–69.4	Moist curing at 28–30 degree for 28 days	[16]
237–284	500–801	55.5–88.5	–	–	–	0.3–0.4	0.36–0.73	0.96–1.48	–	0.43–2.79	S	1	1258–1552	48.0–70.1	–	[30]
273	738	81.2	–	–	–	0.3	0.48	0.90	P	0.90	–	–	1450	52.2	Moist curing at 20 degree for 28 days	[31]
250–256	7120–735	62.6–63.9	–	–	–	0.3	0.46	2.5	P	–	PVA	0–0.5	1361–1390	56.9–59.2	Moist curing at 28–30 degree for 28 days	[41]
258	742	65.0	–	–	–	0.3	0.45	2.5	–	0.66	PVA	0.5	1380	57.5	–	[33]

265	419–874	97.1	–	–	35,50,55	0.3	0.33	1.0	P	0.44–0.99	PE	0, 1.0	1524	45.0–63.5	Moist curing at 25 degree for 28 days	[34]
287	754	65.5	–	–	–	0.4	0.46	–	–	–	PVA	0.9	1450	61.7	Moist curing at 28–30 degree for 14 days	[35]
262	741	65.0	–	–	–	0.3	0.45	2.5	–	1.40	–	0.16	1440	62.0	Moist curing for 28 days	[36]
235–275	585–777	51.5–67.6	–	–	–	0.3–0.4-	0.46	0–2.5	P	–	PE, S	0–0.5	1125–1474	27.1–68.2	Moist curing at 28–30 degree for 28 days	[37]
334–376	429–483	48.0–54.0	–	–	–	0.7	0.78	–	P	–	PVA	0.88–0.99	1072–1201	18.6–23.5	Moist curing at 25 degree for 28 days	[38]
246	705	61.3	–	–	–	0.3	0.45	2.5	P	0.65	PVA	0.5	1361	56.1	–	[39]
289–325	560–905	70–79	0,15,30	–	–	0.3	0.17	1.16	P	0.90	PP, S	0–1.0	1302–1464	46.9–65.8	Moist curing at 28–30 degree for 28 days	[22]
265	775	41.0	–	–	–	0.3	0.43	2.55	P	0.67	PVA	0.5	1450	58	Moist curing at 28–30 degree for 7 or 28 days	[40]

(continued)

Table 1.2 (continued)

Water	OPC	SF	RL of FA	RL of GGBS	RL of LC2	*w/b*	FAC/ binder	SRA/ binder	SP	SP/ binder	Type of fiber	Fiber	Density	28-day compressive strength	Curing regime	Ref.
kg/m^3	kg/m^3	kg/m^3	%	%	%			%		%		% in volume	kg/m^3	MPa		
295	602	66.9	–	–	–	0.4	0.53	1.43	–	0.90	–	–	1293	43.8	Moist curing at 28–30 degree for 28 days	[41]
282–302	606–837	53–73	–	–	–	0.3–0.4	0.42–0.73	1.15–1.47	P	0.54–1.02	–	–	1196–1471	40.9–69.4	Moist curing at 28–30 degree for 28 days	[23]
258–265	742–874	64.5–97.1	–	–	–	0.3	0.33–0.45	0.96–2.49	P	0.78	PVA, S	0–1.0	1329–1691	52.0–87.3	–	[42]
320	681	0	–	–	–	0.5	0.52	–	P	0.37–1.04	–	–	–	–	Air-curing in ambient temperature	[20]
258	742	65	–	–	–	0.3	0.45	2.48	–	0.66	PVA	0.50	1380	64.0	–	[43]
265	775	41	–	–	–	0.3	0.43	2.55	–	0.67	PVA	0.50	1450	55.7	–	[44]
246	705	41.3	–	–	–	0.3	0.45	2.49	–	0.65	PVA	0.50	1361	56.0	–	[45]
244–258	297–742	37–65	0,32,58	–	–	0.3	0.45–1.13	2.46–2.57	P	0.74–0.78	PVA	0.50	1322–1399	30–60	–	[46]
260–283	536–874	59.5–97.1	–	–	–	0.3–0.5	0.33–0.65	0.96–2.49	P	0.78–1.34	PVA, S	0.5–2.0	1289–1573	54.5—93.3	–	[47]
350	450	50	–	–	–	0.7	0.78	–	–	–	–	–	1273	35.5	Moist curing at 20 degree for 28 days	[48]

240	700	230	–	21	–	0.2	0.31	–	P	0.36	PE	2.06	1550	48.0	Moist curing at 23 degree	[49]
293	769	67	–	–	–	0.4	0.46	–	–	–	PVA	0.90	1480	62.0	Moist curing at 28–30 degree for 28 days	[50]

Note: *SF* silica fume, *RL* replacement level, *FA* fly ash, *GGBS* ground granulated blast-furnace slag, *LC2* limestone and calcined clay, *w/b* water/binder, *FAC* fly ash cenosphere, *SRA* shrinkage reducing admixture, *SP* superplasticizer, *P* polycarbonxylated-based, *N* naphthalene-based, *PVA* polyvinyl alcohol, *PE* polyethylene, *S* steel, *PP* polypropylene

SP. Four types of fibers have been used in ULCC, namely polyvinyl alcohol (PVA), polyethylene (PE), polypropylene (PP), and steel with volumetric ratios from 0.07% to 2.0%. The density of ULCC ranges from 1072 kg/m^3 to 1691 kg/m^3, with 28-day compressive strength ranging from 18.6 MPa to 87.3 MPa.

Wang et al. [41] proposed a design method for determining the mix proportion of workable ULCC with a desired density from 1150 kg/m^3 to 1450 kg/m^3. Their method employed a fixed silica fume content of 10% of the total cementitious materials. A shrinkage-reducing admixture (SRA), Eclipse (manufactured by W. R. Grace), was used to reduce the potential shrinkage cracking and entrapped air content at a dosage of 3% of the cement. High range water reducer admixture was used in a range of 0.8–1.0 L/100 kg of the total cementitious materials to achieve a good workability of ULCC. Figure 1.4 presents the flow chart of the proposed method. At the beginning of ULCC mix design, a target density ranging from 1150 kg/m^3 to 1450 kg/m^3 should be specified. The particle size distribution

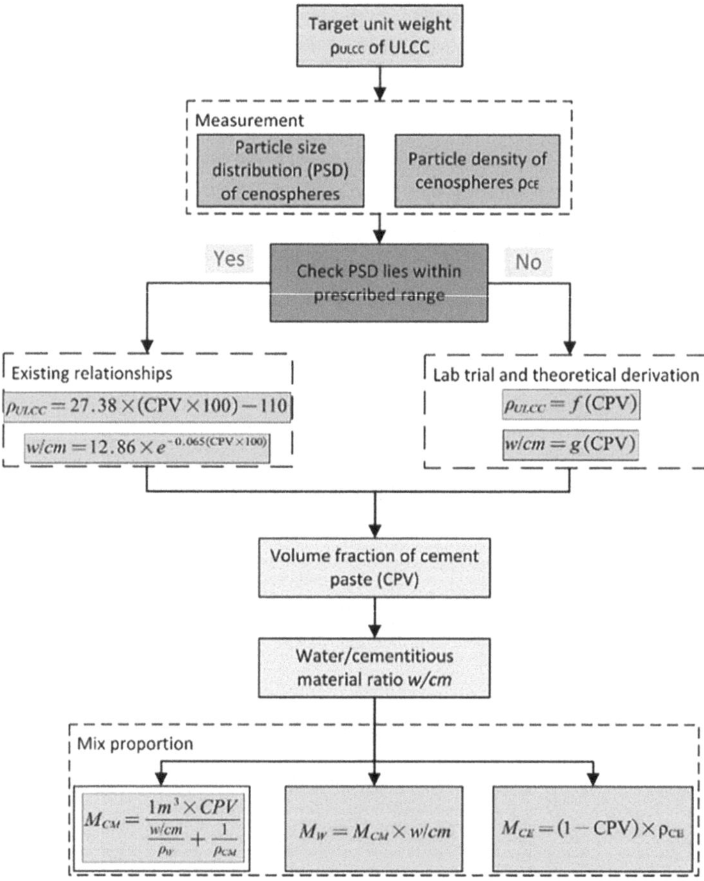

Fig. 1.4 Flow chart of ULCC mix proportion design. Reworked from [41]

(PSD) curve and particle density of FACs should be determined. The PSD curve of the FACs shall lie within the prescribed range so that the existing relationships in [41], which correlate the density of ULCC and water/cementitious materials ratio (*w/cm*) with the volume fraction of cement paste (CPV), can be used without the need of experimental testing. Otherwise, lab trial and analytical derivation as required in [41] should be conducted in order to correlate CPV with the density of ULCC and *w/cm*, by which, one can calculate CPV as well as *w/cm*. The mix proportion of ULCC can then be determined by a volume-based method.

The proposed mix proportion design method promotes the use of ULCC, especially considering the difficulties in determining mix proportion of ULCC, which is significantly different from that of traditional lightweight concrete. Further studies are required to extend the allowable desired density of ULCC beyond the density range of 1150–1450 kg/m^3 as specified in [41] and, more importantly, to enable more general mix to be designed and validated with full consideration of effects of the admixtures.

1.4 Mechanical Properties of ULCC

1.4.1 Compressive Strength

Figure 1.5a illustrates the correlation between density and compressive strength for various cementitious materials, including ULCC [16, 30, 32, 35–38, 40, 42], lightweight cementitious composites (LWCC) [51], lightweight aggregate concrete (LWAC) [35, 36, 40, 52–61], lightweight aggregate foamed concrete (LWAFC) [62], foamed concrete (FC) [63–65], and normal weight aggregate concrete (NWAC) [40, 55]. The data for ULCC demonstrates a strong linear relationship (R^2 = 0.838), indicating a reliable predictive model for its compressive strength based on density. However, some data points fall below the fitted curve, primarily

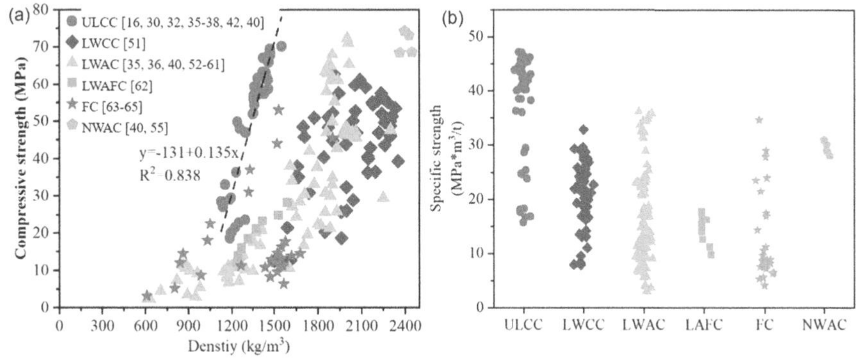

Fig. 1.5 Relationship between density and compressive strength of various concretes (**a**) compressive strength versus density for ULCC, LWCC, LWAC, LWAFC, FC, and NWAC; (**b**) specific strength for the same concrete types

due to the incorporation of aerogel materials in ULCC, which enhance thermal properties but reduce compressive strength [38].

ULCC exhibits superior performance compared to other lightweight concretes, as evidenced by its concentration in the upper-left region of Fig. 1.5a, signifying higher specific strength. This advantage is further emphasized in Fig. 1.5b, where ULCC achieves a maximum specific strength of 47.2 MPa·m^3/t—30.7% higher than that of LWAC (36.1 MPa·m^3/t). The exceptional specific strength of ULCC can be attributed to several factors: the rigid shell structure of FACs, their hollow nature (introducing non-interconnected porosity), and the enhanced ITZ resulting from pozzolanic reactions between FACs and silica fume. In contrast, conventional lightweight aggregates typically exhibit weaker ITZs.

The water-to-binder (*w/b*) ratio significantly impacts the 28-day compressive strength of ULCC, as shown in Fig. 1.6a. Similar to conventional concrete, lower w/b ratios yield higher strengths, with a maximum of 70 MPa achieved at a *w/b* ratio of 0.32. Conversely, increasing the *w/b* ratio to 0.70 reduces strength to approximately 20 MPa. Additionally, the absence of a SRA can substantially weaken ULCC due to entrapped air voids, which compromise the fiber-matrix interface. Research indicates that SRA contributes more to strength enhancement than PE fibers by improving matrix densification [37].

Figure 1.6b highlights the effect of the FAC-to-binder ratio on compressive strength. Exceeding an FAC/binder ratio of 0.5 leads to strength deterioration, as the intrinsic strength of FACs becomes insufficient compared to the cement paste. The optimal FAC dosage ranges between 0.36 and 0.45 for balanced performance.

The influence of SCMs on ULCC strength is depicted in Fig. 1.7. Partial replacement of OPC with LC2 enhances compressive strength at a 35% replacement level, though this benefit diminishes with higher replacements. In contrast, FA replacement consistently reduces strength, with the decline being linearly proportional to the replacement percentage.

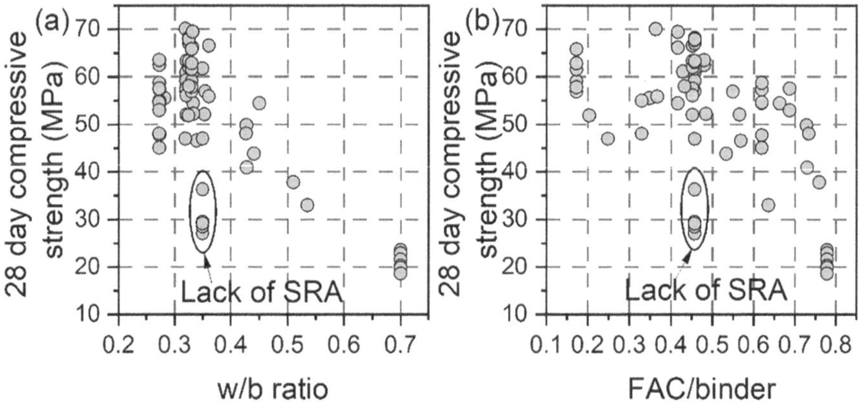

Fig. 1.6 Effects of *w/b* and FAC/binder ratio on 28-day compressive strength of ULCC (**a**) 28-day compressive strength versus *w/b* ratio; (**b**) 28-day compressive strength versus FAC/binder ratio

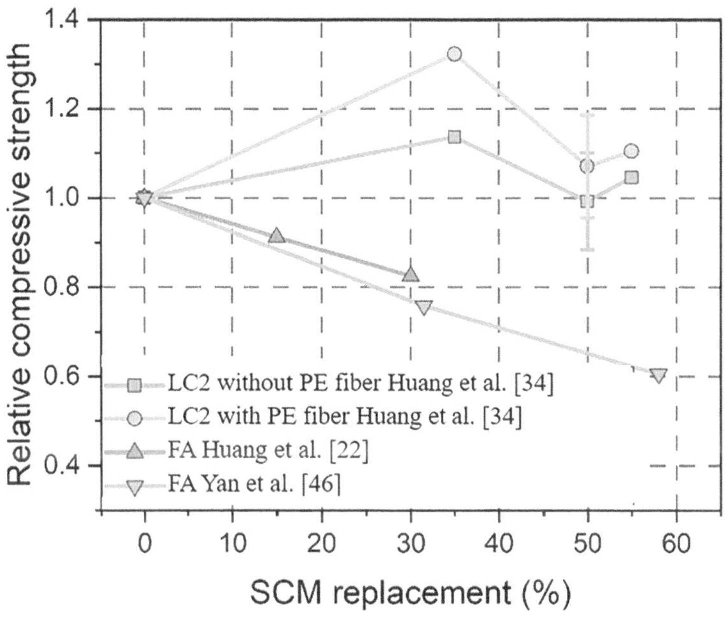

Fig. 1.7 Effects of SCM replacement on 28-day compressive strength of ULCC

1.4.2 Elastic Modulus

The elastic modulus is a critical parameter in assessing the deformation behavior of cementitious materials. Like other lightweight concretes, ULCC exhibits a significantly lower elastic modulus compared to normal-weight concrete due to the inherently low stiffness and density of FACs.

Wu et al. [16] observed that the elastic modulus of ULCC increases with both density and compressive strength. Their study compared experimental results with predictions from the ACI 318 [66] equation, revealing that the ACI model overestimates the elastic modulus for ULCC with densities above 1230 kg/m^3 and compressive strengths exceeding 43 MPa, while underestimating it for lower-density mixes. Additionally, Liu et al. [40] reported that the elastic modulus of ULCC increases by approximately 12% at −60 °C compared to room temperature (30 °C), suggesting temperature-dependent stiffness behavior.

Figure 1.8 illustrates the correlation between elastic modulus, compressive strength, and density for ULCC [16, 30, 32, 35, 36, 38, 40, 42], LWAC [35, 36, 40, 53, 58–60], and NWAC [40, 55]. Despite ULCC's superior compressive strength (as discussed in Sect. 1.3), its elastic modulus remains lower than that of LWAC and NWAC at equivalent strength levels. This indicates that while FACs enhance strength, they do not contribute significantly to stiffness improvement. A strong linear relationship exists between compressive strength and elastic modulus for ULCC (Fig. 1.8a). Furthermore, Fig. 1.8b demonstrates that elastic modulus

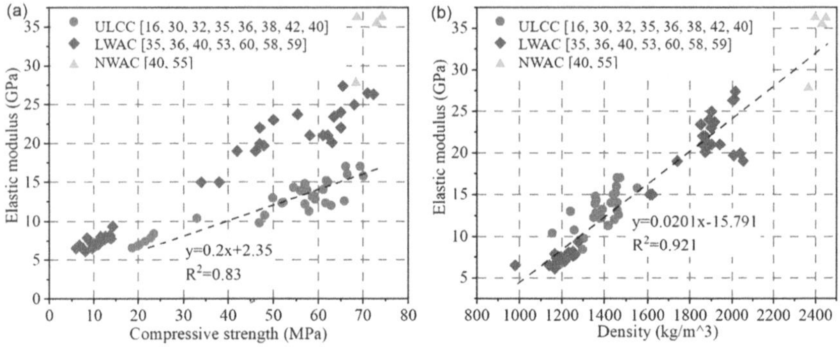

Fig. 1.8 Relationship between elastic modulus and compressive strength, elastic modulus and density of various concretes (**a**) Elastic modulus versus compressive strength; (**b**) Elastic modulus versus density

decreases linearly with density across all cementitious materials, suggesting that lightweight aggregates (whether FACs or conventional LWAC aggregates) reduce stiffness proportionally with density. However, ULCC maintains an advantage over traditional LWAC of similar density due to its higher specific strength and compressive performance. In summary, while ULCC excels in strength-to-weight ratio, its elastic modulus remains constrained by the low stiffness of FACs. The linear dependencies between elastic modulus, strength, and density provide valuable insights for structural design, particularly in applications where deformation resistance must be balanced with lightweight properties.

1.4.3 Tensile Strength and Flexural Strength

The replacement of conventional aggregates with FACs in ULCC results in a more brittle material compared to normal-weight concrete [67]. To mitigate this issue, fiber reinforcement has proven effective in reducing brittleness by promoting multi-cracking behavior under tensile stress. Various fibers, including PVA fibers [32, 35, 39, 43, 68], PE fibers [16, 37], PP fibers [22], and steel fibers [37], have been incorporated into ULCC to enhance its mechanical properties.

Figure 1.9 demonstrates the influence of different fiber types on the tensile strength of ULCC. At comparable dosages (0.5–1.0%), PVA and PE fibers outperform steel fibers in improving tensile strength, with ULCC achieving a maximum tensile strength of 6.2 MPa. However, in flexural applications, steel fibers exhibit superior performance due to their higher elastic modulus (200 GPa vs. 79 GPa for PE fibers), as observed by Wang et al. [37] (vid. Fig. 1.10). The inclusion of SRAs further enhances flexural toughness by reducing solution surface tension and minimizing air voids in the composite. SRAs lower the contact angle between fibers and the matrix, improving interfacial bonding and increasing fiber pull-out

Fig. 1.9 Effect of fibers on tensile strength of ULCC

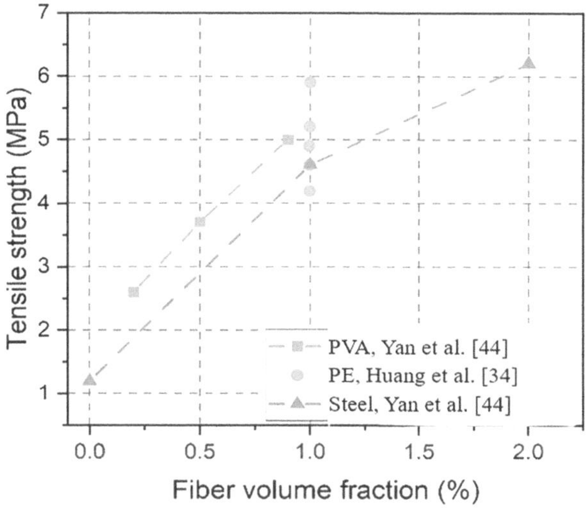

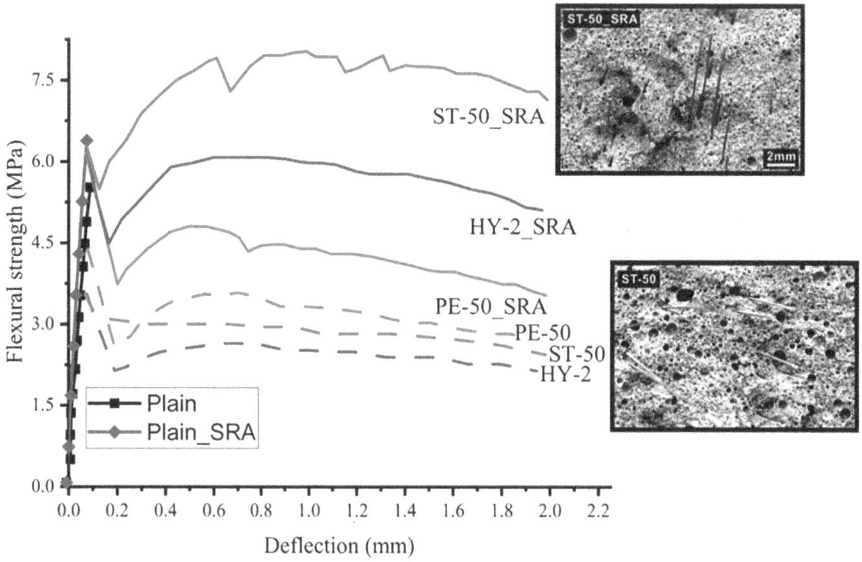

Fig. 1.10 Average load–deflection curves of fiber reinforced ULCC with SRA. Reworked from [37]. Plain: plain ULCC without fibers, plain_SRA: plain ULCC without fibers with shrinkage reduce admixutre (SRA), ST-50_SRA: 0.5% steel fiber (ST) reinforced ULCC with SRA, PE-50_SRA: 0.5% polyethylene fiber (PE) reinforced ULCC with SRA, HY-2_SRA: 0.25% ST + 0.25% PE fibers reinforced ULCC with SRA, PE-50, ST-50, HY-2 with same mix designs albeit without SRA

resistance. As a result, even a modest fiber content of 0.5% combined with SRA can produce ULCC with deflection-hardening behavior.

Liu et al. [40] investigated the effects of low temperatures (down to −60 °C) and curing conditions on ULCC's flexural behavior. Their findings revealed that 28-day moist-cured specimens exhibited 25–38% higher flexural strength than those cured for only 7 days, regardless of testing temperature. While flexural toughness remained largely unaffected by temperature, extended curing significantly improved both strength and toughness.

The tensile and flexural properties of ULCC are primarily governed by fiber type and fiber-matrix interfacial bonding. Enhancing fiber pull-out resistance can further optimize flexural performance. However, research in this area remains limited, highlighting the need for future studies to explore the effects of fiber type, dosage, and interfacial modifications on ULCC's tensile, flexural, and ductile characteristics.

1.4.4 Constitutive Laws

The development of accurate constitutive models for concrete is essential for predicting structural behavior, as these models characterize the material's stress-strain response under various loading conditions, including elastic deformation, plastic yielding, and ultimate failure. For fiber-reinforced ULCC, the stress-strain relationship varies significantly depending on fiber type, influencing both compressive and tensile performance.

Figure 1.11a, c compare the compressive stress-strain relationships of ULCC reinforced with PE and steel fibers [67], respectively. The PE fiber-reinforced composite exhibits a gradual post-peak softening response (vid. Fig. 1.11a), indicating enhanced energy absorption and toughness. In contrast, steel fiber-reinforced ULCC (Fig. 1.11c) demonstrates a more abrupt stress reduction after peak load [30], suggesting less ductile failure. This difference highlights PE fibers' superior ability to mitigate brittle fracture under compression. The initial elastic response in tension is similar for both fiber types, with tensile strengths ranging from 2 MPa to 4 MPa, elastic moduli of 10 GPa to 20 GPa, and strain capacities of 0.01% to 0.03%. However, post-peak behavior diverges markedly (Fig. 1.11b, d). PE fiber reinforcement induces pronounced strain hardening, where stress increases with strain up to approximately 5% deformation, demonstrating effective crack-bridging and load redistribution. Steel fibers, while maintaining a post-peak stress plateau, exhibit limited ductility (~0.3% strain) before rapid strength degradation. These findings underscore PE fibers' advantages in applications demanding high tensile deformation capacity, such as seismic-resistant or impact-resistant structures. The contrasting post-peak behaviors emphasize that fiber choice critically governs ULCC's mechanical performance. PE fibers optimize ductility and toughness, whereas steel fibers may be preferable where higher initial stiffness is prioritized.

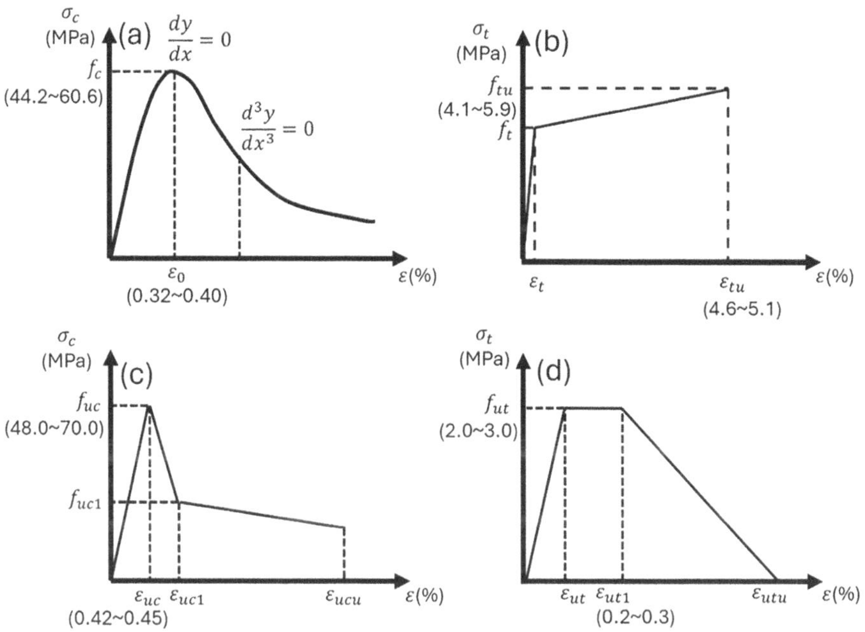

Fig. 1.11 Proposed stress-strain relationship under uniaxial (**a**) compression and (**b**) tension for ULCC reinforced by PE fibers in [69]; proposed stress-strain relationship under uniaxial (**c**) compression and (**d**) tension for ULCC reinforced by steel fibers in [30]. (Reprinted from Liang et al. (2023) and Wang et al. (2018), copyright 2025, with permission from Elsevier)

1.5 Ultra-Lightweight High Ductility Cement Composite

Literature suggests that while most traditional high-ductility concretes exhibit strong tensile strain capacity, their self-weight remains relatively high. Although concrete prepared with fly ash cenospheres or glass microspheres shows a significant reduction in density, its compressive strength is generally low, and its tensile strain capacity does not meet the standard for axial tensile strain of high-ductility concrete, thereby limiting its application in engineering. Therefore, developing lightweight cement-based materials that simultaneously possess light weight, high strength, and high ductility and applying them to practical engineering is highly significant. In this study, an ultra-lightweight and highly ductile cement-based material (ULHDCC) was developed using optimized glass microspheres as fillers and hydrophilic-treated PE fibers based on strength and energy criteria. Its mechanical properties and microstructure were systematically investigated through comprehensive experiments. A novel multifunctional ULHDCC was formulated using three types of hollow microspheres. The incorporation of hollow microspheres as fillers effectively reduced the density of the concrete. Based on strength and energy criteria, the ductility of ULHDCC was enhanced by modification with PE.

ULHDCC was prepared using ordinary Portland cement (PO52.5R), densified silica fume with SiO_2 content exceeding 94%, and three types of hollow microspheres (D38, D42, and D46). The particle size distribution of the microspheres is presented in Fig. 1.12 below.

The basic performance indicators of hydrophilically modified PE fibers are summarized in Table 1.3. A polycarboxylate-type high-efficiency water-reducing agent was employed. The mix ratio design for ULHDCC testing is detailed in Table 1.4.

1.5.1 Apparent Density and Compressive Strength

The fundamental properties of the specimens, summarized in Table 1.5, reveal key relationships between microsphere composition, density, and mechanical performance. The apparent density, measured 24 h after casting and demolding, varied

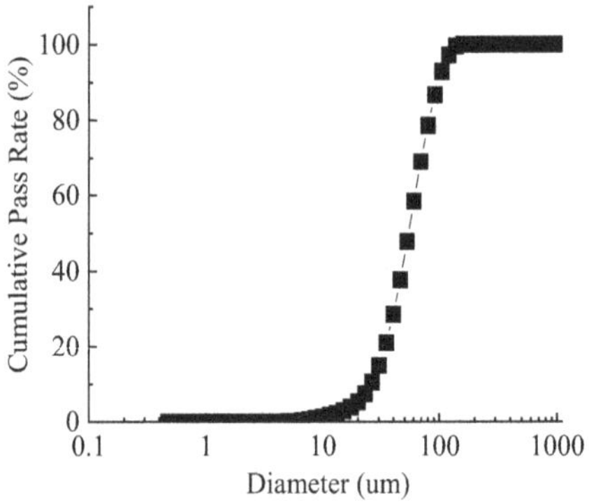

Fig. 1.12 Particle size distribution of microspheres

Table 1.3 Properties of modified PE fibers

Diameter (μm)	Length (mm)	Density (g/cm^3)	Strength (MPa)	Elastic modulus (GPa)	Elongation (%)
24	12	0.97	3000	120	2–3

Table 1.4 Mixture proportion

ID	Cement	Silica fume	Hollowsphere	Water-to-binder ratio	SP	Fiber (%)
D38-PE1%	1	0.18	0.50	0.70	0.03	1
D42-PE1%	1	0.20	0.60	0.70	0.03	1
D46-PE0	1	0.17	0.70	0.70	0.03	0
D46- PE1%	1	0.17	0.70	0.70	0.03	1

Note: The fibers represent the volume proportion. D38-PE1% means that the density of D38 microspheres is added with PE fibers of a volume proportion of 1%. The rest is the proportion of quality

Table 1.5 Properties of ULHDCC

ID	Apparent density (kg/m³)	Compressive strength (MPa)	Tensile strength (MPa)	Thermal conductivity (W/mK)
D38-PE1%	900	27.5	2.5	/
D42-PE1%	909	28.1	2.9	/
D46-PE0	917	32.9	/	0.152
D46-PE1%	842	20.8	2.7	/

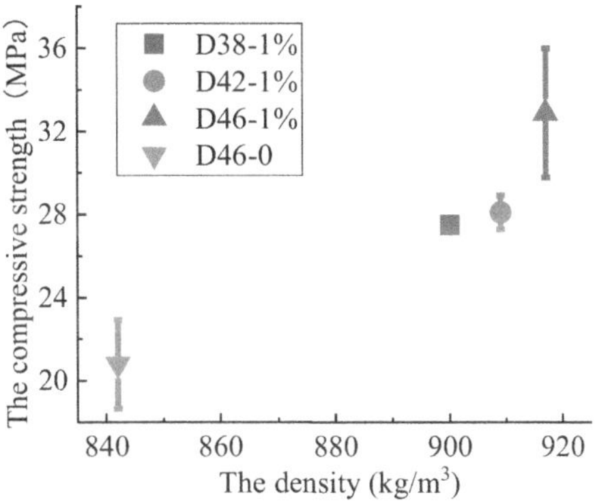

Fig. 1.13 Relationship between compressive strength and density

between 840 kg/m³ and 920 kg/m³ depending on the type of microspheres used. Notably, under consistent volumetric dosages, higher-density microspheres resulted in increased concrete density, demonstrating a direct correlation between microsphere density and bulk material properties. As shown in Table 1.5 and Fig. 1.13, compressive strength exhibited a proportional increase with apparent density, rising from an average of 20.8 MPa to 32.9 MPa. This trend can be attributed to the structural advantages of denser microspheres, which typically feature thicker walls and superior crushing resistance. Consequently, the enhanced microsphere integrity contributed to both the elevated density and improved compressive strength of the lightweight concrete. These findings underscore the critical role of microsphere selection in optimizing the performance of lightweight cementitious composites.

The experimental results demonstrate a significant improvement in compressive performance through fiber reinforcement. For instance, specimens incorporating D46 microspheres exhibited a remarkable 59% strength enhancement (from 20.8 MPa to 32.9 MPa) with the addition of just 1% fibers. This substantial improvement likely stems from multiple mechanisms: the hydrophilic treatment of fibers enhances fiber-matrix bonding, reduces void formation during mixing, and increases overall matrix density. However, these proposed mechanisms require further experimental validation to establish definitive correlations. Beyond strength enhancement, fiber incorporation dramatically alters failure behavior under compression. Unreinforced specimens exhibited characteristic brittle failure patterns, with pronounced surface spalling and delamination similar to conventional high-strength and lightweight concretes. In contrast, fiber-reinforced specimens maintained structural integrity despite crack formation, demonstrating the fibers' effective crack-bridging capability. The restrained crack propagation ultimately resulted in a ductile failure mode where specimens deformed into flattened configurations without disintegration. These findings highlight the dual benefits of fiber reinforcement—both as a strength-enhancing additive and as a toughness-improving component that modifies failure modes. The transition from brittle to ductile failure patterns suggests potential applications where both high strength and damage tolerance are required.

The tensile performance of the material exhibits remarkable improvement with fiber reinforcement. Unreinforced specimens demonstrate minimal tensile strength (approximately 1 MPa), with early-age fragility making them susceptible to cracking and premature failure within 24 h of molding. However, the incorporation of just 1% fibers produces a dramatic transformation in mechanical behavior, elevating ultimate tensile strength to 2.9 MPa while achieving an exceptional 6% strain capacity. Particularly, noteworthy is the D38-PE1% formulation, which attains 8% ultimate tensile strain—a value several orders of magnitude greater than conventional concrete (typically 0.01–0.017%) [70] and exceeds even engineered cementitious composites (ECC) [71] in ductility performance. This extraordinary deformation capacity, illustrated in Fig. 1.14, is accompanied by the development of characteristic multiple microcracks under tension (vid. Fig. 1.15), demonstrating the material's ability to maintain structural integrity while accommodating significant deformation. The transition from brittle fracture to pseudo-ductile behavior underscores the effectiveness of fiber reinforcement in creating a new class of ultra-lightweight, high-ductility cementitious materials. These findings reveal the material's potential for applications requiring both lightweight characteristics and exceptional deformation capacity, though further investigation is needed to optimize the fiber-matrix interface for consistent performance. The observed multi-crack pattern suggests effective stress redistribution mechanisms that could be leveraged in seismic-resistant and impact-resistant structural applications.

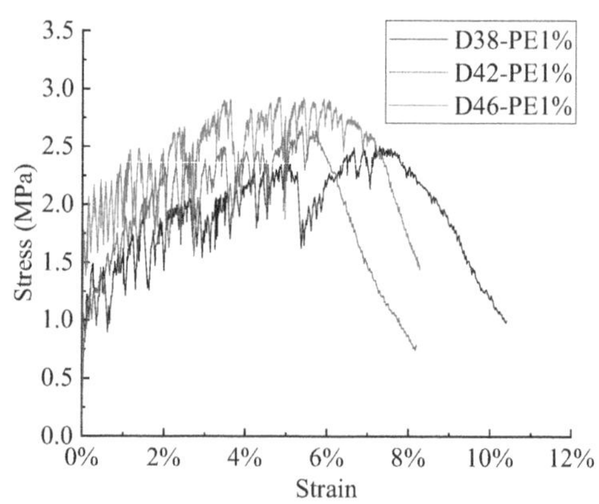

Fig. 1.14 Stress-strain curves under direct tensile. (Reprinted from Huang et al. (2019), copyright 2025, with permission from J. Xi'an Univ. of Arch. & Tech. (Natural Science Edition) (In Chinese))

Fig. 1.15 Multiple-cracks failure mode. (Reprinted from Huang et al. (2019), copyright 2025, with permission from J. Xi'an Univ. of Arch. & Tech. (Natural Science Edition) (In Chinese))

1.5.2 Microstructure and Micromechanical Analysis

The microstructural features of ultra-lightweight high-ductility cementitious composite (ULHDCC) are revealed through scanning electron microscopy (SEM) analysis in Fig. 1.16. At lower magnification (400×), the microstructural homogeneity becomes apparent, with uniformly distributed microspheres embedded in a dense cementitious matrix (see Fig. 1.16a). These spherical particles, primarily composed of SiO_2, maintain their structural integrity due to their high crushing strength (37.9 MPa) [75], as evidenced by the limited microsphere damage observed during mixing and loading processes. Higher magnification images (1600×) in Fig. 1.16b, c) demonstrate two key characteristics: the smooth surface morphology of intact microspheres and the preferential crack formation at ITZs. The observed cracking patterns suggest that the microsphere-matrix interface represents the most vulnerable region in the composite system under the current mix design, despite the microspheres' inherent strength.

Fiber reinforcement plays a pivotal role in the material's performance enhancement. Figure 1.16d illustrates effective fiber anchoring within the matrix, which facilitates the development of multiple microcracks under tensile loading—the fundamental mechanism behind the material's exceptional ductility. The absence of significant fiber breakage in tested specimens confirms optimal fiber-matrix compatibility.

From a design perspective, high-ductility materials must satisfy dual performance criteria: strength equilibrium and energy dissipation capacity [71]. The strength criterion, expressed in Eq. (1.1), establishes the fundamental relationship between initial cracking strength (σ_{cr}) and fiber bridging capacity (σ_0), requiring $\sigma_{cr} \leq \sigma_0$ for proper multiple cracking behavior. This microstructural evidence correlates directly with the macroscopic mechanical performance discussed in previous sections, providing a comprehensive understanding of the material's structure-property relationships.

$$\sigma_{cr} < \sigma_0 \tag{1.1}$$

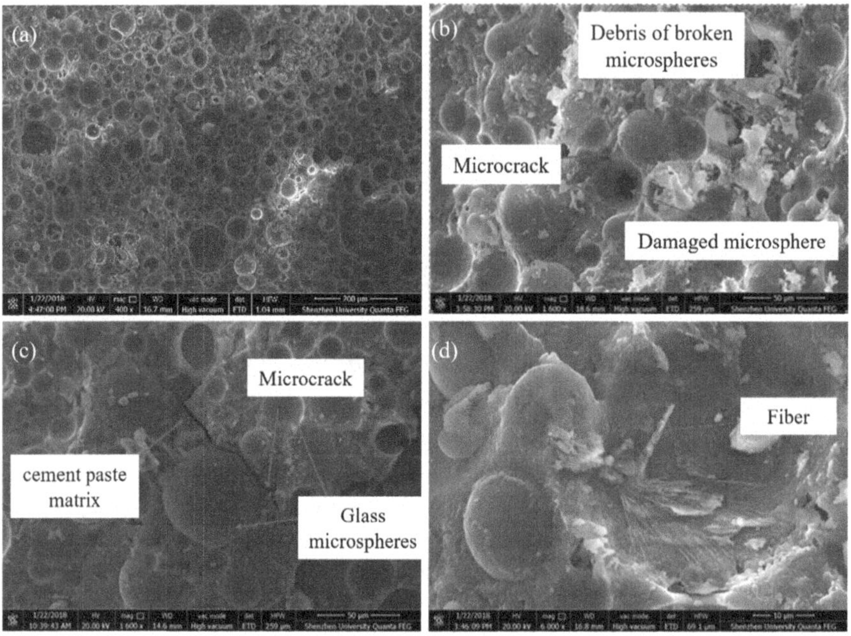

Fig. 1.16 Microscopic morphology (**a**) ULHDCC microstructure at 400× showing uniformly distributed microspheres embedded in a dense cementitious matrix; (**b**) ULHDCC microstructure at 1600× showing microcracks at the microsphere–matrix interface with locally damaged microspheres and debris; (**c**) ULHDCC microstructure at 1600× showing intact glass microspheres and preferential microcrack development along the ITZ; (**d**) ULHDCC microstructure showing fiber anchorage/pull-out within the matrix, supporting multiple microcracking under tensile loading. (Reprinted from Huang et al. (2019), copyright 2025, with permission from J. Xi'an Univ. of Arch. & Tech. (Natural Science Edition) (In Chinese))

The cementitious matrix can avoid direct fiber pull-out failure after cracking when the initial cracking strength is lower than the fiber bridging strength. This enables the fibers to effectively perform their bridging function, transferring stresses to uncracked regions in a cyclic manner, thereby facilitating the development of multiple cracking mechanisms (as illustrated in Fig. 1.15). Experimental results demonstrate that the cementitious matrix strength is indeed lower than the fiber bridging strength, as evidenced in Fig. 1.16b–d. The initial cracking strength (σ_{cr}) is primarily governed by: (1) the specimen's fracture toughness (K_m), (2) internal defect characteristics [70, 71], (3) while the fiber bridging strength (σ_0) depends mainly on fiber type selection and fiber/matrix interfacial properties.

A lower σ_{cr}/σ_0 ratio promotes more favorable crack development. However, excessive reduction of σ_{cr} may compromise other material properties. To optimize this balance, surface-modified PE fibers were specifically selected in this study to: (1) enhance fiber-matrix bonding and (2) significantly improve σ_0 through interfacial strengthening. This strategic fiber modification ensures proper stress transfer while maintaining the composite's overall mechanical performance, achieving the desired multiple cracking behavior characteristic of high-ductility cementitious materials.

1.5.3 Energy Criterion for Crack propagation

The energy criterion ensures stable crack development in cementitious composites. This principle requires that the total energy dissipation—comprising both the crack-tip energy consumption (J_{tip}, i.e., the supplementary energy required to overcome material resistance) and the energy absorbed by fiber bridging—must balance the strain energy generated by the tensile stress (σ_{ss}) under the corresponding displacement (δ_{ss}). Fig. 1.17 schematically illustrates this energy equilibrium.

For controlled cracking behavior, J_{tip} must remain below the maximum supplementary energy capacity, as governed by Eqs. (1.2)–(1.4):

$$J_{\text{tip}} + \int_0^{\delta_{\text{ss}}} \sigma(\delta)\text{d}\delta = \sigma_{\text{ss}}\delta_{\text{ss}} \tag{1.2}$$

$$J_{\text{b}}' + \int_0^{\delta_0} \sigma(\delta)\text{d}\delta = \sigma_0\delta_0 \tag{1.3}$$

$$J_{\text{tip}} < J_{\text{b}}' \tag{1.4}$$

Among these parameters, $J_{\text{tip}} = K_{\text{m}}^2/E_{\text{c}}$, E_{c} represents the elastic modulus of the material, and K_m denotes the fracture toughness, which can be determined using the testing method described in reference [72]. Both the micromechanical model and experimental results suggest that a smaller value of $J_{\text{tip}}/J_{\text{b}}'$ is preferable [73, 74]. The bonding strength between the fiber and matrix significantly influences the magnitude of the supplementary energy. If the bonding strength is excessively low, fibers are prone to being pulled out, leading to an insufficient σ_0. Conversely, if the bonding strength is too high, fibers may break, resulting in a reduced δ_0 [71]. Therefore, it is crucial to select an appropriate fiber/matrix bonding strength rather than indiscriminately increasing it to meet the strength requirements. The energy criterion was

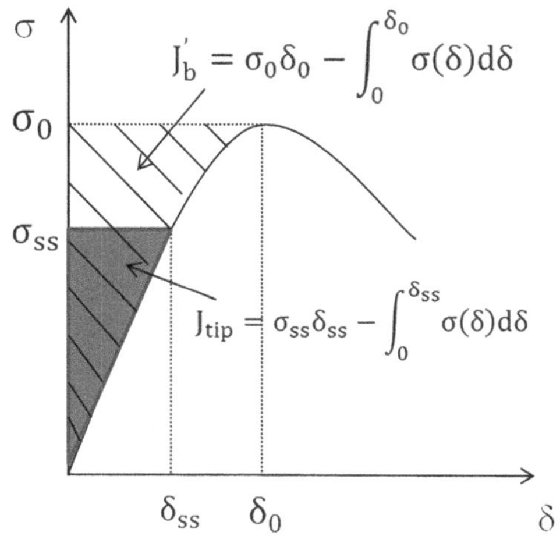

Fig. 1.17 Schematic diagram of energy criterion. (Reprinted from Huang et al. (2019), copyright 2025, with permission from J. Xi'an Univ. of Arch. & Tech. (Natural Science Edition) (In Chinese))

Table 1.6 Calculated fracture parameters

α	$f(\alpha)$	K_m (MPa.m$^{1/2}$)	J_{tip} (J/m^2)	J'_b (J/m^2)
0.4	2.089	9.20×10^{-4}	2.24×10^{-7}	3.45×10^{-2}

validated by considering the D46-PE1% group as an example. The fundamental parameters were derived from reference [72], and a three-point bending test was performed on a notched prism with dimensions of 40 mm × 40 mm × 160 mm. The results obtained using Eqs. (1.5) and (1.6) are presented in Table 1.6, demonstrating that the ratio of J_{tip}/J'_b satisfies the energy criterion outlined in Eq. (1.4).

$$K_m = \frac{1.5\left(FQ + \frac{mg}{2} \times 10^{-2}\right) \times 10^{-3} S\sqrt{a_0}}{th^2} f(\alpha) \tag{1.5}$$

$$f(\alpha) = \frac{1.99 - \alpha(1-\alpha)(2.15 - 3.93\alpha + 2.7\alpha^2)}{(1+2\alpha)(1-\alpha)^{3/2}}, \alpha = \frac{a_0}{h}(6) \tag{1.6}$$

1.6 LC3 Based Ultra-Lightweight Cement Composite

This part focuses on the development of a novel ultra-lightweight cement composite incorporating low-carbon limestone calcined clay cement (ULCC-LC3) and evaluates its resistance to chloride ingress and carbonation. The chloride resistance was assessed using rapid chloride migration (RCM) tests, while accelerated carbonation tests (ACT) were employed to examine carbonation behavior. The hydration characteristics, including heat release and product formation, were analyzed through isothermal calorimetry (IC) and thermogravimetric analysis (TGA), respectively. To further understand the microstructural interactions between ULCC and LC3, investigations were carried out using a mercury intrusion porosimeter (MIP) and scanning electron microscopy (SEM). Additionally, the environmental sustainability of the composite was assessed via life cycle assessment (LCA). The raw materials for ULCC-LC3 included ordinary Portland cement (OPC), cenospheres, silica fume, calcined clay, limestone, gypsum, a polycarboxylate-based superplasticizer, a shrinkage-reducing admixture (SRA), tap water, and polyethylene (PE) fibers. The PE fibers used had properties consistent with those described in Sect. 1.5. Specifically, QK300 cenospheres with particle sizes between 20 and 300 μm were used. The calcined clay was derived from industrial kaolinite tailings, with kaolinite content around 50%, and was calcined at 700–800 °C. The superplasticizer reduced water demand by up to 40%, and the SRA effectively decreased drying shrinkage by 80% at 28 days. The fresh mixtures exhibited excellent workability, with slump values between 180 and 205 mm. As shown in Fig. 1.18, the raw materials displayed a well-graded particle size distribution, contributing to good compaction of the resulting cementitious composite.

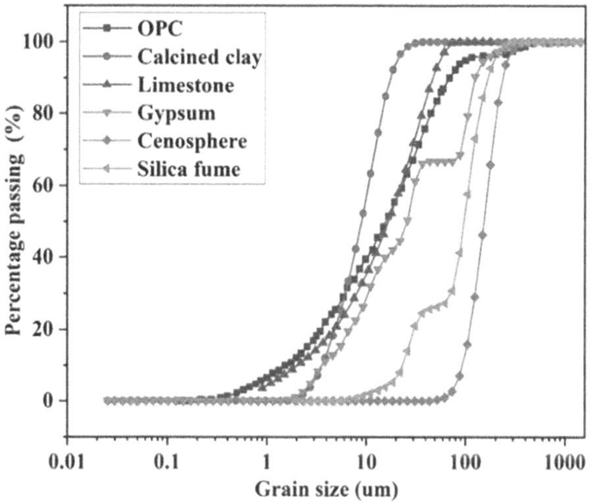

Fig. 1.18 Particle size distribution curves for raw solid materials. (Reprinted from Liang et al. (2023), copyright 2025, with permission from Elsevier)

The mix proportions for ULCC-LC3 are detailed in Table 1.7, with a water-to-binder ratio of 0.27 and a total of twelve mix combinations. The first six mixes were prepared without the addition of polyethylene (PE) fibers, while the latter six incorporated 1% PE fibers by volume. Among the initial six mixes, the first served as a reference, representing a conventional ultra-lightweight cement composite (ULCC) formulation without LC3 or PE fibers. The remaining five mixes in this set were defined based on the clinker content (expressed as a percentage of the total binder) and the mass ratio of calcined clay to limestone. For instance, the mix labeled "50–2:1" contained 50% clinker in the binder (calculated by deducting the calcium sulfate content from ordinary Portland cement as indicated in Table 1.1) and a calcined clay to limestone mass ratio of 2:1. The final six mixes were similar in composition to the first six but included 1% PE fibers by volume along with a higher dosage of superplasticizer. These mixes were distinguished by the suffix "−1" in their group names, which denotes the 1% volumetric proportion of PE fibers.

1.6.1 Mechanical Performances

Figure 1.19 presents the compressive strength results of the mixtures without PE fibers at 1, 3, 7, and 28 days. The rate and pattern of compressive strength gain over time are influenced by the incorporation of LC3 into the ULCC matrix. At 1 day, the compressive strength of the standard ULCC mix reached 48.5% of its 28-day value, whereas the strength of the mixes containing LC3 ranged from 22.6% to 31.9% of their respective 28-day strengths. By day 3, the ULCC mix achieved 53.1% of its 28-day strength, compared to 55.6% for the 50–1:1 mix. By day 7, these two mixes further increased to 91.5% and 95.1% of their 28-day strengths, respectively, while the other mixes in this group showed 7-day to 28-day strength ratios between 72.1%

Table 1.7 Mix design of ULCC-LC3

Group	OPC (kg/m^3)	Calcined clay (kg/m^3)	Limestone (kg/m^3)	Gypsum (kg/m^3)	Cenosphere (kg/m^3)	Silica fume (kg/m^3)	Water (kg/m^3)	Superplasticizier (kg/m^3)	SRA (kg/m^3)	PE fiber (%)
ULCC	873.9	0	0	0	287.6	97.1	264.9	4.3	9.3	0
50–1:1	464.0	196.6	196.6	16.6	287.6	97.1	264.9	6.5	9.3	0
50–2:1	464.0	262.2	131.1	16.6	287.6	97.1	264.9	8.3	9.3	0
50–3:1	464.0	294.9	98.3	16.6	287.6	97.1	264.9	8.5	9.3	0
45–2:1	418.6	291.3	145.7	18.4	287.6	97.1	264.9	9.0	9.3	0
65–2:1	600.4	174.8	87.4	11.4	287.6	97.1	264.9	6.7	9.3	0
ULCC-1	873.9	0	0	0	287.6	97.1	264.9	4.6	9.3	1
50–1:1–1	464.0	196.6	196.6	16.6	287.6	97.1	264.9	6.9	9.3	1
50–2:1–1	464.0	262.2	131.1	16.6	287.6	97.1	264.9	8.7	9.3	1
50–3:1–1	464.0	294.9	98.3	16.6	287.6	97.1	264.9	9.1	9.3	1
45–2:1–1	418.6	291.3	145.7	18.4	287.6	97.1	264.9	9.6	9.3	1
65–2:1–1	600.4	174.8	87.4	11.4	287.6	97.1	264.9	7.0	9.3	1

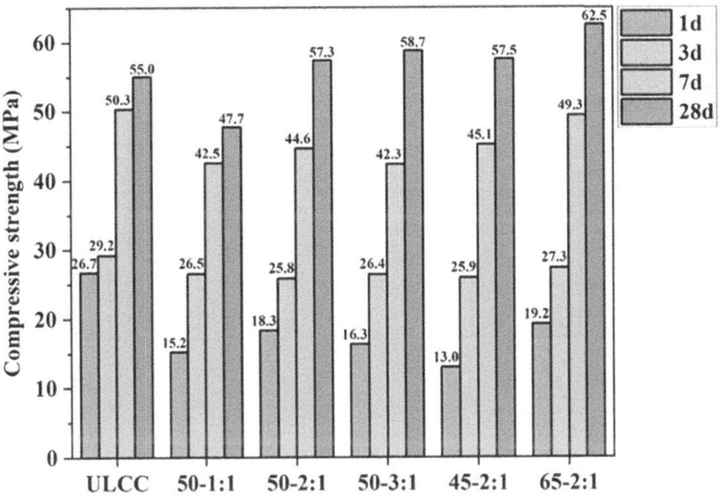

Fig. 1.19 Development of compressive strengths at different hydration ages without PE fibers. (Reprinted from Huang et al. (2023), copyright 2025, with permission from Elsevier)

and 78.9%. This suggests that for ULCC and those mixes with relatively low calcined clay content, the majority of the hydration reactions are largely completed by day 7. In contrast, mixes with higher calcined clay contents continue to exhibit notable strength development beyond day 7, which can be attributed to the ongoing pozzolanic reaction of the calcined clay. This reaction involves the interaction of reactive silica and alumina present in the calcined clay with portlandite, a calcium hydroxide byproduct of cement hydration. Since the pozzolanic reaction occurs at a slower pace compared to the primary cement hydration, it leads to a delayed but progressive gain in compressive strength for mixes with greater amounts of calcined clay.

Investigating the influence of LC3 on the compressive strength of ULCC reveals some noteworthy trends. As illustrated in Fig. 1.20, the most significant drop in early-age compressive strength due to the introduction of LC3 occurs solely on the first day. Beyond that, replacing a portion of OPC with limestone and calcined clay does not lead to any substantial reduction in strength. In fact, by the 28th day, most LC3-containing mixes exhibit higher compressive strength than the standard ULCC, with the exception of the 50–1:1 group. Figure 1.20 also compares the effects of calcined clay and limestone on both the 28-day compressive strength and elastic modulus, distinguishing between mixes with and without PE fibers. A comparison between Fig. 1.20a, b highlights that the addition of PE fibers leads to a notable increase in elastic modulus, rising from approximately 10 GPa to around 17 GPa. It is also evident that omitting PE fibers and substituting OPC with calcined clay and limestone results in a considerable reduction in elastic modulus, as seen in Fig. 1.20a. However, this negative impact on elastic modulus is much less pronounced in mixes that include PE fibers, as shown in Fig. 1.20b. These findings support the beneficial role of PE fibers in enhancing the elastic modulus of

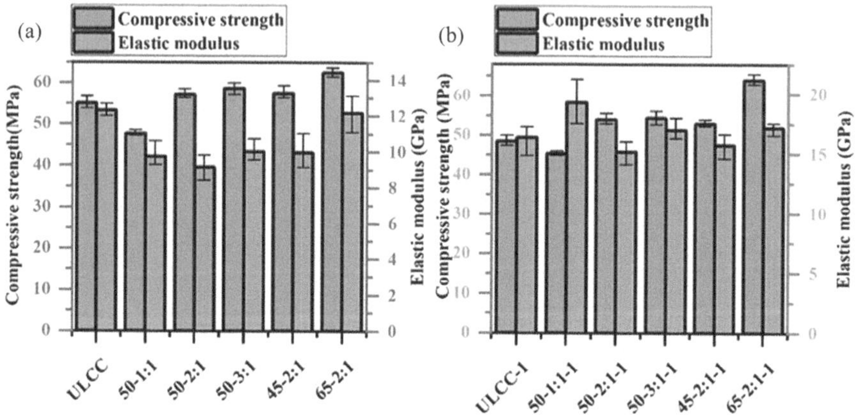

Fig. 1.20 28-day compressive strength and elastic modulus of ULCC-LC3 mixes (**a**) 28-day compressive strength and elastic modulus of ULCC-LC3 mixes without PE fibers (effect of calcined clay–limestone replacement); (**b**) 28-day compressive strength and elastic modulus of ULCC-LC3 mixes with PE fibers (showing the fiber effect on elastic modulus and strength). (Reprinted from Huang et al. (2023), copyright 2025, with permission from Elsevier)

ULCC-LC3 composites. Furthermore, the 28-day compressive strength of the PE fiber-reinforced mixes generally improves with the inclusion of LC3, except for the 50–1:1–1 group, aligning with the strength trends observed in the non-fiber-reinforced counterparts.

The tensile performance of the PE fiber-reinforced mixes, including parameters such as cracked stress, tensile strength, tensile strain, and strain energy, is illustrated in Fig. 1.21. As depicted in Fig. 1.21a, the incorporation of LC3 contributes to an increase in cracked stress, primarily due to its influence on refining the pore structure of the cementitious matrix. Since cracked strength is closely related to pore size, the presence of calcined clay, limestone, and the pozzolanic reaction product—C-A-S-H gel—helps to reduce pore size and enhance the overall structural integrity, thereby improving cracked stress. Figure 1.21b further demonstrates that replacing OPC with calcined clay and limestone leads to a notable rise in tensile strength. In addition, the use of calcined clay results in a slight reduction in tensile strain, decreasing from 5.3% to 4.5%. Moreover, the strain energy of the ULCC-1 mix shows an increase ranging from 12.9% to 34.1% when OPC is substituted with calcined clay and limestone. Overall, the inclusion of LC3 in ULCC mixtures is shown to have a positive impact on key mechanical properties, including compressive strength, elastic modulus, and various aspects of tensile behavior.

1.6.2 RCM Test

As shown in Table 1.8, the chloride ion migration coefficients measured on day 28 ranged from 0.5 to 1.5 × 10^{-12} m^2/s, representing a significant reduction of 86.1%

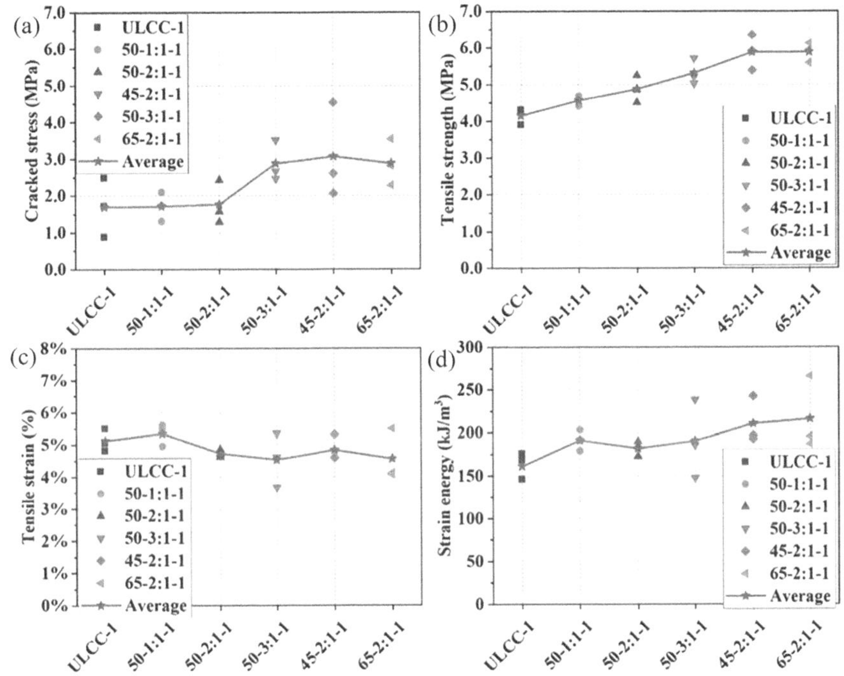

Fig. 1.21 Tensile properties of ULCC-LC3 mixes with PE fibers (**a**) cracked stress of different mix groups; (**b**) tensile strength of different mix groups; (**c**) tensile strain of different mix groups; (**d**) strain energy of different mix groups. (Reprinted from Huang et al. (2023), copyright 2025, with permission from Elsevier)

to 95.4% compared to the migration coefficient of 10.8×10^{-12} m^2/s observed in conventional concrete with comparable compressive strength [76]. Furthermore, when compared to the ULCC mix without LC3, the inclusion of LC3 led to an additional decrease in chloride migration coefficients ranging from 33.3% to 66.7%. Based on the data in Table 1.9, the ULCC mix was classified as having a medium chloride migration resistance, whereas the mixes incorporating LC3 were categorized as having low chloride migration, indicating improved resistance. These findings suggest that the addition of calcined clay and limestone effectively reduces chloride ion penetration. The porous structure of ULCC-LC3 is largely influenced by cenospheres, which are non-interconnected hollow particles. The pozzolanic interaction between cenospheres and silica fume contributes to the further densification of the cement matrix, thereby enhancing its resistance to chloride ingress. This observation aligns with findings from previous research [77]. The use of LC3 in ULCC offers dual advantages for improving chloride resistance: firstly, the refinement of the pore structure, as evidenced by the MIP results, and secondly, the pozzolanic activity of calcined clay, which produces C-A-S-H gel capable of binding chloride ions and further limiting their movement within the matrix [78].

Table 1.8 RCM test results

Specimen ID	No.	Temperature (°C)	Concrete thickness (mm)	Penetration depth (mm)	Testing period (h)	Voltage (V)	Chloride ion migration coefficient	Average migration coefficient
ULCC	1	24.40	51.3	9.6	24	60	2.2	1.5
	2	24.45	51.3	5.3	24	60	1.2	
	3	24.40	51.5	5.2	24	60	1.1	
50–1:1	1	25.35	50.0	4.4	24	60	0.9	0.8
	2	25.40	50.3	4.0	24	60	0.8	
	3	25.35	50.7	2.9	24	60	0.6	
50–2:1	1	24.60	51.8	2.6	24	60	0.5	0.6
	2	24.40	51.9	3.3	24	60	0.7	
	3	24.15	50.1	3.5	24	60	0.7	
50–3:1	1	24.50	51.7	2.1	24	60	0.4	0.5
	2	24.45	51.5	2.9	24	60	0.6	
	3	24.50	51.2	2.2	24	60	0.4	
45–2:1	1	25.05	50.1	6.1	24	60	1.3	1.0
	2	25.00	50.3	4.5	24	60	1.0	
	3	24.75	50.7	3.7	24	60	0.8	
65–2:1	1	25.10	51.8	2.6	24	60	0.5	0.6
	2	24.90	51.6	2.9	24	60	0.6	
	3	24.90	51.7	2.9	24	60	0.6	

Table 1.9 Evaluation of chloride ion migration coefficients

Chloride ion migration coefficient (10^{-12} m^2/s)	>10	10–5	5–1	1–0.5	0.5–0.1	0.1–0.5	<0.05
Evaluation	Very high	High	Medium	Low	Very low	Extremely low	Negligible

(a) Carbonation depths on day 7 (b) Carbonation depths on day 112

Fig. 1.22 Comparison between carbonation depths on day 7 and day 112 (**a**) Carbonation depths on day 7 (**b**) Carbonation depths on day 112

1.6.3 ACT

Figure 1.22 displays visual images of the split surfaces following the application of phenolphthalein solution on days 7 and 112. The carbonation depths measured from day 7 to day 112 for the mixes without PE fibers are illustrated in Fig. 1.23. The ULCC mix exhibited zero carbonation depth on both testing days, demonstrating outstanding resistance to carbonation. In contrast, the ULCC-LC3 mixes showed considerably greater carbonation depths on day 7, with further progression over time, indicating a notable decline in carbonation resistance compared to ULCC. This deterioration can be mainly attributed to two factors: the reduced proportion of ordinary Portland cement (OPC) and the consumption of calcium hydroxide (CH) during the pozzolanic reaction of calcined clay. Consequently, the ULCC-LC3 matrix contains significantly less CH, which plays a critical role in maintaining a high pH environment in the pore solution. Although MIP analysis revealed that the use of LC3 helps reduce the volume of larger pores (those coarser than gel pores), which typically contributes to a lower CO_2 diffusion coefficient and thus improved carbonation resistance, the overall carbonation performance of ULCC-LC3 remains inferior. This suggests that the loss of CH due to OPC replacement and its involvement in the pozzolanic reaction has a more dominant effect on the carbonation behavior of ULCC-LC3. To evaluate and compare the carbonation resistance of ULCC-LC3 with established literature data, Eqs. (1.7) and (1.8) were applied to

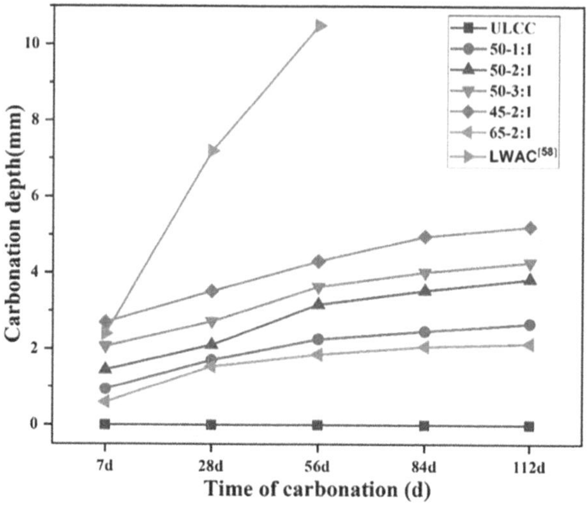

Fig. 1.23 Evolution of carbonation depths under different ages. (Reprinted from Huang et al. (2023), copyright 2025, with permission from Elsevier)

convert the measured accelerated carbonation depths into equivalent real-world carbonation coefficients [79–81]:

$$K_{acc} = D_c/\sqrt{t} \tag{1.7}$$

$$K_a = \frac{K_{acc}}{K_e}\sqrt{\frac{C_0}{C_{acc}}} \tag{1.8}$$

where, K_{acc} represents the accelerated carbonation coefficient, expressed in mm/y$^{0.5}$, D_c represents the depth of carbonation in mm, and t represents the exposure time in years. K_a denotes the actual carbonation coefficient under real environmental conditions, K_e denotes the environmental factor that considers the effect of moisture content on the carbonation coefficient. C_0 denotes the concentration of CO_2 in actual exposure, and C_{acc} represents the concentration of CO_2 in the accelerated carbonation test (20% in this study).

The actual carbonation coefficients were determined by calculating the accelerated carbonation coefficients from the measured carbonation depths in Fig. 1.23 using Eq. (1.7). These accelerated values were then averaged and input into Eq. (1.8) to estimate the real-world carbonation coefficients under moderate to high humidity conditions, corresponding to the XC3 exposure class. For Eq. (1.8), the environmental factor K_e and concentration C_0 were assigned values of 0.6 and 0.1%, respectively, based on findings from previous research [82–85]. The resulting actual carbonation coefficients are presented in Fig. 1.24. Notably, the ULCC mix exhibited an actual carbonation coefficient of 0 mm/y$^{0.5}$, which does not imply the absence of carbonation but rather reflects the extremely low CO_2 diffusion due to its highly dense microstructure, a result of the very low water-to-binder ratio of 0.27. Figure 1.24a illustrates how variations in the calcined clay to limestone ratio affect

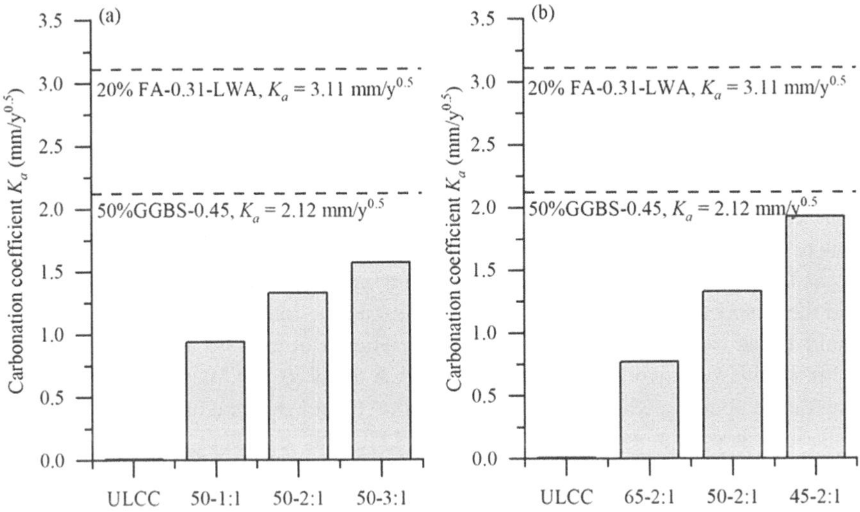

Fig. 1.24 Actual carbonation coefficients and comparison with existing data (**a**) Effect of calcined clay proportion; (**b**) effect of OPC content. (Reprinted from Huang et al. (2023), copyright 2025, with permission from Elsevier)

carbonation resistance, showing that as the proportion of calcined clay increases—while keeping the total content of calcined clay and limestone constant—the resistance to carbonation tends to decrease. Meanwhile, Fig. 1.24b highlights that reducing the OPC content in the mix leads to a lower carbonation resistance, as the cementitious matrix becomes increasingly vulnerable to CO_2 penetration with decreasing OPC levels.

For comparative purposes, two reference concrete mixes with compressive strengths similar to that of ULCC-LC3 are included in Fig. 1.24 as dashed lines. The first reference mix, with a carbonation coefficient of 2.12 mm/y$^{0.5}$, is a conventional concrete with a water-to-binder ratio of 0.45, where 50% of the ordinary Portland cement (OPC) was replaced by ground granulated blast furnace slag (GGBS) [84]. The second reference, having a carbonation coefficient of 3.11 mm/y$^{0.5}$, is a lightweight aggregate concrete (LWAC) with a water-to-binder ratio of 0.31 [86], in which 20% of the OPC was substituted with grade II fly ash and shale haydite served as the lightweight aggregate. Although the inclusion of LC3 in ULCC results in a higher carbonation coefficient compared to the ULCC mix alone, Fig. 1.24 demonstrates that ULCC-LC3 still exhibits superior carbonation resistance relative to both the normal and lightweight concrete mixes with comparable strength levels. On day 112, concrete samples were taken from the carbonation chamber and categorized based on phenolphthalein testing into carbonated and non-carbonated groups. For the ULCC mix, only non-carbonated samples were selected, as no significant carbonation was detected by that stage, as previously indicated in Fig. 1.23.

1.7 High Temperature Effect

Fire is one of the main risks to reinforced concrete structure. Compared with metal, concrete is a good refractory material in construction engineering, but fire could cause deterioration, spalling, and even explosive failure of concrete. The chemical ingredient and physical properties of concrete may significantly change when the concrete is exposed to high temperatures. For example, high temperature may cause concrete spalling and lower its compressive strength and elastic modulus.

As temperature increases in concrete, more water vapor pressure is accumulated and the evaporation of free water causes the excessive internal pore pressure, which could cause the concrete spalling [87, 88]. Huang et al. [32] found that ULCC cylindrical samples with a small content of PVA fibers (0.2–0.5% in volume) did not experience spalling when exposed up to 1000 °C and remained structurally intact. This indicated that the fiber reinforced ULCC has excellent high temperature resistance. Figure 1.25 shows PVA fibers melted at a temperature of exceeding

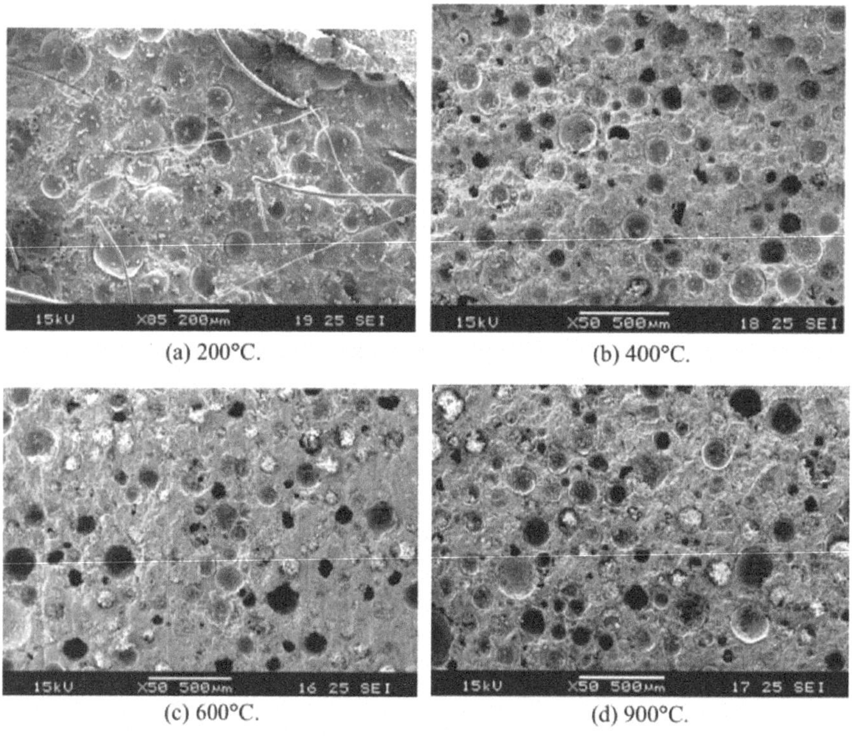

Fig. 1.25 Morphologies of ULCC-02PP subjected to varying heat treatment conditions [22] (**a**) 200 °C, (**b**) 400 °C, (**c**) 600 °C, (**d**) 900 °C. (Reprinted from Huang et al. (2018), copyright 2025, with permission from Elsevier)

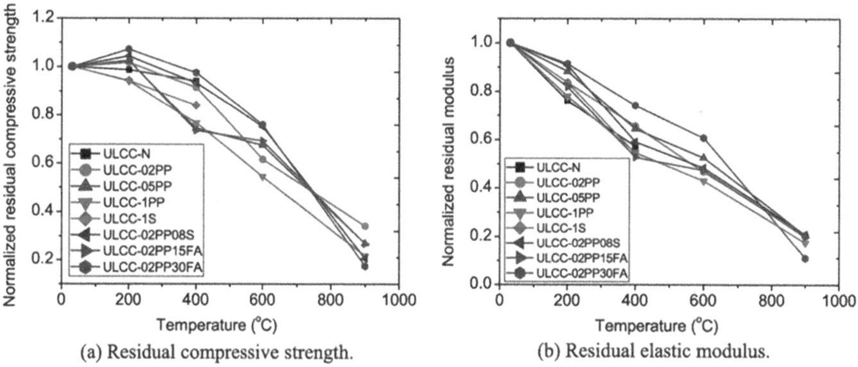

Fig. 1.26 Influence of elevated temperatures on the residual compressive strength and elastic modulus [22] (**a**) Residual compressive strength; (**b**) Residual elastic modulus. (Reprinted from Huang et al. (2018), copyright 2025, with permission from Elsevier)

400 °C, thereby leaving interconnected pores and small channels in the cement matrix, and releasing the pore pressure inside the ULCC. Therefore, the use of PVA fibers significantly inhibit the spalling behavior of ULCC after exposure to high temperature. PVA fiber reinforced ULCC shows promise as a potential material for enhancing fire resistance in structural applications.

Huang et al. [22] later conducted a series of tests on cylindrical and prismatic samples with different types and dosages of fibers, and different amounts of FA to investigate the mechanical properties and microstructure of ULCC exposed up to high temperature of 900 °C. Figure 1.26 compares the changes in elastic modulus and residual compressive strength of ULCC exposed to high temperatures. The test results found that the loss rate of elastic modulus is more pronounced than that of residual compressive strength. A small amount (0.2–0.5% by volume) of PP fiber incorporated in ULCC can effectively prevent the concrete spalling at temperature up to 900 °C, which shows that ULCC with PP fibers can be applied to building structures in which fire hazard is the most critical loading case. The steel fiber will be oxidized when the specimens is heated to 900 °C. The steel fibers failed to bridge the cracks, resulting in lower concrete strength. This suggests that the steel fiber reinforced lightweight concrete slabs without any rebar reinforcement may not meet fire protection requirements. Therefore, further investigation is needed to improve the mechanical performance of steel fibers reinforced lightweight concrete slabs under fire scenario. Moreover, the morphology of the cenospheres exposed to temperature of 900 °C did not change significantly (vid. Fig. 1.25). This phenomenon is similar to the results observed in the literature [89]. The small size pore near the steel fiber hinders the evaporation of water, thereby increasing the water vapor content in the matrix, which causes corrosion of the steel fiber under high temperature and high humidity.

1.8 Sustainability and Economic Performance of ULCC

Assessing the carbon emissions and energy consumption of cement-based materials is crucial, as the cement industry is responsible for approximately 7–9% of global CO_2 emissions. Huang et al. [34] demonstrated that incorporating LC3 into ULCC can effectively lower both carbon emissions and energy usage. Their analysis revealed that the production of cement was the main source of environmental impact in both conventional and modified mixes. By substituting a portion of cement with a 2:1 blend of calcined clay and limestone, they achieved a notable reduction of 41.9% in carbon emissions and 26.7% in energy consumption. While the use of a superplasticizer—particularly in the LC3 blend—led to a slight increase in energy consumption due to potential reductions in workability, the overall advantages of LC3 inclusion were deemed more significant. Despite the need for a higher dosage of superplasticizer, the environmental benefits of reduced emissions and energy use, coupled with maintained durability against chloride ingress and carbonation, highlight the potential of LC3 as a sustainable alternative in ULCC formulations.

The same research team further investigated the carbon emissions, energy consumption as well as cost of different concrete mixes (OPC with SCM replacement, ULCC with LC3 and foam concrete (FC)) at different stages in prefabricated modular construction [90]. Figure 1.27 shows that the overall carbon emissions, energy consumption, and economic performances by considering manufacturing, transportation, construction, installation, deconstruction, and operation stages of ULCC. The lightweight and superior thermal performances improve the performances of ULCC mix during transportation and operation stages, although ULCC may have higher carbon emissions, energy consumption, and costs during manufacturing. Figure 1.28 reveals the advantage of adopting ULCC in scenarios where a long service life is required, e.g., circular economy. In this case, the sustainability and economic performances of ULCC become more pronounced as compared with OPC concrete due to the prolonged operational stage during the whole life cycle of RC structures. It was concluded that ULCC owns the best overall performance considering mechanical, sustainability, economic, and durability index, with 25% improvement as compared with OPC. However, the low elastic modulus of ULCC compared to normal concrete stems from its lightweight, porous structure, which reduces stiffness and increases flexibility. To optimize structural performance, weight savings, and cost-effectiveness, it is highly advantageous to combine metal and Fiber-Reinforced Polymer (FRP) into a single hybrid composite component. This approach synergistically leverages the high strength and impact resistance of metals with the exceptional lightweight, corrosion-resistant, and customizable properties of FRPs, achieving performance superior to the sum of its parts. When combined with metals or FRPs, this necessitates careful design to manage deformations, stress transfer, and serviceability. Strategies like material hybridization, selective FRP use, confinement, and innovative composites can help mitigate these challenges. Understanding these principles allows engineers to leverage ULCC's advantages (e.g., reduced weight) while ensuring structural performance and durability.

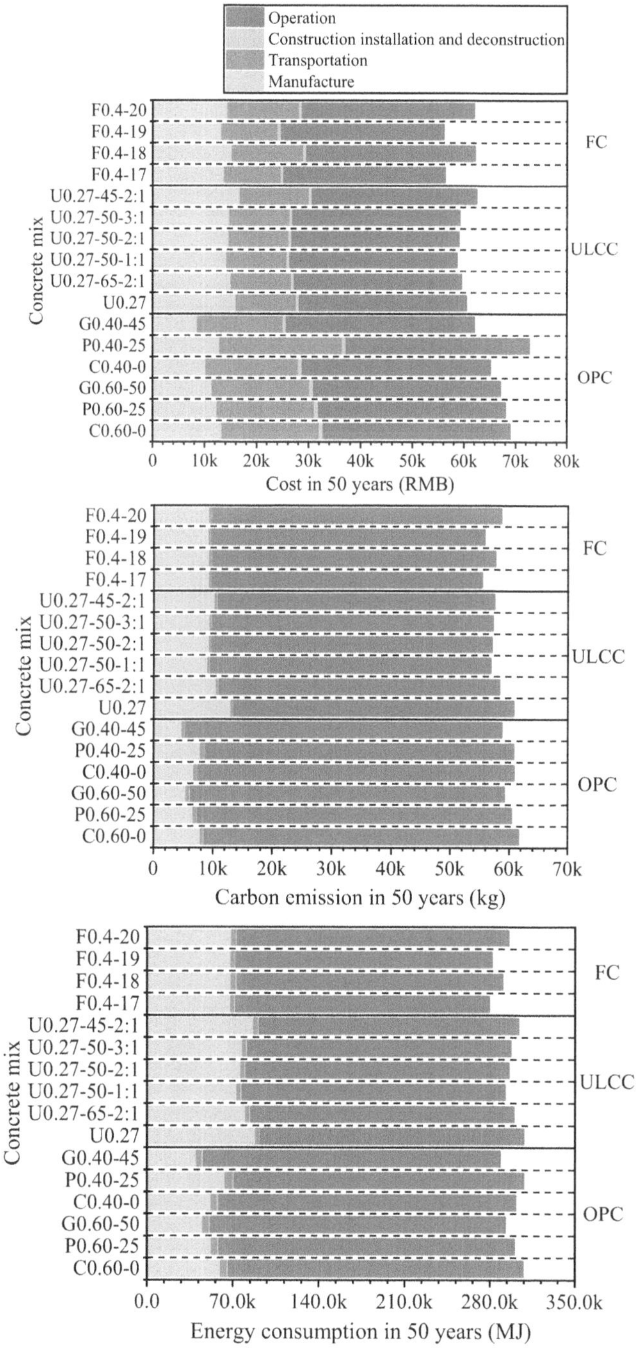

Fig. 1.27 Evaluation of the cost, carbon emissions, and energy usage involved in producing prefabricated modules with different concrete mix designs. Expected service life: 50 years

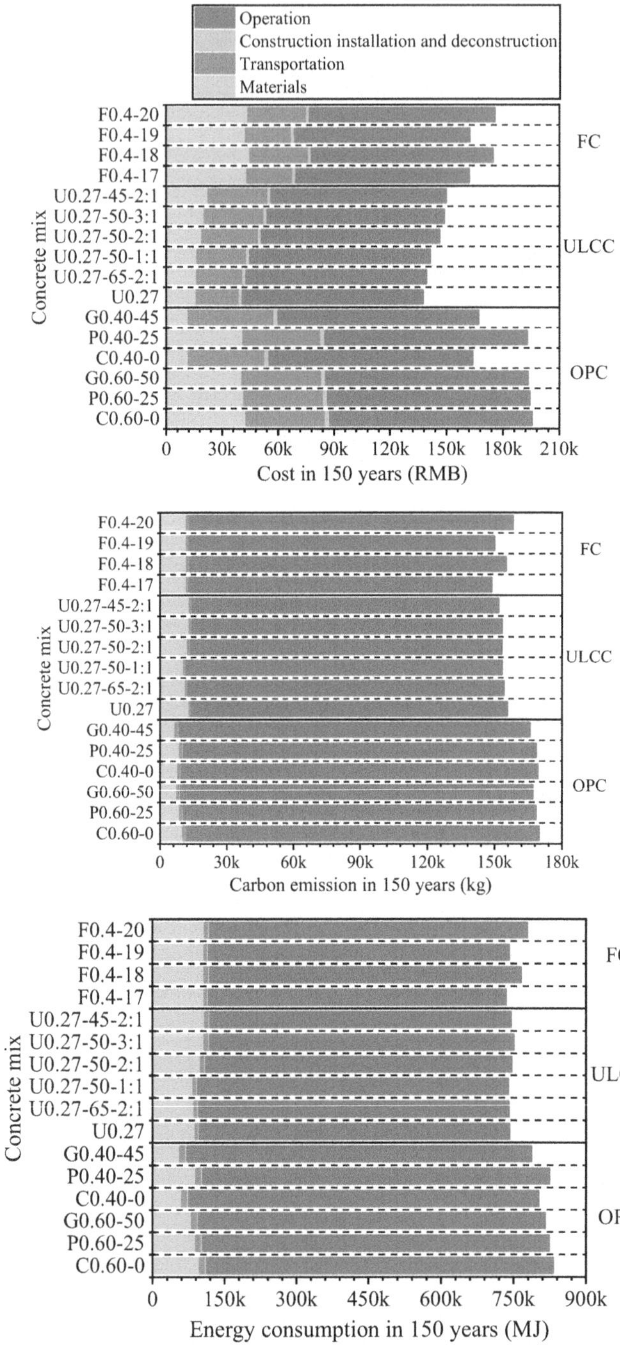

Fig. 1.28 Evaluation of the cost, carbon emissions, and energy usage involved in producing prefabricated modules with different concrete mix designs. Expected service life: 150 years

References

1. S. Chandra, B. Leif, Lightweight aggregate concrete, Elsevier, 2002.
2. ASTM, Guide for structural lightweight-aggregate concrete, American Society for Testing and Materials, West Conshohocken, 2014.
3. K.M.A. Sohel, J.Y.R. Liew, J.B. Yan, M.H. Zhang, K.S. Chia, Behavior of steel–concrete–steel sandwich structures with lightweight cement composite and novel shear connectors, Compos. Struct. 94(12) (2012) 3500–3509.
4. A. Lotfy, K.M.A. Hossain, M. Lachemi, Lightweight self-consolidating concrete with expanded shale aggregates: modelling and optimization, Int. J. Concr. Struct. Mater. 9(2) (2015) 185–206.
5. W.-J. Long, X.-W. Tan, B.-X. Xiao, N.-X. Han, F. Xing, Effective use of ground waste expanded perlite as green supplementary cementitious material in eco-friendly alkali activated slag composites, J. Clean. Prod. 213 (2019) 406–414.
6. D.S. Babu, K.G. Babu, T.H. Wee, Properties of lightweight expanded polystyrene aggregate concretes containing fly ash, Cem. Concr. Res. 35(6) (2005) 1218–1223.
7. A. Lotfy, K.M.A. Hossain, M. Lachemi, Mix design and properties of lightweight self-consolidating concretes developed with furnace slag, expanded clay and expanded shale aggregates, J. Sustain. Cem.Based Mater. 5(5) (2016) 297–323.
8. D. Kramar, V. Bindiganavile, Impact response of lightweight mortars containing expanded perlite, Cem. Concr. Compos. 37 (2013) 205–214.
9. R. Yu, D.V. van Onna, P. Spiesz, Q.L. Yu, H.J.H. Brouwers, Development of ultra-lightweight fibre reinforced concrete applying expanded waste glass, J. Clean. Prod. 112 (2016) 690–701.
10. B. Işıkdağ, Characterization of lightweight ferrocement panels containing expanded perlite-based mortar, Constr. Build. Mater. 81 (2015) 15–23.
11. S.P. McBride, A. Shukla, A. Bose, Processing and characterization of a lightweight concrete using cenospheres, J. Mater. Sci. 37(19) (2002) 4217–4225.
12. K.S. Chia, M.H. Zhang, J.Y.R. Liew, High-strength ultra-lightweight cement composite – material properties, in: Proc. 9th Int. Symp. High Performance Concrete: Design, Verification & Utilization, Rotorua, New Zealand, 2011.
13. Z. Huang, A. Palmer, P. Marshall, J.Y.R. Liew, Arctic platform construction using steel–concrete–steel sandwich composite, in: Proc. Int. Conf. Port and Ocean Engineering Under Arctic Conditions, 2015.
14. D. Montgomery, S. Diamond, The influence of fly ash cenospheres on the details of cracking in fly ash-bearing cement pastes, Cem. Concr. Res. 14(6) (1984) 767–775.
15. A.K.H. Kwan, J.J. Chen, Adding fly ash microsphere to improve packing density, flowability and strength of cement paste, Powder Technol. 234 (2013) 19–25.
16. Y. Wu, J. Wang, P. Monteiro, M. Zhang, Development of ultra-lightweight cement composites with low thermal conductivity and high specific strength for energy efficient buildings, Constr. Build. Mater. 87 (2015) 100–112.
17. E.V. Fomenko, N.N. Anshits, M.V. Pankova, L.A. Solovyov, A.G. Anshits, Fly ash cenospheres: Composition, morphology, structure, and helium permeability, in: World of Coal Ash Conf., 2011.
18. L. Ngu, H. Wu, D. Zhang, Characterization of ash cenospheres in fly ash from Australian power stations, Energy & Fuels 21(6) (2007) 3437–3445.
19. A. Sarkar, R. Rano, K. K. Mishra, A. Mazumder, Characterization of cenospheres collected from ash-pond of a super thermal power plant, Energy Sources Part A 30(3) (2007) 271–283.
20. J. Wang, M. Zhang, W. Li, K.S. Chia, J.Y.R. Liew, Stability of cenospheres in lightweight cement composites in terms of alkali–silica reaction, Cem. Concr. Res. 42(5) (2012) 721–727.
21. F. Blanco, P. García, P. Mateos, J. Ayala, Characteristics and properties of lightweight concrete manufactured with cenospheres, Cem. Concr. Res. 30(11) (2000) 1715–1722.
22. Z. Huang, P. Krishnan, L. Sui, J.Y.R. Liew, Mechanical properties and microstructure of ultra-lightweight cement composites with fly ash cenospheres after exposure to high temperatures, Constr. Build. Mater. 164 (2018) 760–774.

23. V. Rheinheimer, Y. Wu, T. Wu, K. Celik, J. Wang, L. De Lorenzis, P. Wriggers, M.-H. Zhang, P.J.M. Monteiro, Multi-scale study of high-strength low-thermal-conductivity cement composites containing cenospheres, Cem. Concr. Compos. 80 (2017) 91–103.
24. M. Lanzón, V. Cnudde, T. De Kock, J. Dewanckele, X-ray microtomography (μ-CT) to evaluate microstructure of mortars containing low density additions, Cem. Concr. Compos. 34(9) (2012) 993–1000.
25. N. Barbare, A. Shukla, A. Bose, Uptake and loss of water in a cenosphere–concrete composite material, Cem. Concr. Res. 33(10) (2003) 1681–1686.
26. N. Zanjad, S. Pawar, C. Nayak, Review on Application of Lightweight Cenosphere in Construction Material, Macromol. Symp. 413 (2024) 2300006.
27. J. Yan, J. Wang, J.Y.R. Liew, X. Qian, Applications of ultra-lightweight cement composite in flat slabs and double skin composite structures, Constr. Build. Mater. 111 (2016) 774–793.
28. Y. Zhou, Y. Zheng, L. Sui, F. Xing, J. Hu, P. Li, Behavior and modeling of FRP-confined ultra-lightweight cement composites under monotonic axial compression, Compos. B Eng. 162 (2019) 289–302.
29. Z. Huang, J.Y.R. Liew, Compressive resistance of steel–concrete–steel sandwich composite walls with J-hook connectors, J. Constr. Steel Res. 124 (2016) 142–162.
30. J. Wang, X. Gao, J. Yan, Developments and mechanical behaviors of steel fiber reinforced ultra-lightweight cement composite with different densities, Constr. Build. Mater. 171 (2018) 643–653.
31. Z. Huang, L. Sui, F. Wang, S. Du, Y. Zhou, J. Ye, Dynamic compressive behavior of a novel ultra-lightweight cement composite incorporated with rubber powder, Compos. Struct. 244 (2020) 112300.
32. J. Wang, Y. Yang, J.Y.R. Liew, M. Zhang, Method to determine mixture proportions of workable ultra-lightweight cement composites to achieve target unit weights, Cem. Concr. Compos. 53 (2014) 178–186.
33. Z. Huang, J.Y.R. Liew, Experimental and analytical studies of curved steel–concrete–steel sandwich panels under patch loads, Mater. Des. 93 (2016) 104–117.
34. Z. Huang, T. Liang, L. Chen, Experimental studies on durability performances of ultra-lightweight low-carbon LC3 cement composites against chloride ingression and carbonation, Constr. Build. Mater. 395 (2023) 132340.
35. K.-S. Chia, X. Liu, J.Y.R. Liew, M.-H. Zhang, Experimental study on creep and shrinkage of high-performance ultra-lightweight cement composite of 60 MPa, Struct. Eng. Mech. 50(5) (2014) 635–652.
36. K.M.A. Sohel, K. Al-Jabri, M.-H. Zhang, J.Y.R. Liew, Flexural fatigue behavior of ultra-lightweight cement composite and high strength lightweight aggregate concrete, Constr. Build. Mater. 173 (2018) 90–100.
37. J.Y. Wang, K.S. Chia, J.Y.R. Liew, M.H. Zhang, Flexural performance of fiber-reinforced ultra-lightweight cement composites with low fiber content, Cem. Concr. Compos. 43 (2013) 39–47.
38. A. Hanif, S. Diao, Z. Lu, T. Fan, Z. Li, Green lightweight cementitious composite incorporating aerogels and fly ash cenospheres – Mechanical and thermal insulating properties, Constr. Build. Mater. 116 (2016) 422–430.
39. Z. Huang, J. Wang, J.Y.R. Liew, P. Marshall, Lightweight steel–concrete–steel sandwich composite shell subject to punching shear, Ocean Eng. 102 (2015) 146–161.
40. X. Liu, M.-H. Zhang, K.S. Chia, J. Yan, J.Y.R. Liew, Mechanical properties of ultra-lightweight cement composite at low temperatures of 0 to −60 °C, Cem. Concr. Compos. 73 (2016) 289–298.
41. J. Yan, J. Wang, J.Y.R. Liew, X. Qian, W. Zhang, Reinforced ultra-lightweight cement composite flat slabs: Experiments and analysis, Mater. Des. 95 (2016) 148–158.
42. Z. Huang, J.Y.R. Liew, Steel–concrete–steel sandwich composite structures subjected to extreme loads, Int. J. Steel Struct. 16(4) (2016) 1009–1028.

43. J. Yan, X. Liu, J.Y.R. Liew, X. Qian, M. Zhang, Steel–concrete–steel sandwich system in arctic offshore structure: Materials, experiments, and design, Mater. Des. 91 (2016) 111–121.
44. Z. Huang, J.Y.R. Liew, M. Xiong, J. Wang, Structural behaviour of double skin composite system using ultra-lightweight cement composite, Constr. Build. Mater. 86 (2015) 51–63.
45. J. Yan, J.Y.R. Liew, X. Qian, J. Wang, Ultimate strength behavior of curved steel–concrete–steel sandwich composite beams, J. Constr. Steel Res. 115 (2015) 316–328.
46. J.-B. Yan, J.-Y. Wang, J.Y.R. Liew, X. Qian, L. Zong, Ultimate strength behaviour of steel–concrete–steel sandwich plate under concentrated loads, Ocean Eng. 118 (2016) 41–57.
47. Q. Liu, W. Liu, Z. Li, S. Guo, G. Sun, Ultra-lightweight cement composites with excellent flexural strength, thermal insulation and water resistance achieved by establishing interpenetrating network, Constr. Build. Mater. 250 (2020) 118923.
48. B. Xi, Y. Zhou, K. Yu, B. Hu, X. Huang, L. Sui, F. Xing, Use of nano-SiO_2 to develop a high-performance green lightweight engineered cementitious composites containing fly ash cenospheres, J. Clean. Prod. 262 (2020) 121274.
49. X. Liu, K.-S. Chia, M.-H. Zhang, J.Y.R. Liew, Water and chloride ion penetration resistance of high-strength ultra-lightweight cement composite, in: Proc. ICDC 2012 Int. Congr. on Durability of Concrete, Norwegian Concrete Association, 2012.
50. Z. Huang, J.Y.R. Liew, W. Li, Evaluation of compressive behavior of ultra-lightweight cement composite after elevated temperature exposure, Constr. Build. Mater. 148 (2017) 579–589.
51. A.L. Brooks, H. Zhou, D. Hanna, Comparative study of the mechanical and thermal properties of lightweight cementitious composites, Constr. Build. Mater. 159 (2018) 316–328.
52. P. Shafigh, M.A. Nomeli, U.J. Alengaram, H.B. Mahmud, M.Z. Jumaat, Engineering properties of lightweight aggregate concrete containing limestone powder and high volume fly ash, J. Clean. Prod. 135 (2016) 148–157.
53. X. Liu, K.S. Chia, M.-H. Zhang, Water absorption, permeability, and resistance to chloride-ion penetration of lightweight aggregate concrete, Constr. Build. Mater. 25(1) (2011) 335–343.
54. J.N. Farghani, P. Shafigh, B. Alsalhi, S. Shahnazari, H.B. Mahmud, Engineering properties of lightweight aggregate concrete containing binary and ternary blended cement, J. Clean. Prod. 149 (2017) 976–988.
55. K.-S. Youm, J. Moon, J.-Y. Cho, J.J. Kim, Experimental study on strength and durability of lightweight aggregate concrete containing silica fume, Constr. Build. Mater. 114 (2016) 517–527.
56. O. Kayali, M. Haque, B. Zhu, Some characteristics of high strength fiber reinforced lightweight aggregate concrete, Cem. Concr. Compos. 25(2) (2003) 207–213.
57. P. Shafigh, M.Z. Jumaat, H. Mahmud, Oil palm shell as a lightweight aggregate for production of high strength lightweight concrete, Constr. Build. Mater. 25(4) (2011) 1848–1853.
58. J. Xie, J. Liu, F. Liu, J. Wang, P. Huang, Investigation of a new lightweight green concrete containing sludge ceramsite and recycled fine aggregates, J. Clean. Prod. 235 (2019) 1240–1254.
59. D. Bouvard, J. Chaix, R. Dendievel, A. Fazekas, J. Létang, G. Peix, D. Quenard, Characterization and simulation of microstructure and properties of EPS lightweight concrete, Cem. Concr. Res. 37(12) (2007) 1666–1673.
60. B. Chen, J. Liu, Properties of lightweight expanded polystyrene concrete reinforced with steel fiber, Cem. Concr. Res. 34(7) (2004) 1259–1263.
61. A. Sadrmomtazi, J. Sobhani, M.A. Mirgozar, M. Najimi, Properties of multi-strength grade EPS concrete containing silica fume and rice husk ash, Constr. Build. Mater. 35 (2012) 211–219.
62. M.R. Ahmad, B. Chen, Experimental research on the performance of lightweight concrete containing foam and expanded clay aggregate, Compos. B Eng. 171 (2019) 46–60.
63. W. She, Y. Chen, Y. Zhang, M. Jones, Characterization and simulation of microstructure and thermal properties of foamed concrete, Constr. Build. Mater. 47 (2013) 1278–1291.
64. B. Chen, Z. Wu, N. Liu, Experimental research on properties of high-strength foamed concrete, J. Mater. Civ. Eng. 24(1) (2012) 113–118.

65. E. Eltayeb, X. Ma, Y. Zhuge, O. Youssf, J. Mills, Influence of rubber particles on the properties of foam concrete, J. Build. Eng. 30 (2020) 101217.
66. ACI Committee 318, Building code requirements for structural concrete and commentary, part 3, ACI Manual of Concrete Practice, Farmington Hills, 2010.
67. J. Clarke, Structural lightweight aggregate concrete, CRC Press, 1993.
68. J.-B. Yan, J.Y.R. Liew, M.-H. Zhang, Z.-X. Li, Punching shear resistance of steel–concrete–steel sandwich composite shell structure, Eng. Struct. 117 (2016) 470–485.
69. T. Liang, L. Chen, Z. Huang, Y. Zhong, Y. Zhang, Ultra-lightweight low-carbon LC3 cement composites: Uniaxial mechanical behaviour and constitutive models, Constr. Build. Mater. 404 (2023) 133173.
70. X. Liang, Q. Shi, Design Principles of concrete structures, 2nd ed., China Architecture and Building Press, Beijing, 2011 (in Chinese).
71. V.C. Li, Engineered cementitious composites (ECC) – material, structural, and durability performance, in: E.G. Nawy (Ed.), Concrete construction engineering handbook, 2nd ed., CRC Press, 2008, Chapter 24.
72. Z. Huang, F. Wang, Y. Zhou, L. Sui, P. Krishnan, J.Y.R. Liew, A novel, multifunctional, floatable, lightweight cement composite: Development and properties, Constr. Build. Mater. 197 (2019) 75–87.
73. S. Xu, H.W. Reinhardt, Determination of double-k criterion for crack-propagation in quasi-brittle fracture part I: Experimental investigation of crack propagation, Int. J. Fract. 98(1–2) (1999) 151–177.
74. X. Huang, R. Ranade, Q. Zhang, W. Ni, V.C. Li, Mechanical and thermal properties of green lightweight engineered cementitious composites, Constr. Build. Mater. 48 (2013) 954–960.
75. T. Kanda, V.C. Li, Practical design criteria for saturated pseudo strain hardening behavior in ECC, J. Adv. Concr. Technol. 4(1) (2006) 59–72.
76. L. Fu, Application prospect of novel low-carbon cement LC3, Chinese Cement 9 (2020) 98–102 (in Chinese).
77. X. Liu, K.S. Chia, M.-H. Zhang, Development of lightweight concrete with high resistance to water and chloride-ion penetration, Cem. Concr. Compos. 32(10) (2010) 757–766.
78. M.D.A. Thomas, R.D. Hooton, A. Scott, H. Zibara, The effect of supplementary cementitious materials on chloride binding in hardened cement paste, Cem. Concr. Res. 42(1) (2012) 1–7.
79. M. Guo, G. Gong, Y. Yue, F. Xing, Y. Zhou, B. Hu, Performance evaluation of recycled aggregate concrete incorporating limestone calcined clay cement (LC3), J. Clean. Prod. 366 (2022) 132820.
80. L. Chen, R.K.L. Su, Experimental and numerical investigations on macrocell corrosion of partially carbonated reinforced concrete with supplementary cementitious materials, Cem. Concr. Compos. 135 (2023) 104827.
81. R. Neves, F. Branco, J. de Brito, Field assessment of the relationship between natural and accelerated concrete carbonation resistance, Cem. Concr. Compos. 41 (2013) 9–15.
82. A. Leemann, F. Moro, Carbonation of concrete: The role of CO_2 concentration, relative humidity and CO_2 buffer capacity, Mater. Struct. 50 (2017) 30.
83. R.K. Dhir, P.C. Hewlett, Y.N. Chan, Near-surface characteristics of concrete: Prediction of carbonation resistance, Mag. Concr. Res. 41(148) (1989) 137–143.
84. L. Chen, Corrosion rate of macrocell corrosion in reinforced concrete and service life modeling of carbonated concrete containing supplementary cementitious materials, University of Hong Kong, Hong Kong, 2022, Ph.D. thesis.
85. S. Shriram, K. Ramamurthy, S. Ramakrishnan, Effect of occupant-induced indoor CO_2 concentration and bioeffluents on human physiology using a spirometric test, Build. Environ. 149 (2019) 58–67.
86. Y. Gao, L. Cheng, Z. Gao, S. Guo, Effects of different mineral admixtures on carbonation resistance of lightweight aggregate concrete, Constr. Build. Mater. 43 (2013) 506–510.
87. J.-W. Ko, D. Ryu, T. Noguchi, The spalling mechanism of high-strength concrete under fire, Mag. Concr. Res. 63(5) (2011) 357–370.

88. F. Ali, D. O'Connor, A. Abu-Tair, Explosive spalling of high-strength concrete columns in fire, Mag. Concr. Res. 53(3) (2001) 197–204.
89. Q. Li, X. Gao, S. Xu, Microstructure and mechanical properties of high-toughness fiber-reinforced cementitious composites after exposure to elevated temperatures, J. Mater. Civ. Eng. 28(11) (2016) 04016132.
90. L. Chen, Z. Huang, W. Pan, R.K.L. Su, Y. Zhong, Y. Zhang, Low carbon concrete for prefabricated modular construction in circular economy: An integrated approach towards sustainability, durability, cost, and mechanical performances, J. Build. Eng. (2024) 109368.

Chapter 2
Dynamic Behavior of Fiber Reinforced Ultralightweight High Ductility Cement Composite

2.1 Introduction

Marine engineering infrastructures, including offshore platforms, bridges, tunnels, ports, and wind foundations—face severe safety and durability challenges due to dynamic loads. Seismic events impose high inertial forces and deformations, leading to brittle failure in traditional concrete. Ship impacts cause localized or global damage, while repeated wave action and storm loads accelerate fatigue, microcracking, and corrosion. Conventional reinforced concrete performs poorly under such conditions due to low ductility, inadequate crack control, and limited damage tolerance, resulting in costly repairs. Thus, high ductility concrete is urgently needed to enhance resilience in these critical applications.

Current research consistently demonstrates the advantages of rubberized concrete for impact-resistant structures, including high-rise buildings, long-span bridges, and offshore platforms. These massive structures typically require substantial reinforcement, resulting in excessive weight. Lightweight high-strength concrete presents an optimal solution, offering reduced density, enhanced thermal insulation, and improved durability. However, conventional lightweight concrete suffers from inherent limitations of low strength and brittleness, restricting its widespread structural application. This chapter addresses the growing demand for innovative lightweight cementitious materials by developing an ultra-lightweight composite (1450 kg/m^3 density) with exceptional ductility (3–4% direct tensile strain). The material incorporates rubber powder and polyethylene (PE) fibers, combining waste tire recycling benefits with superior mechanical performance. Through comprehensive static and split Hopkinson pressure bar (SHPB) impact testing, the research evaluates key properties including compressive strength, elastic modulus, and strain-rate-dependent failure modes. A modified dynamic increase factor (DIF) prediction model is proposed and validated to characterize the material's rate sensitivity. Advanced finite element analysis using LS-DYNA software successfully reproduces

Z. Huang, Y. Zhou, *Innovative Lightweight High-performance Steel-concrete Sandwich Composite for Marine Structures*,
https://doi.org/10.1007/978-981-95-4681-7_2

the failure mechanisms of ultra-high-performance ductile cementitious composites (ULHDCC). The simulation employs a sophisticated multi-parameter sampling approach (Latin Hypercube method) to establish a representative design domain model. Building on this foundation, the study develops a high-fidelity explicit constitutive model through genetic programming (GP) algorithm, leveraging their iterative optimization capabilities to accurately capture the material's complex behavior.

The research fills critical knowledge gaps regarding the dynamic response of rubberized lightweight cement composites, while simultaneously advancing sustainable construction through waste tire utilization. The developed material system demonstrates potential to overcome traditional trade-offs between weight reduction and mechanical performance in structural applications.

2.2 Dynamic Compressive Behavior

2.2.1 Materials and Test Program

The rubberized ultra-lightweight cement composite (RULCC) was developed using cement, fly ash cenospheres (FAC), silica fume, rubber powder (380 μm), and hydrophilic-coated PE fibers, targeting 35 MPa compressive strength at 1450 kg/m^3 density. The FAC fine aggregate (870 kg/m^3 specific gravity, 0.902 g/cm^3 fineness modulus) was partially replaced with surface-treated rubber powder (5–20% by volume), exhibiting <1% water absorption. The binder system comprised CEM I 52.5R cement (100%), silica fume (11 wt%), and FAC (38.7–48.4 wt %). Seven mix groups were formulated, varying rubber content (0–20%) and fiber dosage (0–0.7 vol%), as shown in Table 2.1. Polycarboxylate-based superplasticizer ensured workability. This optimized design balances lightweight properties with mechanical performance while incorporating recycled rubber, addressing both structural and sustainability requirements.

Table 2.1 Mix proportions of RULCC (by weight)

Mix ID	Cement	SF	FAC	Rubber	Water	SP	SRA	Fiber
R0-0	1	0.11	0.484	0.000	0.37	0.001	0.001	0
R5-0	1	0.11	0.460	0.027	0.37	0.001	0.001	0
R10-0	1	0.11	0.436	0.054	0.37	0.001	0.001	0
R15-0	1	0.11	0.412	0.080	0.37	0.001	0.001	0
R20-0	1	0.11	0.387	0.107	0.37	0.001	0.001	0
R0-0.7PE	1	0.11	0.484	0.000	0.37	0.001	0.001	0.7%
R10-0.7PE	1	0.11	0.436	0.054	0.37	0.001	0.001	0.7%

Note: *SF* silica fume, *FAC* fly ash cenospheres, *SP* superplasticizer, *SRA* shrinkage reducing agent. R10-0.7PE represents the RULCC with 10% rubber powder replacement of FAC and 0.7% PE fiber by volume

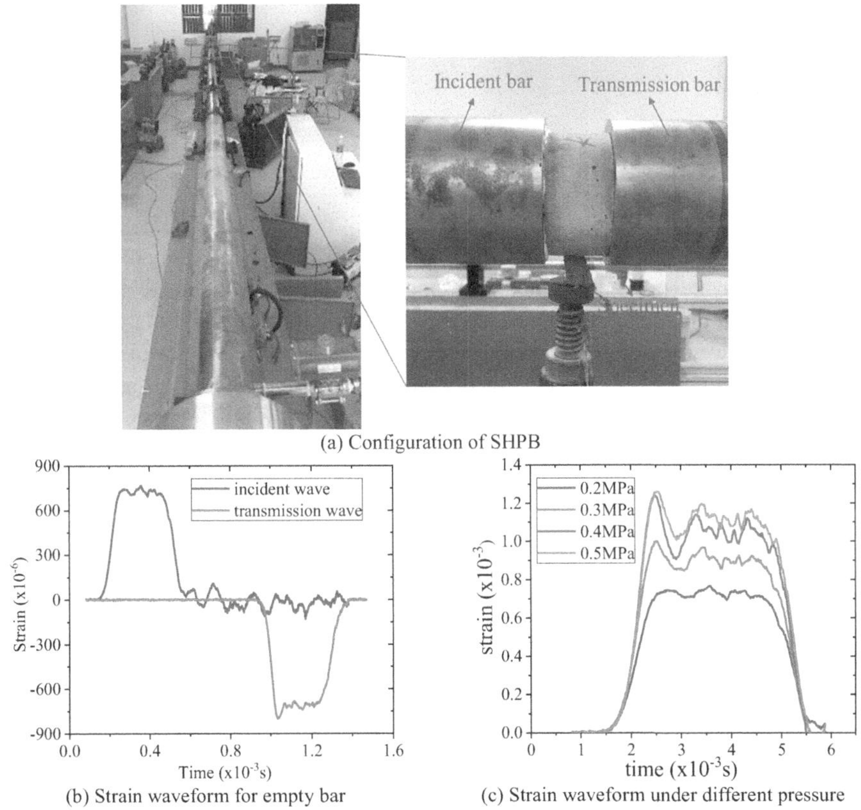

Fig. 2.1 SHPB test and impact waveform input. (**a**) Configuration of SHPB. (**b**) Strain waveform for empty bar. (**c**) Strain waveform under different pressure. (Reprinted from Huang et al. [14], copyright 2025, with permission from Elsevier)

Static compression tests were conducted on Φ100 × 200 mm cylinders using an MTS machine following ASTM C39 standards [1], while uniaxial tensile tests adhered to JSCE recommendations [2]. Dynamic properties were evaluated through SHPB tests [3], with five specimens tested per strain rate condition to obtain averaged stress-strain responses. The SHPB setup (refer to Fig. 2.1a) employs striker, incident, and transmission bars with a damper, operating under two fundamental assumptions: (1) one-dimensional stress wave propagation, and (2) uniform stress distribution across the specimen. These conditions enable calculation of material properties from bar deformations using standard wave equations refer to Eqs. (2.1)–(2.3), with Fig. 2.1b showing typical input waveforms.

$$\sigma = \frac{A_0}{A_s} E_0 \varepsilon_t(t) \tag{2.1}$$

$$\varepsilon = -2\frac{C_0}{L_s}\int_0^t \varepsilon_r(t)\mathrm{d}t \tag{2.2}$$

$$\dot{\varepsilon} = -2\frac{C_0}{L_s}\varepsilon_r(t) \tag{2.3}$$

where A_0 and A_s denote the cross-sectional areas of the bar and specimen, E_0 and C_0 represent the bar's elastic modulus and wave velocity, L_s is specimen length, and $\varepsilon_t(t)$ and $\varepsilon_r(t)$ are transmitted and reflected waves. Key parameters are listed in Table 2.1, and Fig. 2.1c shows a baseline strain wave from an empty test. Specimen end surfaces were precision-ground to minimize friction effects. Seven mix groups were tested under four strain rates (90–190/s), controlled via air pressures (0.2–0.5 MPa) to simulate impact/blast scenarios (e.g., vehicular, marine, or military structural impacts).

2.2.2 *Static Compressive Test Behavior*

The compressive strength of RULCC exhibits a progressive decline with increasing rubber content, showing reductions of 29.5%, 47.7%, 54.8%, and 60.3% at 5%, 10%, 15%, and 20% rubber replacement ratios, respectively (vid. Fig. 2.2). This strength reduction stems from poor interfacial bonding between the organic rubber particles and inorganic cement matrix, creating weak zones that promote microcrack initiation. Elastic modulus similarly decreases by 15.7–33.6% across the tested rubber contents, reflecting the combined effects of rubber's low stiffness and the absence of coarse aggregates. Microstructural analysis (see Fig. 2.3) reveals uniform FAC distribution without segregation [4], though rubber-cement interfaces demonstrate clear debonding failures. The composites' fracture behavior transitions from brittle failure (audible cracking in rubber-free specimens) to more ductile failure with larger fragment sizes in rubber-modified mixes. PE fiber incorporation further reduces compressive strength by 8.8–16.7% due to introduced air voids during

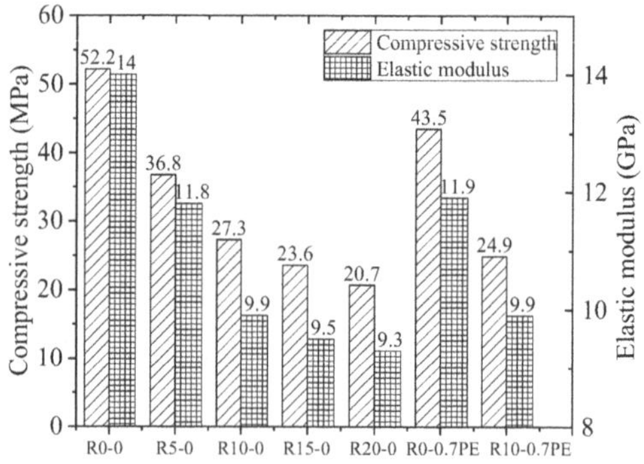

Fig. 2.2 Compressive strength and elastic modulus of different mixture. (Reprinted from Huang et al. [14], copyright 2025, with permission from Elsevier)

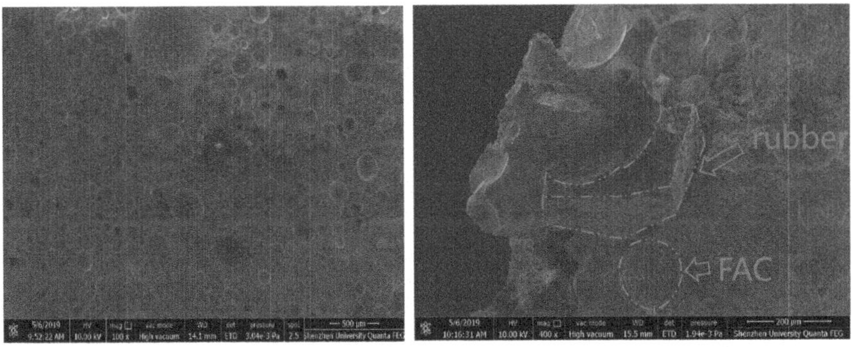

(a) Morphology of the composites

(b) Morphology of rubber powder and FAC in composites

Fig. 2.3 SEM morphology. (**a**) Morphology of the composites. (**b**) Morphology of rubber powder and FAC in composites. (Reprinted from Huang et al. [14], copyright 2025, with permission from Elsevier)

mixing. The RULCC shows greater strength reduction than conventional rubberized concrete [5] due to: (1) higher effective replacement ratios when substituting FAC rather than sand, and (2) smaller particle sizes (≤380 μm) that create less dense microstructures compared to standard aggregates (0–10 mm). These findings highlight the need for future research optimizing rubber particle size distribution to improve mechanical performance.

2.2.3 *Impact Test Behavior-Dynamic Compressive*

2.2.3.1 Failure Modes

Figure 2.4 illustrates the failure patterns across different mix groups, revealing two key trends: (1) increasing strain rates consistently produced more severe damage, and (2) higher rubber content resulted in larger but fewer fragments, particularly under high strain-rate loading. This contrasts with static loading conditions where failure typically initiates from void defects, propagating through a limited number of dominant cracks. The transition from single-crack to multiple-crack failure under impact loading stems from the material's inability to rapidly dissipate energy through slow crack propagation. Rubber modification significantly alters this failure mechanism through its inherent elasticity and energy dissipation capacity. The viscoelastic properties of rubber particles enable more efficient kinetic energy absorption, reducing crack multiplicity and fragmentation—an effect that becomes increasingly pronounced at higher rubber contents.

Fiber reinforcement demonstrates strain-rate-dependent effectiveness. At low strain rates, 0.7% PE fibers effectively restrain matrix cracking without fragmentation (refer to Fig. 2.4f, g). However, under extreme loading conditions (146.8–185.1/s),

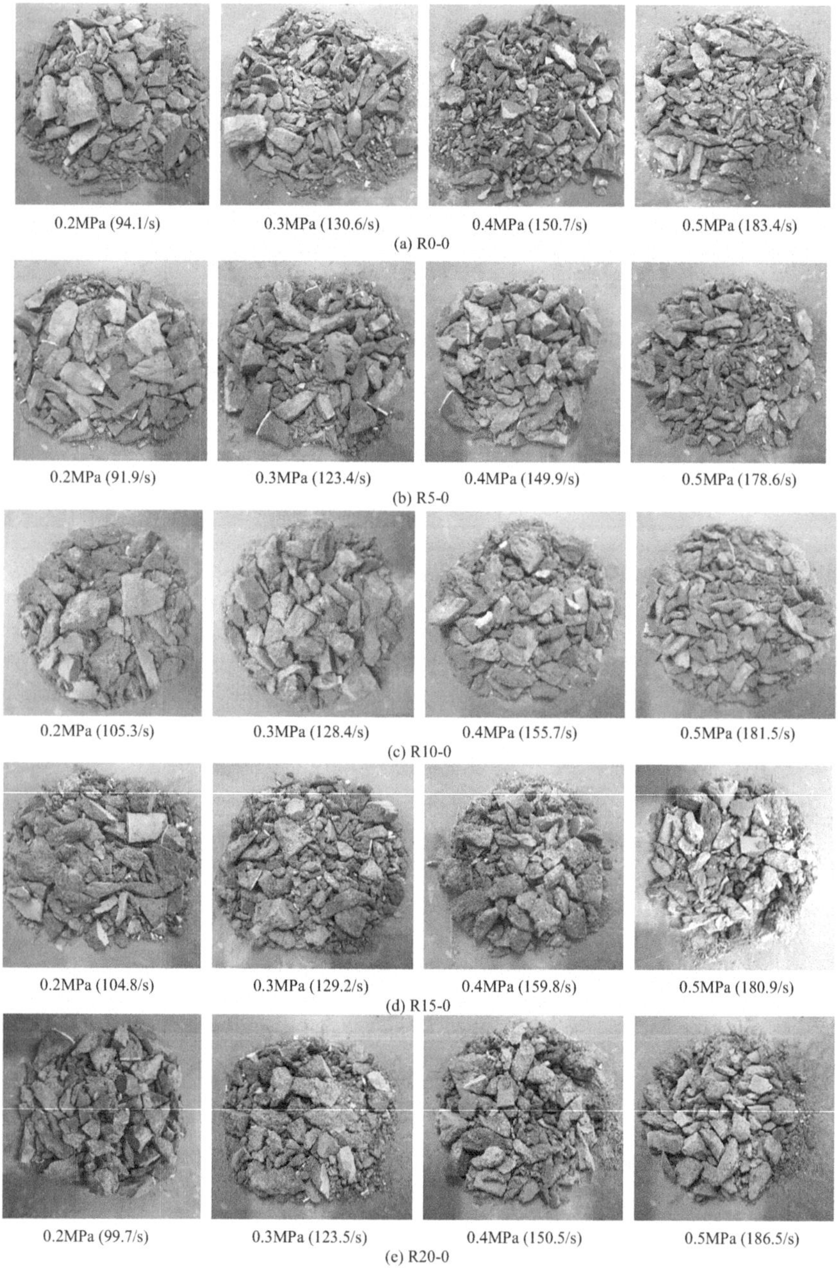

Fig. 2.4 Failure modes of different mixture under different strain rate. (**a**) R0-0. (**b**) R5-0. (**c**) R10-0. (**d**) R15-0. (**e**) R20-0. (**f**) R0-0.7PE. (**g**) R10-0.7PE. (Reprinted from Huang et al. [14], copyright 2025, with permission from Elsevier)

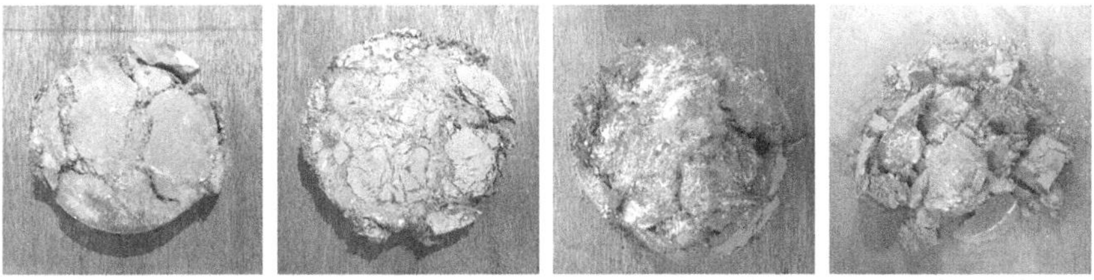

0.2MPa (102.1/s) 0.3MPa (128.3/s) 0.4MPa (146.8/s) 0.5MPa (166.4/s)

(f) R0-0.7PE

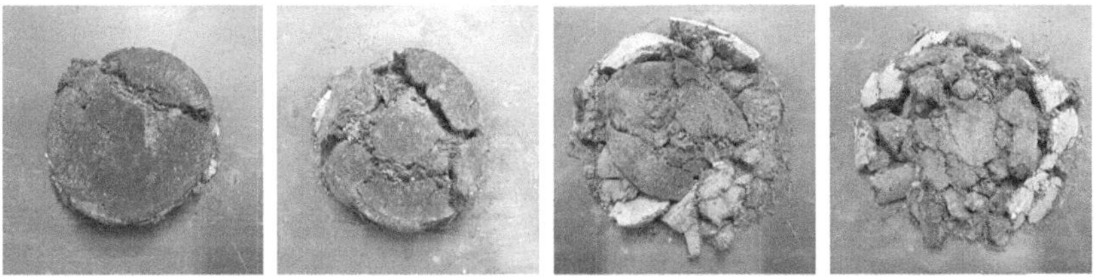

0.2MPa (93.1/s) 0.3MPa (138.2/s) 0.4MPa (158.7/s) 0.5MPa (185.1/s)

(g) R10-0.7PE

Fig. 2.4 (continued)

both rubberized (R10-0.7PE) and non-rubberized (R0-0.7PE) fiber-reinforced specimens exhibited comparable damage levels, losing structural integrity through combined cracking and fragmentation. Interestingly, the R5-0 mix (5% rubber without fibers) matched the energy absorption capacity of fiber-reinforced specimens (R0-0.7PE), as evidenced by stress-strain curve analysis. These findings suggest that optimal impact resistance may be achieved through balanced rubber and fiber incorporation, though further investigation is needed to quantify their synergistic effects.

2.2.3.2 Dynamic Compressive Stress-Strain Curves

Figure 2.5 presents the dynamic stress-strain responses of various mix designs under different strain rates. The incorporation of PE fibers notably smooths the stress-strain curves (Fig. 2.5a, f), particularly in the post-peak phase, where the increased area under the curves indicates enhanced ductility and energy absorption capacity. This behavior mirrors the effect of steel fibers in concrete under impact loading, as reported by Nili et al. [6]. Table 2.2 presents the dynamic peak strain and the enclosed area of the dynamic stress-strain curves. The enclosed area is defined as the integration of the normalized peak stress $\left(\frac{\sigma_d}{\sigma_s}\right)$ with regard to the strain. Key observations from Table 2.2 reveal that: (1) a 5% rubber content without fibers achieves comparable energy absorption to 0.7% PE fiber reinforcement without rubber; (2) rubber modification increases peak stress by 71.6–92.5% while significantly boosting peak strain (19–246% increase); (3) the optimal rubber content for energy dissipation appears between 5–10%, beyond which performance declines;

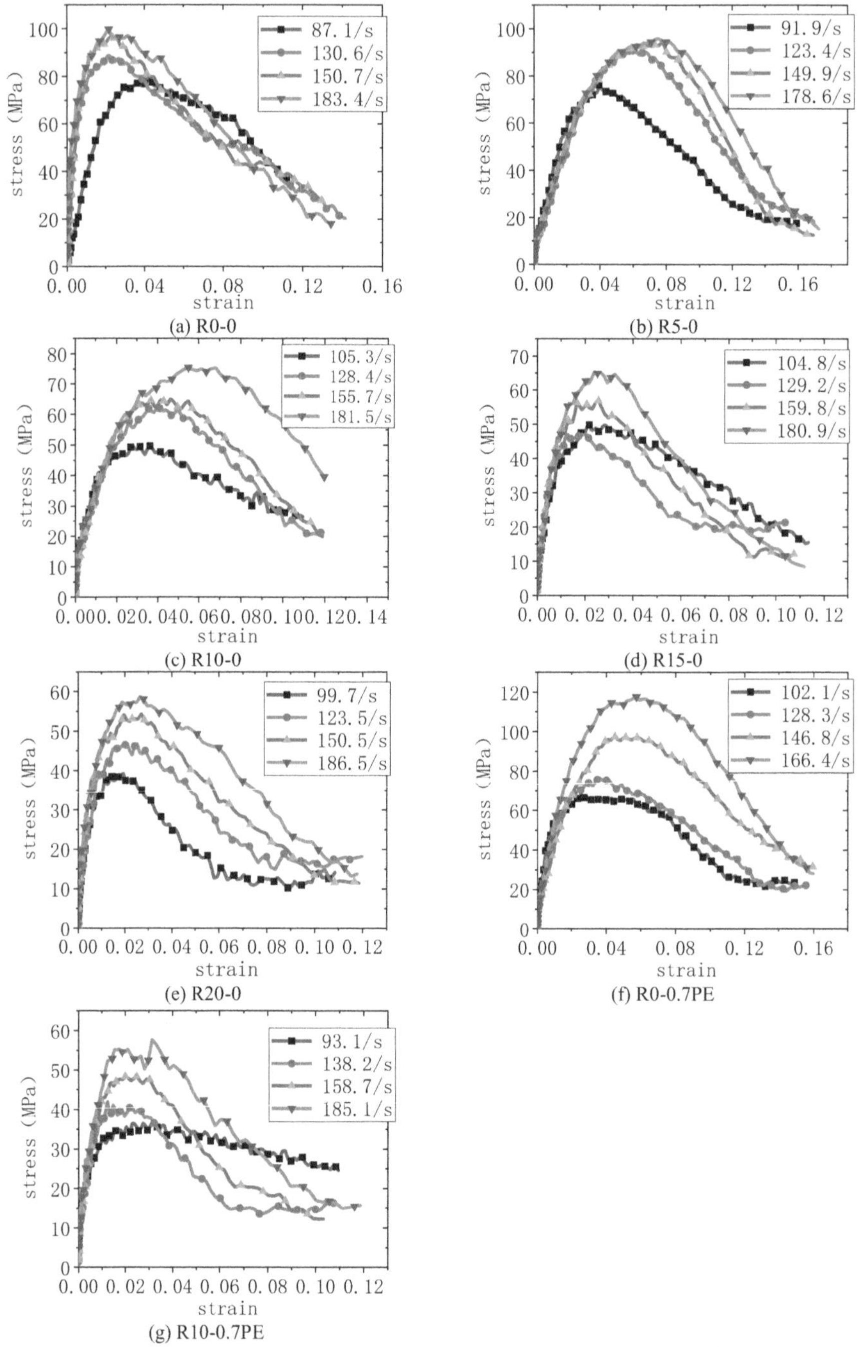

Fig. 2.5 Dynamic compressive stress-strain curves. (**a**) R0-0. (**b**) R5-0. (**c**) R10-0. (**d**) R15-0. (**e**) R20-0. (**f**) R0-0.7PE. (**g**) R10-0.7PE. (Reprinted from Huang et al. [14], copyright 2025, with permission from Elsevier)

Table 2.2 SHPB test results

Mix ID	Static compressive strength (MPa)	Impact velocity (m/s)	Strain rate (s)	Dynamic peak stress (MPa)	Dynamic peak strain	Enclosed area of stress-strain curve
R0-0	52.2	8.3	94.1	79.7 (52.7%[a])	4.298% (8.6[a])	1.29
		10.3	130.6	88.7 (69.9%)	2.112% (3.7)	1.41
		12.3	150.7	96.4 (84.7%)	2.267% (4.0)	1.47
		13.0	183.4	100.8 (93.1%)	6.320% (13.0)	2.09
R0-0.7PE	43.5	8.7	102.1	68.0 (56.3%)	2.527% (4.6)	1.58
		10.9	128.3	76.3 (75.4%)	3.601% (7.0)	1.74
		12.5	146.8	97.3 (123.7%)	4.510% (9.0)	2.44
		14.0	166.4	117.5 (170.1%)	5.734% (11.7)	2.97
R5-0	36.8	8.4	91.9	76.3 (107.3%)	3.948% (7.8)	1.93
		10.5	123.4	91.2 (147.8%)	6.104% (12.6)	2.51
		12.2	149.9	93.7 (154.6%)	7.838% (16.4)	2.60
		13.6	178.6	95.8 (160.3%)	7.527% (15.7)	2.91
R10-0	27.3	8.2	105.3	49.6 (81.7%)	3.633% (7.1)	1.52
		10.7	128.4	62.1 (127.5%)	3.964% (7.8)	1.84
		12.0	155.7	65.0 (138.1%)	4.631% (9.3)	2.01
		13.7	181.5	74.6 (173.3%)	5.506% (11.2)	3.05
R10-0.7PE	24.9	8.7	93.1	37.4 (50.2%)	3.236% (6.2)	1.35
		10.7	138.2	41.9 (68.3%)	1.242% (1.8)	1.03
		12.4	158.7	48.3 (94.0%)	2.002% (3.4)	1.23
		13.8	185.1	57.6 (131.3%)	3.140% (6.0)	1.67
R15-0	23.6	8.5	104.8	49.4 (109.3%)	2.644% (4.9)	1.70
		10.6	129.2	50.3 (113.1%)	4.567% (9.1)	1.76
		12.3	159.8	56.8 (140.7%)	2.561% (4.7)	1.35
		13.6	180.9	65.3 (176.7%)	2.391% (4.3)	1.70
R20-0	20.7	8.0	99.7	40.8 (97.1%)	1.938% (3.3)	1.08
		10.4	123.5	46.8 (126.1%)	2.107% (3.7)	1.60
		12.1	150.5	53.7 (159.4%)	2.696% (5.0)	1.83
		13.6	186.5	57.4 (177.3%)	2.650% (4.9)	2.18

[a] The value in the bracket represents the increasement ratio compared to the static value

(4) PE fibers improve energy absorption by 22.3–66.0% across tested strain rates; and (5) combined rubber and fiber use shows reduced effectiveness, likely due to compromised fiber-matrix bonding at higher rubber contents. These findings demonstrate that moderate rubber incorporation (5–10%) effectively balances strength and energy dissipation, while PE fibers primarily enhance post-peak behavior. The results suggest potential for optimizing hybrid reinforcement systems in impact-resistant applications.

2.2.3.3 Effect of Strain Rate

Figure 2.6 illustrates the relationship between strain rate and the DIF for various concrete types, including normal concrete, conventional rubberized concrete, ULCC, and RULCC. The results demonstrate that RULCC exhibits greater strain rate sensitivity in dynamic compressive strength compared to other materials. Similar to previous studies [7–9], the strength of concrete increases with higher strain rates. In this study, different loading pressures (0.2–0.5 MPa) corresponding to strain rates of 91.9–186.5/s were applied. As shown in Fig. 2.6, the DIF increases approximately linearly with strain rate. This behavior stems from distinct failure mechanisms under static and dynamic loading. Static failure typically initiates from localized microcracks that propagate into dominant fractures, whereas dynamic loading induces simultaneous microcrack formation across the material. Since crack development consumes significant energy, higher strain rates—which allow less time for energy dissipation—result in increased stress resistance. Figure 2.7a

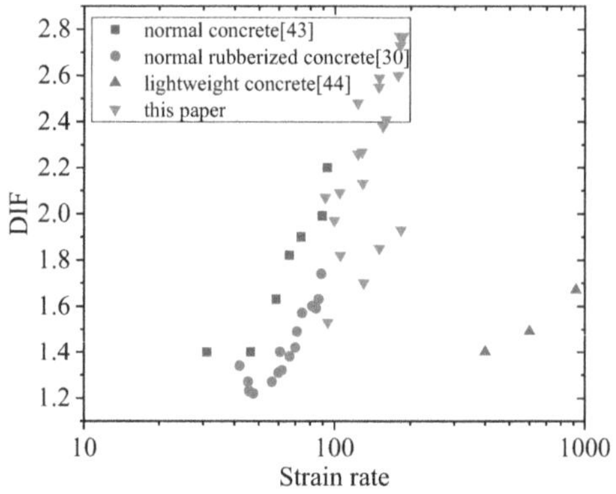

Fig. 2.6 DIF for different types of concrete. (Reprinted from Huang et al. [14], copyright 2025, with permission from Elsevier)

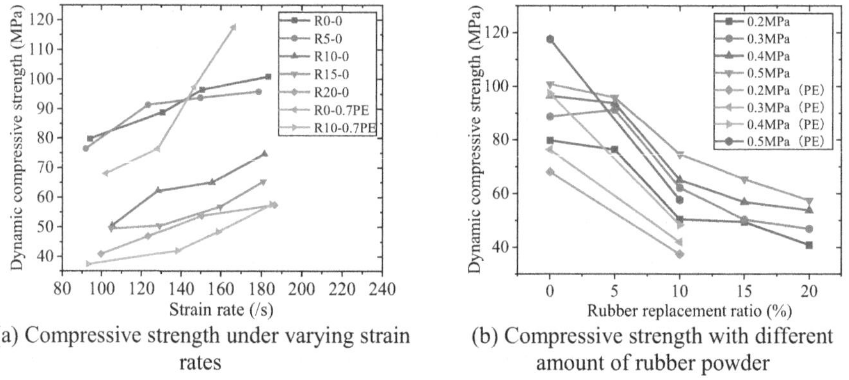

Fig. 2.7 Effect of strain rate and rubber replacement ratio. (**a**) Compressive strength under varying strain rates. (**b**) Compressive strength with different amount of rubber powder. (Reprinted from Huang et al. [14], copyright 2025, with permission from Elsevier)

highlights the strain rate effect on RULCC's dynamic compressive strength. While all mixes exhibit near-linear strength increases with strain rate, the PE fiber-reinforced mix (R0-0.7PE) shows a pronounced strength gain beyond 146.8/s. This suggests that PE fibers enhance fracture resistance through crack bridging and strain hardening, absorbing more impact energy. In contrast, rubber-modified composites lack this reinforcing mechanism, emphasizing the role of fiber reinforcement in high-rate loading scenarios. These findings underscore the importance of material composition in dynamic performance, with fiber reinforcement proving, particularly effective at elevated strain rates.

2.2.3.4 Effect of Rubber Content

As shown in Fig. 2.7b, the dynamic compressive strength of RULCC exhibits a clear decreasing trend with higher rubber content, mirroring behavior observed under static loading conditions. This reduction stems from rubber's inherent material limitations—its hydrophobic nature, weak interfacial bonding with cementitious matrices, and low elastic modulus collectively compromise structural integrity. The strength degradation occurs consistently across all tested formulations, including both fiber-reinforced and non-reinforced specimens. The parallel performance trends under static and dynamic loading confirm that rubber's detrimental effects on mechanical properties persist regardless of loading conditions. These findings align with established understanding of rubber-cement composites, where increased rubber content systematically diminishes load-bearing capacity while potentially improving other material characteristics such as energy absorption and ductility. The consistent response across different strain rates suggests that rubber's influence on strength remains dominant even when considering strain-rate effects, highlighting the need for careful material design when incorporating rubber in structural applications.

2.2.4 Analytical Modeling

2.2.4.1 Modified Equation for DIF

The dynamic increase factor (DIF), representing the ratio of dynamic to static compressive strength, serves as a key metric for evaluating strain-rate effects. Various predictive models exist for different concrete types: CEB-FIP [10] for conventional concrete, Chen [7] for rubberized concrete, and Ngo et al. [11] for high-strength concrete. These formulations, summarized in Table 2.3, provide standardized approaches for quantifying strength enhancement under dynamic loading conditions.

Table 2.4 compares experimental DIF results with predictions from three established models: CEB-FIP [10], Chen [7], and Ngo et al. [11]. The analysis

Table 2.3 Existing DIF models in the literature[a]

References	Concrete type	DIF formulae
CEB-FIP [10]	Normal concrete	$\mathrm{DIF}=\frac{\sigma_d}{\sigma_s}=\begin{cases}\left(\frac{\dot{\varepsilon}_c}{\dot{\varepsilon}_s}\right)^{0.014} & \dot{\varepsilon}_d\le 30\mathrm{s}^{-1}\\ 0.012\left(\frac{\dot{\varepsilon}_c}{\dot{\varepsilon}_s}\right)^{\frac{1}{3}} & \dot{\varepsilon}_d> 30\mathrm{s}^{-1}\end{cases}$
Chen [7]	Conventional rubberized concrete	$\mathrm{DIF}=\frac{\sigma_d}{\sigma_s}=0.9079(\dot{\varepsilon}_d)^{0.3158}\gamma \quad \dot{\varepsilon}_d>30s^{-1}$ $\log\gamma=6.156\alpha-0.492$ $\alpha=(5+3(101.6-1.85R)\,\%\,f_{cu}/4)^{-1}$
Ngo et al. [11]	High strength concrete	$\mathrm{DIF}=\frac{\sigma_d}{\sigma_s}=\left(\frac{\dot{\varepsilon}_d}{\dot{\varepsilon}_s}\right)^{1.026\alpha} \quad \dot{\varepsilon}_c\le\dot{\varepsilon}_1$ $\mathrm{DIF}=\frac{\sigma_d}{\sigma_s}=A\times\ln\dot{\varepsilon}_d-B \quad \dot{\varepsilon}_c>\dot{\varepsilon}_1$ $A=-0.0044\sigma_s+0.9866,$ $B=-0.0128\sigma_s+2.1396$ $\alpha=1/(20+\sigma_s)/2,$ $\dot{\varepsilon}_1=0.0022\sigma_s^2-0.1989\sigma_s+46.137$

[a] σ_d and σ_s denote dynamic and static compressive strength, respectively; $\dot{\varepsilon}_d$ and $\dot{\varepsilon}_s$ are the respective dynamic and static strain rates; f_{cu} is the compressive strength of concrete in MPa; R is the rubber content by volume fraction

reveals significant discrepancies between these models and the test data for rubber-modified ULCC: (1) The CEB-FIP model overestimates DIF for non-rubberized ULCC, while Ngo et al.'s high-strength concrete model shows better agreement due to similar brittle behavior; and (2) Chen's rubberized concrete model produces inconsistent predictions—underestimating DIF at low rubber content (5%), but overestimating at higher contents (10–20%). These limitations prompted the development of modified DIF equations specifically tailored for ULCC and RULCC, building upon the CEB-FIP equation structure while incorporating material-specific parameters.

The new DIF formula of ULCC follows the structure of the CEB-FIP's equation,

$$\mathrm{DIF}=\frac{\sigma_d}{\sigma_s}=0.0106\left(\frac{\dot{\varepsilon}_c}{\dot{\varepsilon}_s}\right)^{\frac{1}{3}} \quad \dot{\varepsilon}_d>30s^{-1} \tag{2.4}$$

The proposed dynamic increase factor (DIF) model for rubberized ULCC adopts the framework of Ngo et al.'s [11] high-strength concrete formulation, recognizing similar mechanical behavior between these materials. Unlike the CEB-FIP model, this approach incorporates the rubber replacement ratio (R) as a key parameter through functional coefficients. This is different from the CEB-FIP equation that can not consider the rubber content. While Chen's model also considers rubber content and matrix strength, its dimensional inconsistencies and computational complexity limit its practical application. Building on these considerations and experimental results, we propose a refined DIF equation that maintains dimensional consistency while effectively capturing the influence of rubber content on strain-rate

Table 2.4 Verification of DIFs using existing models

Mix ID	σ_s (MPa)	σ_d (MPa)	DIF_{test}	DIF_{CEB}	DIF_{test}/DIF_{CEB}	DIF [7]	DIF_{test}/DIF [7]	DIF [11]	DIF_{test}/DIF [11]	DIF_{pre}	DIF_{test}/DIF_{pre}
R0-0	52.2	79.7	1.53	1.76	0.87	1.69	0.91	1.64	0.93	1.55	0.99
		88.7	1.70	1.96	0.87	1.87	0.91	1.81	0.94	1.73	0.98
		96.4	1.85	2.06	0.90	1.96	0.94	1.89	0.98	1.82	1.02
		100.8	1.93	2.19	0.88	2.08	0.93	2.00	0.97	1.94	0.99
R5-0	36.8	76.3	2.07	1.74	1.19	1.87	1.11	1.63	1.27	2.11	0.98
		91.2	2.48	1.92	1.29	2.06	1.20	1.78	1.39	2.35	1.06
		93.7	2.55	2.05	1.24	2.19	1.16	1.88	1.36	2.51	1.02
		95.8	2.60	2.17	1.20	2.31	1.13	1.98	1.31	2.65	0.98
R10-0	27.3	49.6	1.82	1.82	1.00	2.21	0.82	1.70	1.07	2.15	0.85
		62.1	2.27	1.95	1.16	2.35	0.97	1.80	1.26	2.34	0.97
		65.0	2.38	2.08	1.14	2.50	0.95	1.90	1.25	2.54	0.94
		74.6	2.73	2.19	1.25	2.62	1.04	1.99	1.37	2.69	1.01
R15-0	23.6	49.4	2.09	1.82	1.15	2.36	0.89	1.69	1.24	2.07	1.01
		50.3	2.13	1.95	1.09	2.52	0.85	1.80	1.18	2.32	0.92
		56.8	2.41	2.10	1.15	2.69	0.90	1.92	1.26	2.56	0.94
		65.3	2.77	2.18	1.27	2.80	0.99	1.99	1.39	2.71	1.02
R20-0	20.7	40.8	1.97	1.79	1.10	2.48	0.79	1.67	1.18	1.93	1.02
		46.8	2.26	1.92	1.18	2.65	0.85	1.78	1.27	2.22	1.02
		53.7	2.59	2.05	1.26	2.82	0.92	1.89	1.37	2.48	1.04
		57.4	2.77	2.21	1.25	3.02	0.92	2.01	1.38	2.77	1.00
Mean					1.12		0.96		1.22		0.99
Standard deviation					0.14		0.11		0.16		0.05

sensitivity. The modified formulation addresses the limitations of existing models through, which is proposed below.

$$\mathrm{DIF} = \frac{\sigma_\mathrm{d}}{\sigma_\mathrm{s}} = \beta \times \lg\left(\frac{\dot{\varepsilon}_\mathrm{c}}{\dot{\varepsilon}_\mathrm{s}}\right) - \gamma \qquad \dot{\varepsilon}_\mathrm{d} > 30s^{-1} \tag{2.5}$$

where, β and γ are defined as a linear function of rubber replacement ratio R, i.e., $\beta = 8.121R + 1.458$, $\gamma = 54.587R + 7.250$, where the constants are obtained from the test results through regression. The DIF of RULCC calculated from Eq. (2.5) is larger than that of normal concrete, which indicates that RULCC exhibits greater strain-rate sensitivity than normal concrete. This enhanced sensitivity stems from ULCC's characteristic microstructure—the material's higher void content and lower elastic modulus promote superior deformation capacity and energy absorption characteristics under dynamic loading.

2.2.4.2 Verification of the Proposed Model

Table 2.4 compares the accuracy of various DIF prediction models against experimental results. The analysis reveals that while the CEB-FIP and Ngo et al.'s models [7] tend to underestimate DIF values for ULCC and RULCC (with mean ratios of 1.12 and 1.22 respectively), Chen's model [7] shows reasonable accuracy for non-rubberized ULCC (mean ratio = 0.96). The newly proposed model demonstrates superior predictive capability, achieving a mean ratio of 0.99 with the smallest standard deviation (0.05). As illustrated in Fig. 2.8, the test data consistently falls within the bounds predicted by the new model, which uniquely incorporates both strain rate and rubber replacement ratio as parameters. This contrasts with the CEB-FIP model's generic exponential function approach. Additional validation tests across the full range of rubber contents (0–20%) further confirm the model's reliability, suggesting its potential for future dynamic design applications of RULCC.

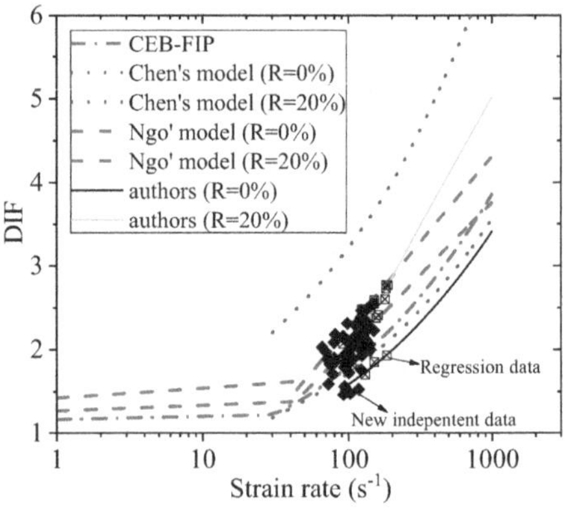

Fig. 2.8 Proposed DIF model vs. existing DIF models. (Reprinted from Huang et al. [14], copyright 2025, with permission from Elsevier)

2.3 Dynamic Splitting Behavior

2.3.1 *Materials and Test Program*

Seven mix designs of ultra-lightweight concrete incorporating rubber powder and PE fibers were formulated, systematically varying rubber content (0–20% by volume of fine aggregates) and fiber reinforcement (0% and 0.7% by volume), as shown in Table 2.5. Existing research suggests that rubberized concrete exhibits enhanced strain rate sensitivity when rubber content remains below 30% [9]. Beyond this threshold, excessive air entrapment occurs, increasing porosity and potentially causing microcrack formation due to incompatible deformation between rubber particles and the cementitious matrix [12]. Previous investigations have demonstrated significant improvements in damping capacity, with a 90% increase observed at 25% rubber replacement [13]. Building on these findings, the current study limits rubber content to 20% while incorporating 0.7% PE fibers to optimize both dynamic response and ductility [14].

The test program employs complementary static and dynamic splitting testing approaches to evaluate the mechanical properties of RULCC. Static compression and splitting tests provide baseline measurements of compressive strength, splitting tensile strength, and elastic modulus. Dynamic characterization utilizes SHPB system (refer to Fig. 2.9) to investigate high strain-rate behavior, with dynamic increasing factors quantified through tension-DIF (dynamic-to-static splitting strength ratio).

For dynamic splitting tests, Brazilian disc specimens (φ36mm × 18 mm) [15] are prepared and tested using a specialized 72 mm diameter SHPB configuration. Rigorous specimen preparation includes precision alignment of incident and transmission bars as shown in Fig. 2.10. The experimental setup incorporates a pulse shaping technique to ensure consistent strain rates, improve incident pulse characteristics, and minimize stress wave oscillations during impact testing. This methodology enables reliable measurement of dynamic stress-strain response while maintaining constant strain-rate conditions throughout the loading duration [16, 17].

Table 2.5 Mix proportions of RULHDCC (kg/m^3)

ID	Water	Cement	FAC	SF	Rubber	SP	SRA	Fiber
R0	259	702	339.9	78	0	7.0	9.0	0
R5	259	702	322.9	78	38.3	7.0	9.0	0
R10	259	702	305.9	78	76.7	7.0	9.0	0
R15	259	702	288.9	78	115	7.0	9.0	0
R20	259	702	271.9	78	153.3	7.0	9.0	0
R0-0.7PE	259	702	339.9	78	0	7.0	9.0	6.8
R10-0.7PE	259	702	305.9	78	76.7	7.0	9.0	6.8

Note: *SF* silica fume, *FAC* fly ash cenospheres, *SP* superplasticizer, *SRA* shrinkage reducing agent

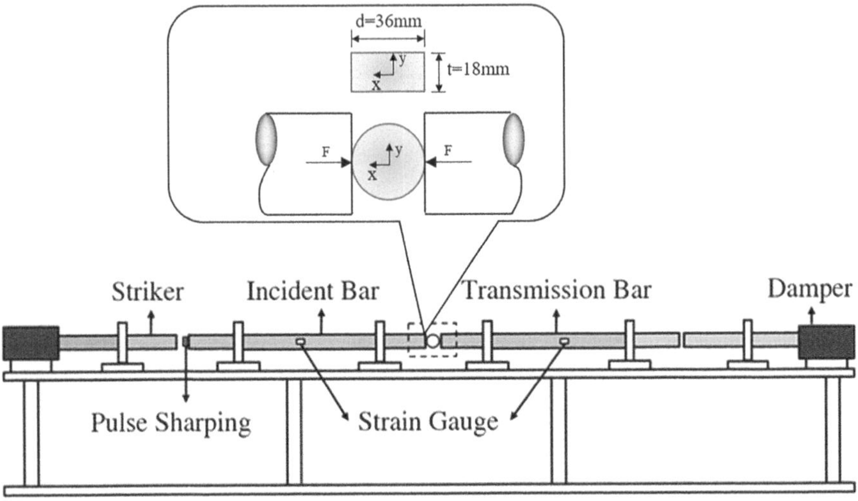

Fig. 2.9 Configuration of typical SHPB. (Reprinted from Huang et al. [45], copyright 2025, with permission from Elsevier)

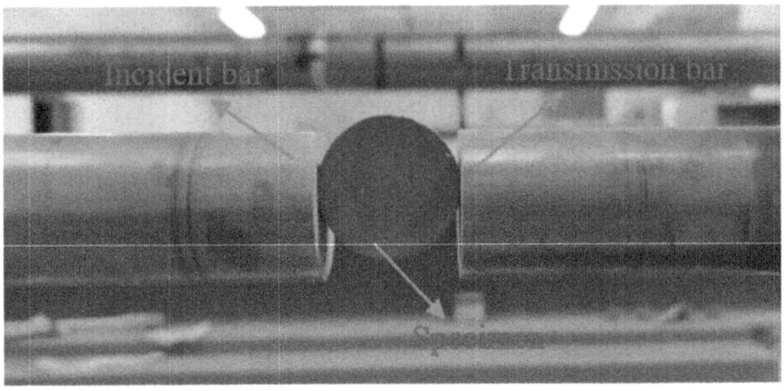

Fig. 2.10 Radial loading mode in dynamic splitting test. (Reprinted from Huang et al. [45], copyright 2025, with permission from Elsevier)

2.3.2 Principle of Splitting Test

2.3.2.1 Static Splitting

The Brazilian disc test was conducted to determine the static splitting tensile strength of concrete specimens following ASTM C39 [1] and ISRM guidelines [18, 19]. This standardized indirect tensile test utilizes a disc-shaped specimen subjected to diametral compression, where the thickness-to-diameter ratio of approximately 0.5 enables simplification of the three-dimensional stress analysis to a two-dimensional plane stress problem. The resulting tensile stress at the disk center was calculated using classical elasticity theory for radial symmetric loading conditions, with the analytical solution and corresponding equations presented in Eqs. (2.6)–(2.9). This approach provides a reliable method for characterizing the tensile properties of quasi-brittle materials such as concrete while maintaining experimental simplicity.

$$\sigma_x = \frac{2F}{\pi h}\left(\frac{\sin^2\theta_1 \cos\theta_1}{r_1} \frac{\sin^2\theta_2 cos\theta_2}{r_2}\right) - \frac{2F}{\pi hd} \tag{2.6}$$

$$\sigma_y = \frac{2F}{\pi h}\left(\frac{\cos^3\theta_1}{r_1} + \frac{\cos^3\theta_2}{r_2}\right) - \frac{2F}{\pi hd} \tag{2.7}$$

$$\tau_{xy} = \frac{2F}{\pi h}\left(\frac{\sin\theta_1 \cos^2\theta_1}{r_1} - \frac{\sin\theta_2 \cos^2\theta_2}{r_2}\right) \tag{2.8}$$

$$r_1 = r_2 = 0.5d, \theta_1 = \theta_2 = 0,$$

$$\sigma_x = -\frac{2F}{\pi hd}, \sigma_y = \frac{6F}{\pi hd} \tag{2.9}$$

According to [20, 21], the formula for calculating the splitting tensile strength of the sample in the Brazilian disc test is given in Eq. (2.10) below:

$$\sigma_T = \frac{2F}{\pi hd} \tag{2.10}$$

where, σ_T is the splitting tensile strength; h is the concrete thickness of the sample; and d is the sample diameter.

2.3.2.2 Dynamic Splitting

The dynamic splitting strength quantifies a material's resistance to rapid tensile failure under impact loading conditions. In this test investigation, Brazilian disk specimens were subjected to high-velocity impacts within a SHPB configuration (Fig. 2.10). The test methodology involved precise measurement of strain waves ($\varepsilon_I(t)$ and $\varepsilon_t(t)$) propagating through the incident and transmission bars during impact events. These strain measurements enabled calculation of the dynamic splitting strength through fundamental wave mechanics relationships provided in Eqs. (2.11)–(2.13), with the peak force corresponding to initial crack formation serving as the critical strength parameter.

$$\varepsilon(t) = \varepsilon_I(t) - \varepsilon_t(t) \tag{2.11}$$

$$F(t) = A_0 E_0 \varepsilon(t) \tag{2.12}$$

$$\sigma = \frac{2F}{\pi hd} \tag{2.13}$$

where, A_0 is the cross-sectional area of the specimen; E_0 is the elastic modulus of the bar.

High-speed imaging revealed a progressive failure mechanism in the splitting tests, where specimen fracture developed through multiple impact events rather than single-impact failure. The initial impact generated primary cracking, while subsequent impacts propagated these fractures, leading to complete failure. The analysis focused on the first crack formation event, using the corresponding peak impact force for strength calculations (Eq. (2.13)), as it represents the material's inherent resistance to crack initiation under dynamic loading conditions.

2.3.3 *Impact Test Behavior-Dynamic Splitting*

2.3.3.1 Failure Mode

Figure 2.11 demonstrates the significant influence of pulse shaping on specimen failure behavior during dynamic splitting tests. The pulse shaper's critical role in modifying incident pulse characteristics produces fundamentally different fracture patterns compared to unmodified impacts. When the pulse shaper is used (see Fig. 2.11a), specimens show more progressive crack propagation and controlled energy dissipation, whereas its absence (see Fig. 2.11b) results in sudden brittle fracture with irregular cracking. Material composition further dictates failure characteristics under dynamic splitting conditions. Rubber-modified specimens develop dominant fractures along the impact axis, separating into distinct fragments within 50 μs. In contrast, fiber-reinforced composites demonstrate distributed microcracking, reflecting the fibers' ability to redistribute tensile stresses across wider material volumes. All specimens show localized crushing at bar contact points, forming characteristic triangular damage zones. Notably, rubber-modified samples exhibit substantially greater deformation capacity than conventional composites, confirming the energy-absorbing properties reported in previous studies [22]. These observations collectively demonstrate how both testing methodology (pulse shaping) and material design (rubber/fiber modification) govern dynamic fracture behavior in cementitious composites.

Fig. 2.11 Failure mode of dynamic splitting test. (**a**) Pulse shaper used. (**b**) Pulse shaper not used. (Reprinted from Huang et al. [45], copyright 2025, with permission from Elsevier)

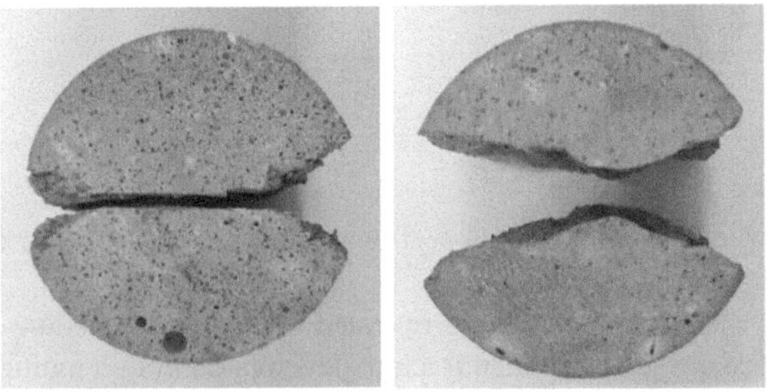

(a) Pulse shaper used (b) Pulse shaper not used

2.3.3.2 Stress-Time Curves

Figure 2.12 presents the stress-time relationships obtained from dynamic impact tests, calculated using Eq. (2.10), which demonstrate the critical influence of pulse shaping on test reliability. The brass pulse shaper (BPS) configuration, following established methodology from prior research [16, 17], effectively maintains constant strain rate conditions during 70–100% of the failure stress range. This test approach, utilizing a 1 mm thick, 10 mm diameter brass shaper, produces significantly more stable stress-time responses compared to unmodified (N) impact conditions at both 0.3 and 0.6 MPa loading pressures. The comparative results validate the pulse shaper's dual functionality: extending incident pulse rise time while ensuring consistent strain rates throughout the loading duration. The modified pulses generate smooth stress histories that confirm test validity across the complete strain rate spectrum. This controlled loading characteristic proves particularly valuable for investigating material behavior at different strain rate regimes, from quasi-static to dynamic conditions. The methodology aligns with previous successful applications in ultra-high performance concrete testing [16], demonstrating its effectiveness for characterizing impact response in cementitious materials.

2.3.3.3 Dynamic Splitting Strength and DIF

Table 2.6 presents a comprehensive comparison of the material's response under both static and dynamic loading conditions, quantifying key parameters including splitting strength values, strain rates, and dynamic increasing factors (DIF). For the splitting test, DIF is defined as the ratio of dynamic to static splitting strength as given by Eq. (2.14).

$$\mathrm{DIF} = \frac{f_{\mathrm{td}}}{f_{\mathrm{ts}}} \tag{2.14}$$

Figure 2.13 illustrates the characteristic stress-time behavior of the composite during dynamic splitting tests. The initial linear phase ($R^2 > 0.95$) prior to peak stress demonstrates consistent elastic response, allowing accurate determination of stress rate ($\dot{\sigma}$) and strain rate ($\dot{\varepsilon}$) through Eqs. (2.15)–(2.16).

$$\dot{\sigma} = \frac{\Delta\sigma}{\Delta t} \tag{2.15}$$

$$\dot{\varepsilon} = \frac{\dot{\sigma}}{E} \tag{2.16}$$

The dynamic splitting strength of RULHDCC exhibits a distinct strain-rate dependency, though the degree of enhancement varies with material composition. Rubber-modified specimens demonstrate particularly pronounced strain-rate sensitivity, likely attributable to the elastic behavior of rubber particles under impact loading.

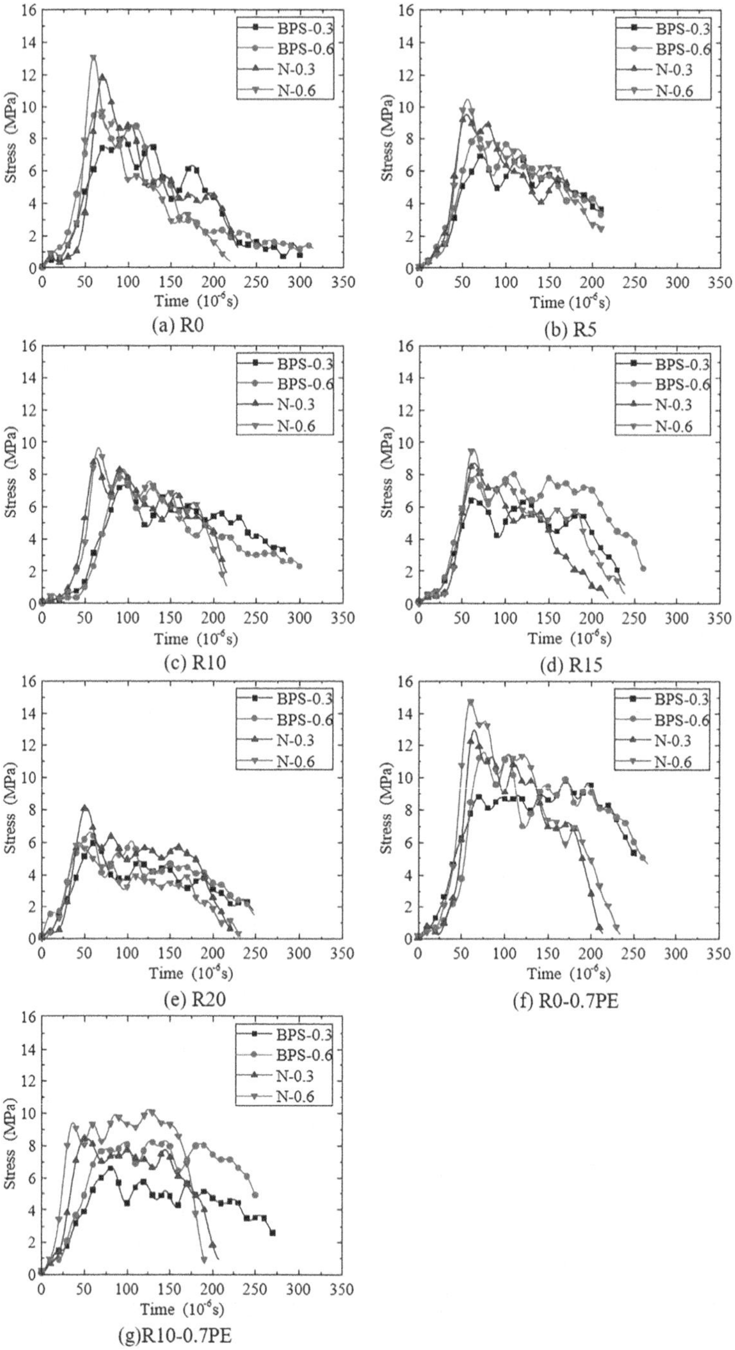

Fig. 2.12 Stress-time curves of samples under different pressures. (**a**) R0. (**b**) R5. (**c**) R10. (**d**) R15. (**e**) R20. (**f**) R0-0.7PE. (**g**) R10-0.7PE. (Reprinted from Huang et al. [45], copyright 2025, with permission from Elsevier)

Table 2.6 Static and dynamic splitting test results of different mix proportions

Specimen	Dynamic tensile strength (MPa)	Static tensile strength (MPa)	DIF	$\Delta\sigma/\Delta t$	Elastic modulus (GPa)	Strain rate (s^{-1})
R0-0.3BPS	8.16	2.62	3.12	14433.00	14.00	20.21
R0-0.6BPS	9.68	2.62	3.70	22302.00	14.00	31.22
R0-0.3N	11.84	2.62	4.52	39628.00	14.00	55.48
R0-0.6N	13.09	2.62	4.99	53197.00	14.00	74.48
R5-0.3BPS	6.97	2.27	3.07	13747.26	11.80	16.22
R5-0.6BPS	8.21	2.27	3.62	18520.00	11.80	21.85
R5-0.3N	9.53	2.27	4.20	31283.64	11.80	36.91
R5-0.6N	10.46	2.27	4.61	38599.09	11.80	45.55
R10-0.3BPS	7.36	2.02	3.64	14027.00	9.90	13.89
R10-0.6BPS	8.34	2.02	4.13	18576.00	9.90	18.39
R10-0.3N	8.99	2.02	4.45	28067.00	9.90	27.79
R10-0.6N	9.64	2.02	4.77	37598.00	9.90	37.22
R15-0.3BPS	6.54	1.82	3.59	15623.00	9.50	14.84
R15-0.6BPS	7.91	1.82	4.35	19516.00	9.50	18.54
R15-0.3N	8.71	1.82	4.79	27458.00	9.50	26.09
R15-0.6N	9.59	1.82	5.27	31211.00	9.50	29.65
R20-0.3BPS	6.00	1.68	3.57	11565.00	9.30	10.76
R20-0.6BPS	6.58	1.68	3.92	14372.00	9.30	13.37
R20-0.3N	8.14	1.68	4.84	26767.00	9.30	24.89
R20-0.6 N	5.96	1.68	3.55	18597.00	9.30	17.30
R10-0.7PE-0.3BPS	6.63	4.87	1.36	8777.00	9.60	8.43
R10-0.7PE-0.6BPS	8.29	4.87	1.70	15175.00	9.60	14.57
R10-0.7PE-0.3N	8.45	4.87	1.74	25039.00	9.60	24.04
R10-0.7PE-0.6N	10.22	4.87	2.10	35515.00	9.60	34.09
R0-0.7PE-0.3BPS	9.83	7.24	1.36	15594.00	11.90	18.56
R0-0.7PE-0.6BPS	11.59	7.24	1.60	36297.00	11.90	43.19
R0-0.7PE-0.3N	12.93	7.24	1.79	49852.00	11.90	59.32
R0-0.7PE-0.6N	14.73	7.24	2.03	58089.00	11.90	69.13

While the dynamic splitting strength generally decreases nonlinearly with higher rubber content—more significantly than under static conditions—this trend plateaus beyond 10% rubber replacement. At this threshold, specimens with 10–15% rubber content show comparable dynamic strength values across high strain rates, suggesting a saturation effect in rubber's influence on impact resistance. Interestingly, specimen R20-4 displayed unexpectedly low dynamic performance, matching the strength of R20-1 tested at lower strain rates. Post-test examination revealed localized rubber agglomeration in R20-4, which likely created weak zones that compromised both splitting strength and energy dissipation capacity. This observation highlights the critical role of particle distribution uniformity in determining the dynamic performance of rubber-modified cementitious composites, particularly at higher replacement ratios. The results collectively demonstrate that while rubber

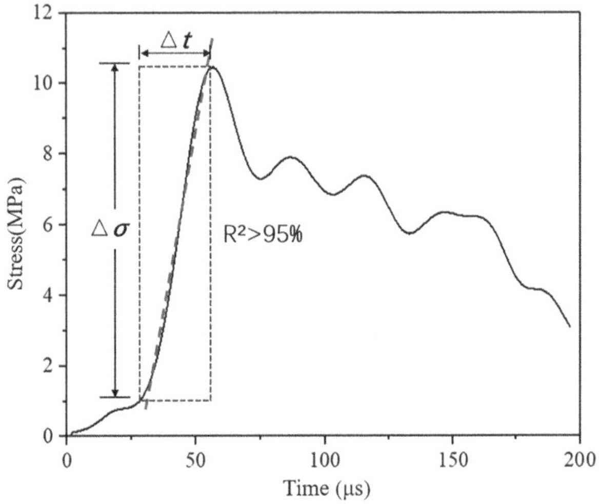

Fig. 2.13 Typical stress-time curve of dynamic splitting test. (Reprinted from Huang et al. [45], copyright 2025, with permission from Elsevier)

enhances strain-rate sensitivity, its effectiveness becomes limited beyond optimal concentrations, and proper dispersion is essential for consistent dynamic behavior.

2.3.3.4 DIF Prediction Model and Validation

Figure 2.14 demonstrates a linear relationship between the logarithmic strain rate and DIF for all material compositions. The analysis reveals significant variations in strain rate sensitivity, with sample R15 showing the most pronounced effect (slope = 5.09), approximately 1.54 times greater than the reference material (R0). However, most rubber-modified specimens exhibited reduced sensitivity compared to the control, contrasting with previous findings for conventional rubberized concrete [22]. This discrepancy likely stems from fundamental differences in material composition and microstructure. The unique combination of FAC and rubber in RULHDCC creates a distinct mechanical response, where FAC contributes structural integrity while rubber enhances energy absorption. The material's inherent porosity and the superior strength-to-weight ratio of FAC further differentiate its behavior from ordinary concrete. Fiber reinforcement appears to moderate strain rate effects, as evidenced by the nearly identical response of fiber-reinforced specimens (R0-0.7PE and R10-0.7PE). These findings highlight the complex interplay between material constituents in determining dynamic performance characteristics.

The stress-time response during initial loading exhibits near-linear characteristics, allowing for reliable strain rate determination using Eqs. (2.15)–(2.16) within acceptable error margins. The CEB-FIP provides an established framework for DIF calculation based on strain rate effects, as given in Eq. (2.17):

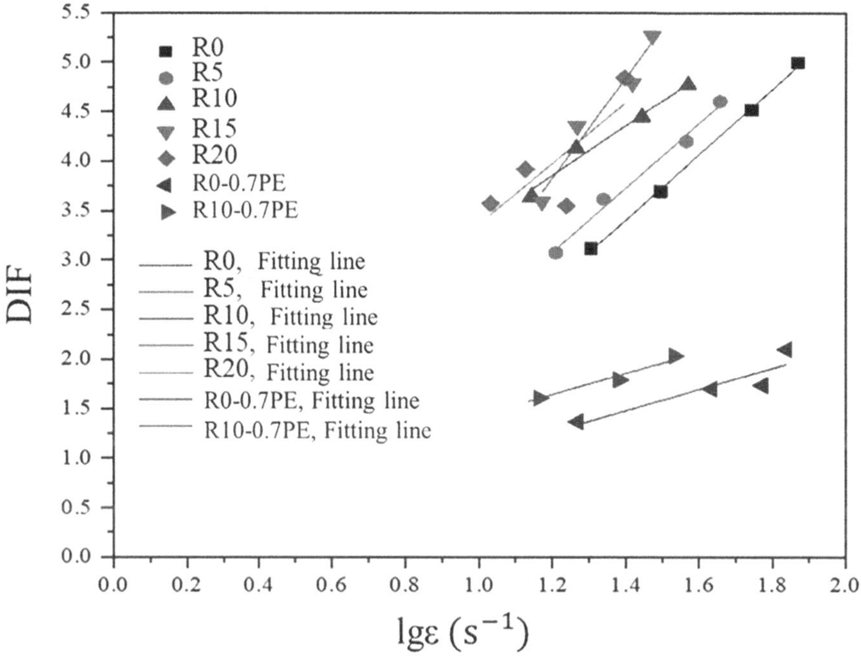

Fig. 2.14 Relationship between strain rate and DIF. (Reprinted from Huang et al. [45], copyright 2025, with permission from Elsevier)

$$\mathrm{DIF} = \begin{cases} (\dot{\varepsilon}/\dot{\varepsilon_0})^{0.018} & \dot{\varepsilon} \leq 10s^{-1} \\ 0.0062(\dot{\varepsilon}/\dot{\varepsilon_0})^{1/3} & \dot{\varepsilon} > 10s^{-1} \end{cases} \tag{2.17}$$

where, $\dot{\varepsilon_0} = 1 \times 10^{-6}s^{-1}$.

Figure 2.15 compares the DIF curves calculated by the CEB-FIP model with those from other models available in the literature [23–27]. It was found that the DIF prediction equation proposed by CEB-FIP underestimates the strength-increasing effect of most cement-based materials under impact, making it conservative for design purposes. The equation does not take into account the influence of strength grade. To address this issue, Malvar and Crawford proposed the Tension-DIF correction formula [21], as shown in Eq. (2.18):

$$\mathrm{DIF} = \begin{cases} (\dot{\varepsilon}/\dot{\varepsilon_0})^{\delta} & \dot{\varepsilon} \leq 1s^{-1} \\ \beta(\dot{\varepsilon}/\dot{\varepsilon_0})^{1/3} & \dot{\varepsilon} > 1s^{-1} \end{cases} \tag{2.18}$$

where, $\dot{\varepsilon_0} = 1 \times 10^{-6}s^{-1}$; $\log\beta = 6\delta - 2$; $\delta = 1/(1 + 8(f_{cs}/f_{co}))$; f_{cs} is the static splitting strength of the material; $f_{co} = 10$ MPa.

The test results demonstrate distinct predictive relationships for different material formulations within the 30–50 MPa compressive strength range. As shown in Fig. 2.15, the dynamic behavior of non-fiber RULCC specimens correlates well

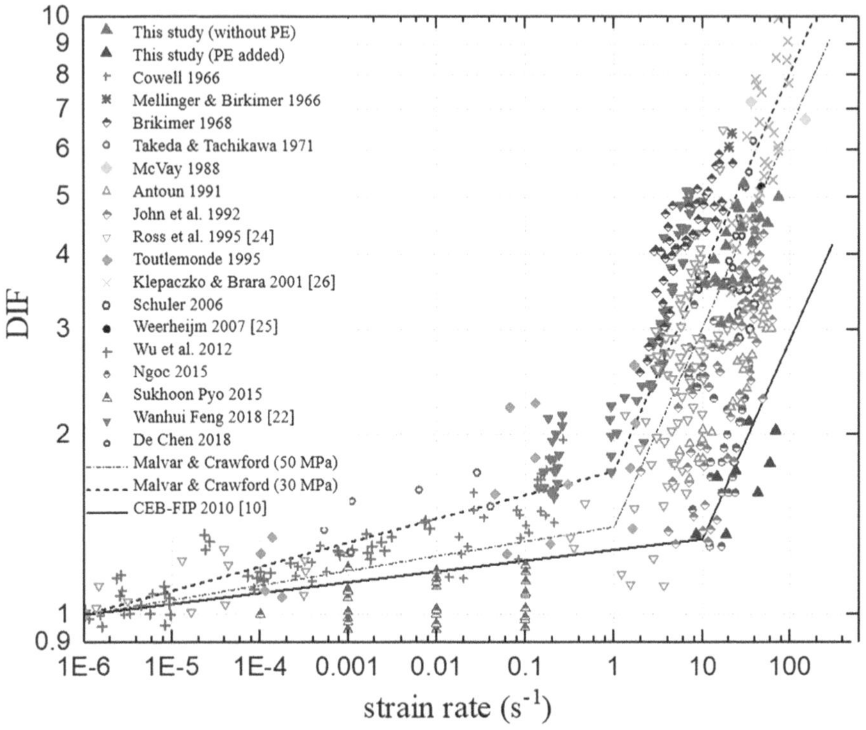

Fig. 2.15 Database of DIFs for dynamic splitting tests. (Reprinted from Huang et al. [45], copyright 2025, with permission from Elsevier)

with Malvar and Crawford's modified model, while fiber-reinforced RULHDCC follows the CEB-FIP predictions. This divergence highlights the significant influence of fiber reinforcement on strain-rate sensitivity, with PE fibers apparently altering the material's dynamic response to align with conventional concrete behavior.

2.4 Numerical Simulation for Dynamic Compressive Behavior

2.4.1 H-J-C Model

The SHPB impact tests were numerically simulated using LS-DYNA's explicit solver [27], employing distinct material models for system components. The steel bars were characterized using the J2 plasticity model (MAT_24), while the RULCC response was captured through the Holmquist-Johnson-Concrete (H-J-C) model—a validated approach for simulating cementitious materials under high strain-rate loading [28]. The H-J-C formulation incorporates three key components: (1) a

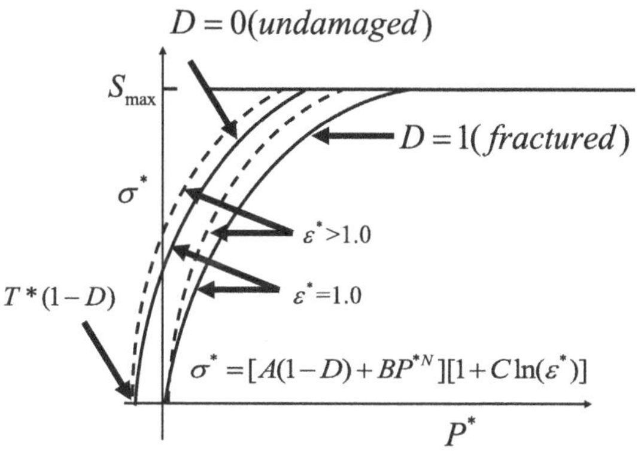

Fig. 2.16 H-J-C model. (Reprinted from Huang et al. [46], copyright 2025, with permission from Elsevier)

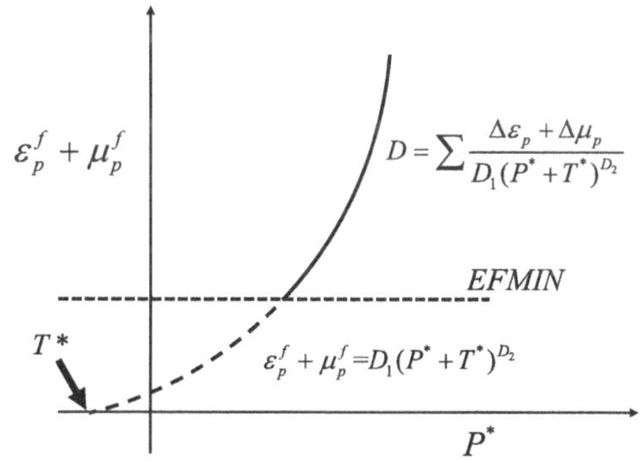

Fig. 2.17 Material damage evolution relationship. (Reprinted from Huang et al. [46], copyright 2025, with permission from Elsevier)

normalized strength model (see Eq. (2.19) and Fig. 2.16); (2) damage evolution equations; and (3) a state equation. It effectively addressing large deformations, strain-rate effects, and high-pressure conditions through combined Lagrangian-Eulerian meshing. This methodology aligns with established practices in dynamic concrete modeling [28–30], providing a robust framework for analyzing the RULCC's impact response.

$$\sigma^* = \left[A(1-D) + \mathrm{BP}^{*N}\right]\left[1 + C\ln(\dot{\varepsilon}^*)\right] \tag{2.19}$$

where, A, B, C and N are material constants; σ^* is the normalized effective strengh, $\sigma^* = \sigma/f_c \le S_{\max}$ and $S_{\max}$ is the normalized maximum strength; $f_c$is the uniaxial compresive strength; $P^* = P/f_c$is the normalized pressure; $\dot{\varepsilon}^* = \dot{\varepsilon}/\dot{\varepsilon}_0$ is the effective strain rate, $\dot{\varepsilon}$ is the true strain rate while $\dot{\varepsilon}_0 = 1.0s^{-1}$ is the reference strain rate; $T^* = T/f_c$ while T is the maximum tensile strength of the concrete.

The damage described by the H-J-C model is attributed to the accumulation of plastic strain, as illustrated in Fig. 2.17. The plastic strain comprises both the equivalent plastic strain and the plastic volume strain. The equation governing the damage evolution is presented below:

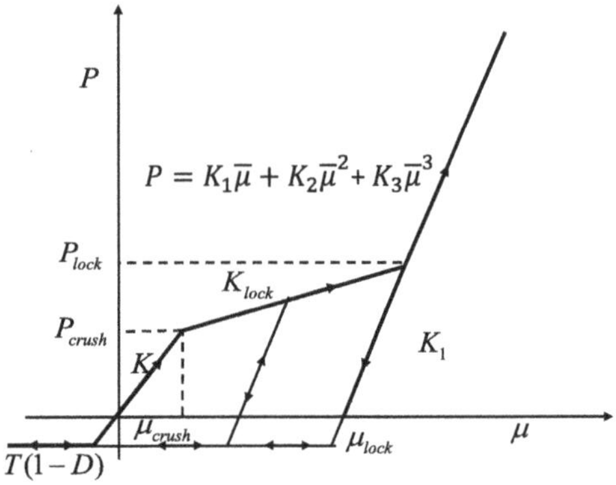

Fig. 2.18 Pressure-volumetric strain curve. (Reprinted from Huang et al. [46], copyright 2025, with permission from Elsevier)

$$D = \sum \frac{\Delta\varepsilon_p + \Delta\mu_p}{\varepsilon_p^f + \mu_p^f} \tag{2.20}$$

$$\varepsilon_p^f + \mu_p^f = D_1 (P^* + T^*)^{D_2} \tag{2.21}$$

where, D_1 and D_2 are damage constants; D denotes damge factor, $D \in [0, 1]$; $D_1(P^* + T^*)^{D_2} \geq \varepsilon_{f,\min}$,$\varepsilon_{f,\min}$is the minimum fracture strain that can be determined by a tensile test; $\Delta\varepsilon_p$and$\Delta\mu_p$ represent the effective plastic strain and plastic volume strain, respectively, for the element at each integration step; $\varepsilon_p^f + \mu_p^f$ is the plastic strain at the current integration step.

The H-J-C model employs a three-phase equation of state characterizing the pressure-volume response. The state equation of the H-J-C model is divided into three stages, i.e., elastic, transition and compaction stages, as shown in Fig. 2.18. The first stage $(-T(1 - D) \leq P \leq P_{\text{crush}})$ exhibits linear elastic, and the relationship between the hydrostatic pressure P and the volume strain is governed by:

$$P = K\mu \tag{2.22}$$

where, P is the hydrostatic pressure; K is the bulk modulus; μ is the volume strain; P_{crush} is the hydrostatic pressure at elastic limit.

The second stage $(P_{\text{crush}} < P \leq P_{\text{lock}})$ is the transition stage, in which the voids in the material is gradually compacted to produce plastic deformation. The loading and unloading equations are as follows:

$$P = \frac{(\mu - \mu_{\text{crush}})(P_{\text{lock}} - P_{\text{crush}})}{\mu_p - \mu_{\text{crush}}} + P_{\text{crush}} \tag{2.23}$$

$$P = P_0 - [(1 - F)K + FK_1](\mu_0 - \mu) \tag{2.24}$$

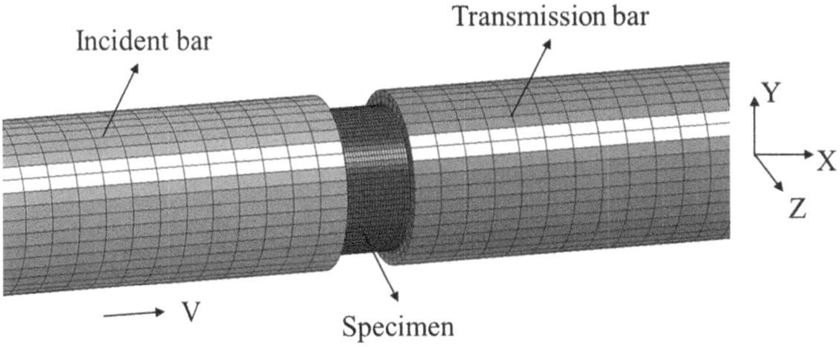

Fig. 2.19 Full-scaled FE model for concrete under SHPB test. (Reprinted from Huang et al. [46], copyright 2025, with permission from Elsevier)

$$F = \frac{\mu_0 - \mu_{\text{crush}}}{\mu_p - \mu_{\text{crush}}} \tag{2.25}$$

where, μ_p is the volume deformation corresponding to the compaction hydrostatic pressure P_{lock}; P_0 is the hydrostatic pressure corresponding to the volume deformation μ_0 before unloading; F is the unloading proportionality coefficient.

The third stage ($P_{\text{lock}} < P$) is the pore free compaction stage, during which the material has been completely damage. The loading and unloading equations are governed by:

$$P = K_1\overline{\mu} + K_2\overline{\mu}^2 + K_3\overline{\mu}^3 \tag{2.26}$$

$$P = K_1\overline{\mu} \tag{2.27}$$

$$\overline{\mu} = \frac{\mu\text{-}\mu_{\text{crush}}}{1 + \mu_{\text{lock}}} \tag{2.28}$$

where, μ_{lock} is the volume strain at the compaction limit; $\overline{\mu}$ is the modified volume strain, which is to prevent the material from softening when it just enters the third stage; K_1, K_2 and K_3 are the pressure constants.

The model simplifies the SHPB setup by excluding the absorption rod while maintaining key components: impact bullet, incident bar, test specimen, and transmission bar. All elements employ 8-node solid discretization with reduced integration and hourglass stabilization. As shown in Fig. 2.19, the sample is positioned between the pressure bars, accurately replicating experimental boundary conditions. This optimized configuration balances computational efficiency with physical fidelity, focusing mesh refinement on critical stress wave interaction zones while eliminating non-essential components from the dynamic simulation.

The numerical simulation employs surface-to-surface contact with penalty formulation, designating the stiffer SHPB surfaces as master contact surfaces. Impact loading is replicated through prescribed bullet velocities, generating stress waves that closely match experimental measurements (see Fig. 2.20). The excellent agreement between simulated and experimental strain histories validates the model's

Fig. 2.20 Validation between FE simulation and test for incident wave. (Reprinted from Huang et al. [46], copyright 2025, with permission from Elsevier)

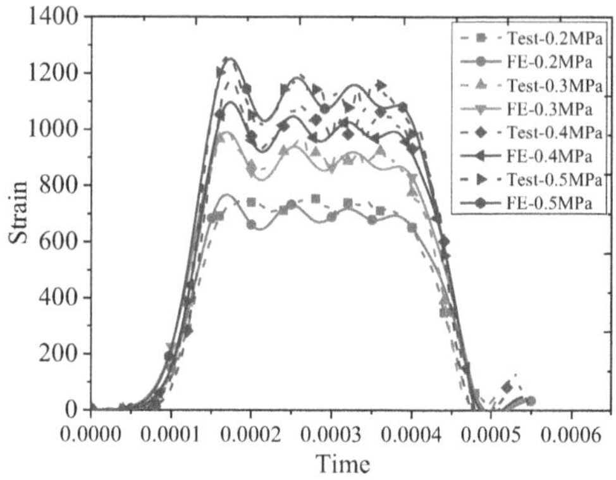

accuracy in capturing wave propagation phenomena. This verification confirms the FE model's capability to reproduce the essential physics of SHPB testing while maintaining computational efficiency through appropriate contact algorithms and master-slave surface assignments.

2.4.2 *Modified H-J-C Model*

To account for the enhanced mechanical properties of fiber-rubber modified RULCC, the conventional HJC concrete model has been adapted through two key modifications: (1) a normalized cohesive strength factor (α) addressing PE fiber contributions, and (2) a pressure hardening coefficient (β) representing rubber particle effects. Both the parameters can be calculated using Eq. (2.29) and Eq. (2.30), respectively. Following established approaches [31–33] for fiber-reinforced concrete, these dimensionless factors linearly modify the yield surface while preserving the original HJC framework's pressure and strain-rate dependencies. The proposed formulation maintains the model's fundamental damage mechanics while capturing the unique strength-toughness balance imparted by the composite's hybrid reinforcement system. This adaptation enables accurate simulation of RULCC's distinctive dynamic response without compromising computational efficiency.

$$\alpha = a\cdot\frac{V_{\mathrm{PE}}l_{\mathrm{PE}}}{d_{\mathrm{PE}}} - b\cdot M_r V_r \tag{2.29}$$

$$\beta = c\cdot\frac{V_{\mathrm{PE}}l_{\mathrm{PE}}}{d_{\mathrm{PE}}} - d\cdot M_r V_r \tag{2.30}$$

where, V_{PE} is the fiber content in volume; l_{PE} is the fiber length; d_{PE} is the fiber diameter; M_{r} is the rubber size (test sieve of $M_{\mathrm{r}} = 40$ mesh according to ASTM E11-01[34], which is 380 μm in size in this study); V_{r} is the rubber content in volume. The four constant coefficients in both the above equations are determined by binary regression on the experimental results presented in [4], which are, respectively, $a = 0.3016$, $b = 0.0239$, $c = 0.2997$, and $d = 0.0247$. The modified strength equation, given in Eq. (2.31), integrates these parameters while retaining the original HJC model's state and damage evolution equations. This formulation maintains the base model's computational framework while accounting for composite-specific reinforcement effects through the introduced α and β factors. The calibration process ensures physical consistency with experimental observations of fiber-rubber interactions in the RULCC matrix.

$$\sigma^* = \left[\alpha A(1-D) + \beta \mathrm{BP}^{*N}\right]\left[1 + C\ln(\dot{\varepsilon}^*)\right] \tag{2.31}$$

2.4.3 Determination of H-J-C Model Parameters

The simulation employs 21 material parameters from the HJC model, with key parameters determined as follows: experimentally calibrated values for measurable properties, calculated values for derived parameters, and established defaults for less sensitive variables. As shown in Table 2.7, $\varepsilon f_{\min}$ adopts the conventional value of 0.01 following Wu et al.'s findings [4] regarding its minimal impact on results, while $S_{\max}$ maintains the standard 7.0 MPa threshold recommended in literature [5, 11, 36]. The crush stress p_{cursh} under uniaxial compression follows the conventional HJC formulation (given in Eq. (2.32)), ensuring consistency with the model's fundamental mechanics while accommodating the RULCC's modified behavior through the adjusted strength parameters. This balanced approach maintains model fidelity while optimizing computational efficiency.

$$p_{\text{cursh}} = f_c/3 \tag{2.32}$$

According to Holmquist et al. [37], the volume strain, μ_{crush}, is related to p_{crush} by:

$$\mu_{\text{cursh}} = p_{\text{cursh}}/K \tag{2.33}$$

in which, the volume modulus K is determined by:

$$K = \frac{E}{3(1-2\nu)} \tag{2.34}$$

p_{cursh} is the compaction stress of the concrete shown in Fig. 2.17, which is between 0.8–1.2 MPa according to previous studies [37, 38].

The volume strain μ_{lock} at full compaction is given by:

Table 2.7 Parameters of the modified H-J-C model

Material	R0-0	R5-0	R10-0	R15-0	R20-0	R0-0.7PE	R10-0.7PE
Density ρ (kg/m^3)	1395	1397	1399	1401	1403	1405	1409
G (GPa)	5.833	4.917	4.125	3.958	3.875	4.958	4.125
A	0.54	0.56	0.50	0.48	0.46	0.57	0.48
B	2.25	2.25	2.20	2.18	2.15	2.36	2.09
C	0.007	0.007	0.007	0.007	0.007	0.007	0.007
N	1.36	1.36	1.36	1.36	1.36	1.36	1.36
f_c (MPa)	52.2	36.8	27.3	23.6	20.7	43.5	24.9
T (MPa)	2.4	1.9	1.6	1.4	1.3	2.1	1.4
$\dot{\varepsilon}_0$	1	1	1	1	1	1	1
εf_{min}	0.01	0.01	0.01	0.01	0.01	0.01	0.01
S_{max}	7	7	7	7	7	7	7
P_{cursh} (MPa)	17.40	12.27	9.10	7.87	6.90	14.50	8.30
μ_{cursh}	0.002237	0.001871	0.001655	0.001491	0.001335	0.002193	0.001509
P_{lock} (GPa)	0.9	0.9	0.9	0.9	0.9	0.9	0.9
μ_{lock}	0.12	0.12	0.12	0.12	0.12	0.12	0.12
D_1	0.04	0.04	0.04	0.04	0.04	0.04	0.04
D_2	1	1	1	1	1	1	1
K_1 (GPa)	85	85	85	85	85	85	85
K_2 (GPa)	−171	−171	−171	−171	−171	−171	−171
K_3 (GPa)	208	208	208	208	208	208	208
f_s	0.004	0.004	0.004	0.004	0.004	0.004	0.004

$$\mu_{lock} = \frac{\rho_{grain}}{\rho_0} - 1 = \frac{V_0}{V_g} - 1 = \frac{w}{1 - w} \tag{2.35}$$

where, ρ_{grain}, ρ_0, V_g, V_0, and w are compaction density, initial density, compaction volume, initial volume, and porosity ratio of the concrete, respectively. For lightweight concrete, the porosity ratio w is high and of 28–50% [39].

For accurate simulation of lightweight concrete behavior, the cement matrix densities (ρ_{grain} and ρ_0) are used rather than compact density due to the material's high porosity (11.21%–12.63% for w/c = 0.34) [40]. From Eq. (2.35), μ_{lock} can be expressed as $\frac{V_0}{V_g} - 1 = \frac{w}{1-w}$, where V_0 and V_g are the initial volume and the volume after compaction of cement matrix, respectively. f_s takes the value in Li [41] or 0.004 as proposed by Liu et al. [8]. T is the tensile strength of lightweight concrete from tensile test or Eurocode 2 [42], which is governed by:

$$T = \eta_1 f_{ctk} \tag{2.36}$$

$$\eta_1 = 0.04 + 0.6\frac{\rho}{2200} \tag{2.37}$$

The elasatic modulus, E_c, of the ULCC was measured by tests [43] based on ASTM C469/C469M [44]. The shear modulus G is obtained by:

$$G = \frac{E_c}{2(1+\nu)} \tag{2.38}$$

where v is taken from the test results in reference [43]. The key parameters of the modifed H-J-C model are shown in Table 2.7.

2.4.4 Validation of FE Analysis

The finite element simulations employing the modified H-J-C constitutive model demonstrate excellent agreement with test results, particularly for low fiber-content specimens (Fig. 2.21a–g). Comparative analysis reveals the model accurately captures both the stress-strain response and failure modes across varying strain rates. As quantified in Table 2.8, the numerical predictions show strong correlation with SHPB test data, exhibiting minimal deviations in peak stress (average $< 2.91\%$) and strain rate (average $< 4.51\%$). This validation confirms the model's effectiveness in simulating the dynamic behavior of rubberized concrete, with particular accuracy in representing the material response before fiber reinforcement dominates the failure mechanisms. The results demonstrate the robustness of the calibrated parameters in adapting the H-J-C framework to this novel composite material.

The modified H-J-C model effectively simulates the impact response of RULCC, as demonstrated through comparative analysis of experimental and numerical results compared in Figs. 2.22 and 2.23. High-speed camera recordings at 0.00025 s intervals reveal that rubber (R10-0) and fiber (R0-0.7PE) modifications significantly mitigate damage severity compared to plain specimens (R0-0). Numerical strain contours show excellent correlation with observed failure patterns, accurately capturing damage progression from intact (blue) to fully failed (red) states. The model successfully predicts key dynamic characteristics, including: (1) peak stress values within 2.91% accuracy; (2) strain rate correlations within 4.51% deviation; and (3) Damage initiation and propagation patterns. Element deletion analysis (refer to Fig. 2.23) confirms the model's capability to replicate crack development when exceeding material strain limits. This validated framework provides a reliable basis for subsequent metamodeling of RULCC's strain-rate-dependent behavior under various impact scenarios.

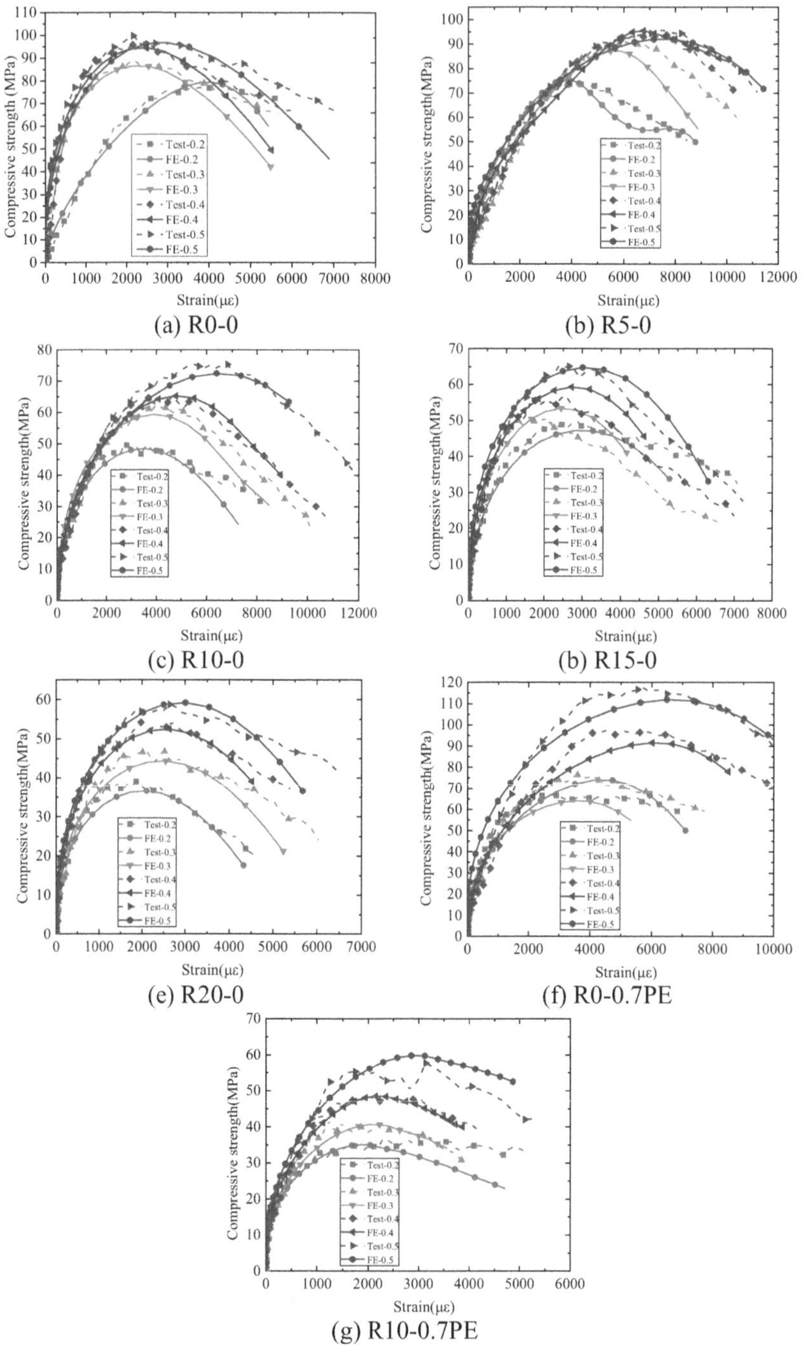

Fig. 2.21 Comparison of stress-strain curves for different strain rates. (**a**) R0-0. (**b**) R5-0. (**c**) R10-0. (**b**) R15-0. (**e**) R20-0. (**f**) R0-0.7PE. (**g**) R10-0.7PE. (Reprinted from Huang et al. [46], copyright 2025, with permission from Elsevier)

Table 2.8 Comparison of test results and numerical simulation

		Test				Numerical simulation				Difference		
Mix ID	SCS (MPa)	DIS (MPa)	DPS	SR (s)	DIF	DIS (MPa)	DPS	SR (s)	DIF	DIS (%)	DPS (%)	SR (%)
R0-0	52.20	79.70	4203	94.10	1.53	80.12	5206	96.70	1.53	0.53	23.86	2.76
		88.70	2112	130.60	1.70	88.16	2030	121.30	1.69	0.61	3.88	7.12
		96.40	2267	150.70	1.85	96.71	2309	154.60	1.85	0.32	1.85	2.59
		100.80	3467	183.40	1.93	102.43	3480	199.60	1.96	1.62	0.37	8.83
R0-0.7PE	43.50	68.00	4312	102.10	1.56	70.26	4627	96.70	1.62	3.32	7.31	5.29
		76.30	3601	128.30	1.75	78.07	3843	117.30	1.79	2.32	6.71	8.57
		97.30	5778	146.80	2.24	97.75	6219	147.70	2.25	0.46	7.63	0.61
		117.50	5734	166.40	2.70	115.86	6370	159.60	2.66	1.40	11.09	4.09
R5-0	36.80	76.30	3948	91.90	2.07	75.35	3849	97.30	2.05	1.25	2.50	5.88
		91.20	5515	123.40	2.48	87.17	5499	114.70	2.37	4.42	0.30	7.05
		93.70	7838	149.90	2.55	94.85	6337	156.70	2.58	1.23	19.15	4.54
		95.80	7527	178.60	2.60	91.96	7167	175.60	2.50	4.01	4.78	1.68
R10-0	27.30	49.60	4532	105.30	1.82	49.50	3264	100.60	1.81	0.20	27.98	4.46
		62.10	3964	128.40	2.27	63.53	4508	134.40	2.33	2.31	13.72	4.67
		65.00	4631	155.70	2.38	65.24	4387	149.30	2.39	0.37	5.27	4.11
		74.60	5506	181.50	2.73	73.00	7124	196.40	2.67	2.14	29.38	8.21
R10-0.7PE	24.90	37.40	3236	93.10	1.50	35.08	1861	87.30	1.41	6.21	42.50	6.23
		41.90	1242	138.20	1.68	40.63	2021	132.10	1.63	3.04	62.73	4.41
		48.30	2002	158.70	1.94	48.45	2152	155.60	1.95	0.32	7.47	1.95
		57.60	3140	185.10	2.31	59.80	2854	192.10	2.40	3.81	9.11	3.78
R15-0	23.60	49.40	2644	104.80	2.09	50.66	1958	100.90	2.15	2.54	25.93	3.72
		50.30	4567	129.20	2.13	46.61	1722	119.70	1.97	7.35	62.29	7.35
		56.80	2561	159.80	2.41	58.58	2198	158.70	2.48	3.13	14.18	0.69
		65.30	2391	180.90	2.77	67.93	2426	177.60	2.88	4.03	1.44	1.82

(continued)

Table 2.8 (continued)

		Test				Numerical simulation				Difference		
Mix ID	SCS (MPa)	DIS (MPa)	DPS	SR (s)	DIF	DIS (MPa)	DPS	SR (s)	DIF	DIS (%)	DPS (%)	SR (%)
R20-0	20.70	40.80	1938	99.70	1.97	35.37	2719	94.30	1.71	13.32	40.29	5.42
		46.80	2107	123.50	2.26	48.16	2794	119.60	2.33	2.90	32.59	3.16
		53.70	2696	150.50	2.59	55.63	2580	144.40	2.69	3.60	4.29	4.05
		57.40	2650	186.50	2.77	60.06	2447	180.60	2.90	4.63	7.68	3.16
Mean										2.91	17.01	4.51
Standard deviation										0.027	0.177	0.023

Note: *SCS* static compressive strength, *DIS* dynamic impact strength, *DPS* dynamic peak strain, *SR* strain rate

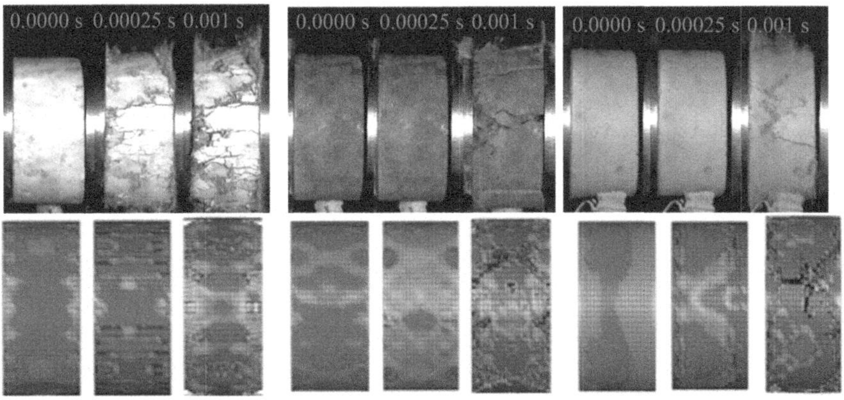

(a)R0-0(0.4MPa) (b)R10-0(0.3MPa) (c)R0-0.7PE(0.2MPa)

Fig. 2.22 Comparison of failure modes between FE and tests. (**a**) R0-0 (0.4 MPa). (**b**) R10-0 (0.3 MPa). (**c**) R0-0.7PE (0.2MPa). (Reprinted from Huang et al. [46], copyright 2025, with permission from Elsevier)

R0-0(No impact) R0-0(0.2MPa) R0-0(0.3MPa) R0-0(0.4MPa) R0-0(0.5MPa)

Time = 0.5ms

Fig. 2.23 Failure of RULCC under different loading pressure. (Reprinted from Huang et al. [46], copyright 2025, with permission from Elsevier)

2.5 Data-Driven Model for Predicting DIF

The limitations of conventional experimental methods for engineering design have become increasingly apparent, particularly when evaluating complex systems with multiple variables. While finite element analysis has emerged as a powerful alternative, its computational demands remain prohibitive for extensive parametric studies and optimization routines. This study addresses these challenges by developing a metamodeling approach to predict the dynamic strength characteristics of RULCC composites. The proposed surrogate model specifically accounts for varied rubber and fiber volume fractions and different impact strain rate conditions. By transforming the validated FE simulations into an efficient algebraic framework, this methodology enables rapid performance evaluation while maintaining predictive accuracy. The metamodel serves as a practical design tool that balances computational efficiency with engineering precision, particularly valuable for optimization scenarios requiring numerous iterations. This approach represents a significant advancement in computational materials design, offering a streamlined pathway for developing advanced cementitious composites with tailored dynamic properties.

2.5.1 Design of Experiments Sampling

The study employs an Audze-Eglais criterion-based Latin hypercube design of experiments (DoE) to efficiently populate the design space while minimizing computational costs. This approach strategically selects sampling points by optimizing their spatial distribution through minimization of a potential energy function, U as defined in Eq. (2.39), which is derived from pairwise Euclidean distances. The methodology offers two key advantages for RULCC characterization: (1) comprehensive coverage of the multi-dimensional design space (incorporating rubber content, fiber reinforcement, and strain rate variables), and (2) significant reduction in required sampling points compared to full factorial designs. This optimized experimental design enables effective metamodel construction while maintaining statistical robustness, particularly valuable for capturing the complex nonlinear relationships between material composition and dynamic performance. The spatial uniformity criterion ensures balanced representation across all design variables, providing reliable data for subsequent surrogate model development.

$$U = \sum_{p=1}^{P} \sum_{q=p+1}^{P} \frac{1}{L_{\mathrm{pq}}^{2}} \tag{2.39}$$

where, P is the total number of points, L_{pq} is the Euclidean distance between points p and q $(p \neq q)$.

The investigation establishes practical bounds for three key variables governing RULCC's impact performance, as summarized in Table 2.9: (1) rubber content (0–20% by volume), (2) fiber reinforcement (0–1% by volume), and (3) applied strain rates (10–200 s^{-1}).

The study employs a space-filling design methodology to select 60 optimal material compositions within the defined parameter space. These sampling points, distributed uniformly across the three-dimensional fiber-rubber-strain rate domain, demonstrate excellent spatial coverage as verified by Euclidean distance analysis. The minimum Euclidean distance between each of the points to their surrounding points is governed by:

Table 2.9 Bounds of design variables

Design variables	Lower bound	Upper bound
Fiber contents (%)	0	0.7
Rubber contents (%)	0	20
Strain rates (/s)	20	200

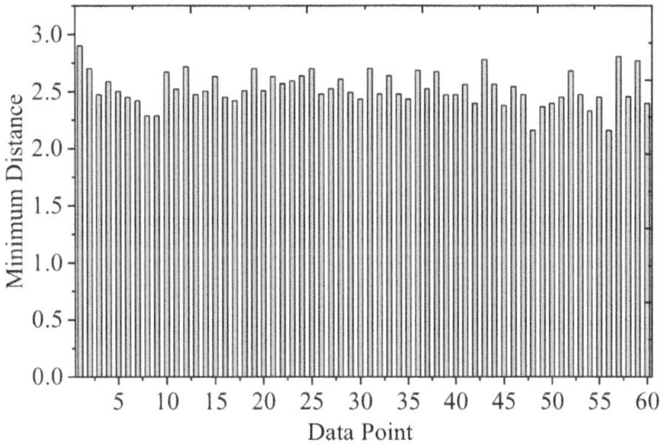

Fig. 2.24 Minimum distances between points generated by OLH within the design space. (Reprinted from Huang et al. [46], copyright 2025, with permission from Elsevier)

$$D_{\min}^{i} = \min\{L_{ij}\}\ j = 1, 2, \ldots i - 1, i + 1, \ldots n \tag{2.40}$$

where $D_{\min}^{i}$ is the minimum distance from point i (=1,... .60) to all the surrounding points. The values L_{ij} are the Euclidean distances calculated from these points, and Eq. (2.40) is minimized to optimize their spatial configuration. Figure 2.24 shows that a reasonably uniform spacing has been achieved.

2.5.2 Data Generation

The study employs a high-performance cloud computing platform to systematically generate the required dataset for metamodel development. Using the validated HJC model, material properties at each of the 60 sampling points are determined through an optimized computational workflow: (1) key mechanical parameters (compressive strength, elastic modulus, and tensile strength) are derived from experimental data [35] using Levenberg-Marquardt nonlinear regression; and (2) remaining HJC parameters (A, B, P_c, u_c) are calculated through the model's governing equations. The cloud-based implementation utilizes elastic scaling and multi-server parallel processing to efficiently complete the 60 simulations (vid. Fig. 2.25). This distributed computing approach demonstrates significant advantages in both computational speed and resource utilization compared to conventional methods. Table 2.10 presents the comprehensive set of simulation parameters, ensuring transparency in the metamodel's foundational data. The framework successfully balances numerical accuracy with practical computational efficiency, enabling large-scale parametric studies that would be prohibitively expensive through physical experiment alone.

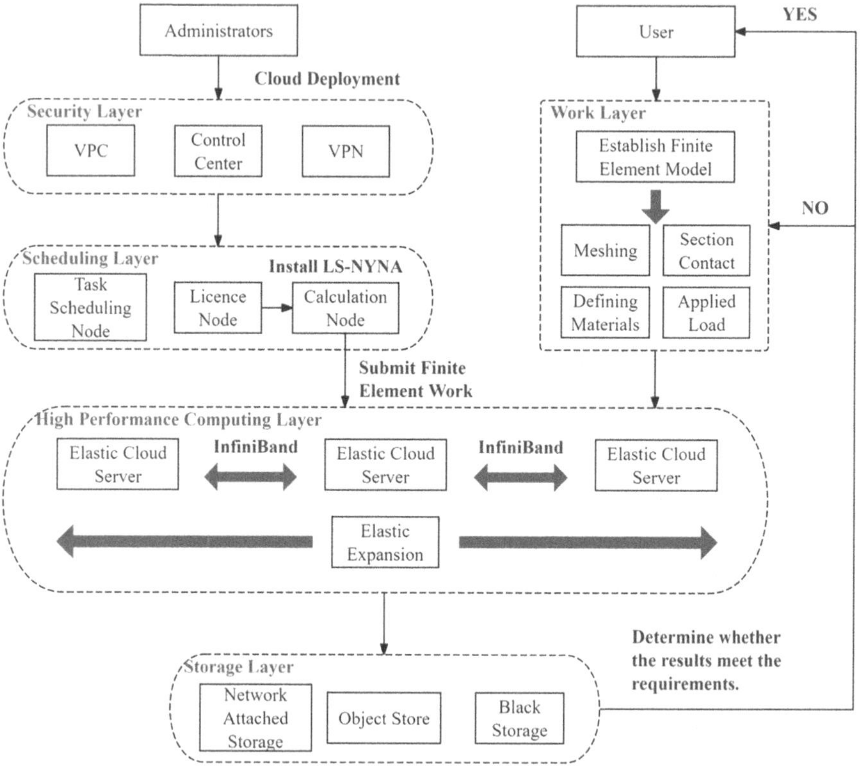

Fig. 2.25 Multi-node cloud simulation framework based on cloud platform

2.5.3 *Model Development*

This study implements genetic programming (GP) to derive optimal functional representations for predicting the dynamic strength of RULCC composites. The algorithm employs an evolutionary optimization framework that systematically explores potential mathematical structures through tree-based representations (see Fig. 2.26a for a typical expression of$(x_1 \times x_2 - x_3)^2$). The GP workflow (see Fig. 2.26b) operates through several key mechanisms: initial population generation with randomized mathematical expressions; fitness evaluation against simulation data; iterative refinement through genetic operators (reproduction, mutation, and crossover); and termination upon achieving satisfactory prediction accuracy. Unlike conventional regression techniques that require predefined functional forms, this approach automatically discovers appropriate nonlinear relationships between material composition, strain rate, and dynamic strength. The tree-structured representation enables efficient exploration of complex mathematical spaces while maintaining interpretability of the resulting metamodels. The evolutionary optimization process continues until the derived expressions demonstrate sufficient accuracy across the

Table 2.10 Parameters of HJC model for extended FE simulation

Material	Density ρ (kg/m^3)	G (GPa)	A	B	C	N	f_c (MPa)	T (MPa)	$\dot{\varepsilon}_0$	εf_{min}	S_{max}	P_c (MPa)	μ_c	P_1 (GPa)	μ_1	D_1	D_2	K_1 (GPa)	K_2 (GPa)	K_3 (GPa)	f_s
No.1	1400	4.848	0.55	2.33	0.007	1.36	41.2	2.0	1	0.01	7	13.73	0.0021246	0.9	0.12	0.04	1	85	−171	208	0.004
No.2	1400	4.332	0.50	2.20	0.007	1.36	25.5	1.72	1	0.01	7	8.51	0.0014737	0.9	0.12	0.04	1	85	−171	208	0.004
No.3	1400	4.421	0.50	2.21	0.007	1.36	27.4	1.76	1	0.01	7	9.12	0.0015475	0.9	0.12	0.04	1	85	−171	208	0.004
No.4	1400	4.477	0.50	2.21	0.007	1.36	28.8	1.79	1	0.01	7	9.61	0.0016092	0.9	0.12	0.04	1	85	−171	208	0.004
No.5	1400	5.050	0.54	2.24	0.007	1.36	40.0	2.06	1	0.01	7	13.33	0.0019796	0.9	0.12	0.04	1	85	−171	208	0.004
No.6	1400	4.945	0.53	2.24	0.007	1.36	38.2	2.01	1	0.01	7	12.73	0.0019302	0.9	0.12	0.04	1	85	−171	208	0.004
No.7	1400	5.098	0.54	2.25	0.007	1.36	41.3	2.08	1	0.01	7	13.77	0.0020256	0.9	0.12	0.04	1	85	−171	208	0.004
No.8	1400	3.832	0.47	2.14	0.007	1.36	21.1	1.48	1	0.01	7	7.02	0.0013742	0.9	0.12	0.04	1	85	−171	208	0.004
No.9	1400	3.889	0.47	2.15	0.007	1.36	21.7	1.51	1	0.01	7	7.23	0.0013950	0.9	0.12	0.04	1	85	−171	208	0.004
No.10	1400	3.558	0.45	2.11	0.007	1.36	17.8	1.35	1	0.01	7	5.94	0.0012522	0.9	0.12	0.04	1	85	−171	208	0.004
No.11	1400	4.711	0.52	2.22	0.007	1.36	34.8	1.90	1	0.01	7	11.60	0.0018460	0.9	0.12	0.04	1	85	−171	208	0.004
No.12	1400	3.994	0.48	2.15	0.007	1.36	23.2	1.56	1	0.01	7	7.72	0.0014494	0.9	0.12	0.04	1	85	−171	208	0.004
No.13	1400	5.275	0.55	2.27	0.007	1.36	44.7	2.16	1	0.01	7	14.89	0.0021170	0.9	0.12	0.04	1	85	−171	208	0.004
No.14	1400	4.622	0.52	2.21	0.007	1.36	33.6	1.85	1	0.01	7	11.19	0.0018156	0.9	0.12	0.04	1	85	−171	208	0.004
No.15	1400	4.743	0.52	2.22	0.007	1.36	35.8	1.91	1	0.01	7	11.92	0.0018853	0.9	0.12	0.04	1	85	−171	208	0.004
No.16	1400	4.252	0.50	2.17	0.007	1.36	27.4	1.68	1	0.01	7	9.14	0.0016131	0.9	0.12	0.04	1	85	−171	208	0.004
No.17	1400	5.340	0.56	2.28	0.007	1.36	45.7	2.19	1	0.01	7	15.25	0.0021415	0.9	0.12	0.04	1	85	−171	208	0.004
No.18	1400	3.687	0.46	2.09	0.007	1.36	18.8	1.41	1	0.01	7	6.25	0.0012721	0.9	0.12	0.04	1	85	−171	208	0.004
No.19	1400	4.098	0.49	2.14	0.007	1.36	25.1	1.60	1	0.01	7	8.37	0.0015314	0.9	0.12	0.04	1	85	−171	208	0.004
No.20	1400	5.122	0.55	2.27	0.007	1.36	42.4	2.09	1	0.01	7	14.14	0.0020704	0.9	0.12	0.04	1	85	−171	208	0.004
No.21	1400	3.566	0.46	2.06	0.007	1.36	16.9	1.35	1	0.01	7	5.65	0.0011874	0.9	0.12	0.04	1	85	−171	208	0.004
No.22	1400	4.106	0.49	2.14	0.007	1.36	25.4	1.61	1	0.01	7	8.46	0.0015456	0.9	0.12	0.04	1	85	−171	208	0.004
No.23	1400	4.937	0.54	2.25	0.007	1.36	39.5	2.00	1	0.01	7	13.17	0.0020011	0.9	0.12	0.04	1	85	−171	208	0.004
No.24	1400	4.026	0.48	2.12	0.007	1.36	24.1	1.57	1	0.01	7	8.05	0.0014991	0.9	0.12	0.04	1	85	−171	208	0.004

(continued)

Table 2.10 (continued)

Material	Density ρ (kg/m^3)	G (GPa)	A	B	C	N	f_c (MPa)	T (MPa)	$\dot{\varepsilon}_0$	εf_{min}	S_{max}	P_c (MPa)	μ_c	P_1 (GPa)	μ_1	D_1	D_2	K_1 (GPa)	K_2 (GPa)	K_3 (GPa)	f_s
No.25	1400	5.050	0.54	2.27	0.007	1.36	41.4	2.05	1	0.01	7	13.81	0.0020507	0.9	0.12	0.04	1	85	−171	208	0.004
No.26	1400	3.558	0.46	2.04	0.007	1.36	16.7	1.35	1	0.01	7	5.57	0.0011733	0.9	0.12	0.04	1	85	−171	208	0.004
No.27	1400	4.389	0.51	2.17	0.007	1.36	30.4	1.74	1	0.01	7	10.15	0.0017342	0.9	0.12	0.04	1	85	−171	208	0.004
No.28	1400	4.477	0.51	2.18	0.007	1.36	32.0	1.78	1	0.01	7	10.67	0.0017875	0.9	0.12	0.04	1	85	−171	208	0.004
No.29	1400	4.534	0.52	2.19	0.007	1.36	33.0	1.81	1	0.01	7	11.01	0.0018211	0.9	0.12	0.04	1	85	−171	208	0.004
No.30	1400	5.203	0.55	2.30	0.007	1.36	43.9	2.12	1	0.01	7	14.65	0.0021112	0.9	0.12	0.04	1	85	−171	208	0.004
No.31	1400	3.421	0.45	1.99	0.007	1.36	14.5	1.28	1	0.01	7	4.84	0.0010600	0.9	0.12	0.04	1	85	−171	208	0.004
No.32	1400	3.929	0.48	2.09	0.007	1.36	22.8	1.52	1	0.01	7	7.61	0.0014522	0.9	0.12	0.04	1	85	−171	208	0.004
No.33	1400	4.759	0.53	2.23	0.007	1.36	37.0	1.91	1	0.01	7	12.33	0.0019431	0.9	0.12	0.04	1	85	−171	208	0.004
No.34	1400	5.235	0.56	2.32	0.007	1.36	44.5	2.14	1	0.01	7	14.84	0.0021255	0.9	0.12	0.04	1	85	−171	208	0.004
No.35	1400	3.776	0.47	2.05	0.007	1.36	20.3	1.45	1	0.01	7	6.78	0.0013458	0.9	0.12	0.04	1	85	−171	208	0.004
No.36	1400	3.703	0.47	2.03	0.007	1.36	19.1	1.41	1	0.01	7	6.38	0.0012913	0.9	0.12	0.04	1	85	−171	208	0.004
No.37	1400	3.824	0.48	2.06	0.007	1.36	21.2	1.47	1	0.01	7	7.07	0.0013865	0.9	0.12	0.04	1	85	−171	208	0.004
No.38	1400	4.751	0.53	2.24	0.007	1.36	37.0	1.91	1	0.01	7	12.35	0.0019492	0.9	0.12	0.04	1	85	−171	208	0.004
No.39	1400	4.227	0.50	2.14	0.007	1.36	28.2	1.66	1	0.01	7	9.41	0.0016697	0.9	0.12	0.04	1	85	−171	208	0.004
No.40	1400	4.993	0.55	2.29	0.007	1.36	41.0	2.02	1	0.01	7	13.66	0.0020516	0.9	0.12	0.04	1	85	−171	208	0.004
No.41	1400	3.276	0.45	1.92	0.007	1.36	12.1	1.21	1	0.01	7	4.02	0.0009210	0.9	0.12	0.04	1	85	−171	208	0.004
No.42	1400	5.074	0.55	2.31	0.007	1.36	42.3	2.06	1	0.01	7	14.09	0.0020829	0.9	0.12	0.04	1	85	−171	208	0.004
No.43	1400	4.550	0.52	2.20	0.007	1.36	33.9	1.81	1	0.01	7	11.30	0.0018622	0.9	0.12	0.04	1	85	−171	208	0.004
No.44	1400	4.606	0.52	2.22	0.007	1.36	34.9	1.84	1	0.01	7	11.62	0.0018925	0.9	0.12	0.04	1	85	−171	208	0.004
No.45	1400	4.308	0.51	2.15	0.007	1.36	29.9	1.69	1	0.01	7	9.96	0.0017334	0.9	0.12	0.04	1	85	−171	208	0.004
No.46	1400	4.106	0.50	2.11	0.007	1.36	26.4	1.60	1	0.01	7	8.81	0.0016091	0.9	0.12	0.04	1	85	−171	208	0.004
No.47	1400	3.647	0.47	1.99	0.007	1.36	18.5	1.38	1	0.01	7	6.15	0.0012648	0.9	0.12	0.04	1	85	−171	208	0.004
No.48	1400	3.478	0.46	1.95	0.007	1.36	15.5	1.30	1	0.01	7	5.18	0.0011168	0.9	0.12	0.04	1	85	−171	208	0.004

No.49	1400	3.308	0.45	1.90	0.007	1.36	12.6	1.22	1	0.01	7	4.21	0.0009545	0.9	0.12	0.04	1	85	−171	208	0.004
No.50	1400	3.752	0.48	2.01	0.007	1.36	20.4	1.43	1	0.01	7	6.79	0.0013580	0.9	0.12	0.04	1	85	−171	208	0.004
No.51	1400	4.163	0.50	2.12	0.007	1.36	27.6	1.62	1	0.01	7	9.20	0.0016580	0.9	0.12	0.04	1	85	−171	208	0.004
No.52	1400	4.413	0.51	2.18	0.007	1.36	31.9	1.74	1	0.01	7	10.64	0.0018087	0.9	0.12	0.04	1	85	−171	208	0.004
No.53	1400	4.921	0.54	2.31	0.007	1.36	40.2	1.98	1	0.01	7	13.40	0.0020429	0.9	0.12	0.04	1	85	−171	208	0.004
No.54	1400	4.751	0.54	2.27	0.007	1.36	37.6	1.90	1	0.01	7	12.52	0.0019769	0.9	0.12	0.04	1	85	−171	208	0.004
No.55	1400	3.357	0.45	1.90	0.007	1.36	13.6	1.24	1	0.01	7	4.52	0.0010106	0.9	0.12	0.04	1	85	−171	208	0.004
No.56	1400	4.767	0.54	2.28	0.007	1.36	37.9	1.91	1	0.01	7	12.63	0.0019869	0.9	0.12	0.04	1	85	−171	208	0.004
No.57	1400	3.470	0.46	1.92	0.007	1.36	15.6	1.29	1	0.01	7	5.20	0.0011250	0.9	0.12	0.04	1	85	−171	208	0.004
No.58	1400	3.945	0.49	2.06	0.007	1.36	24.1	1.52	1	0.01	7	8.03	0.0015259	0.9	0.12	0.04	1	85	−171	208	0.004
No.59	1400	4.163	0.50	2.12	0.007	1.36	27.9	1.62	1	0.01	7	9.31	0.0016766	0.9	0.12	0.04	1	85	−171	208	0.004
No.60	1400	4.445	0.52	2.20	0.007	1.36	32.8	1.75	1	0.01	7	10.92	0.0018423	0.9	0.12	0.04	1	85	−171	208	0.004

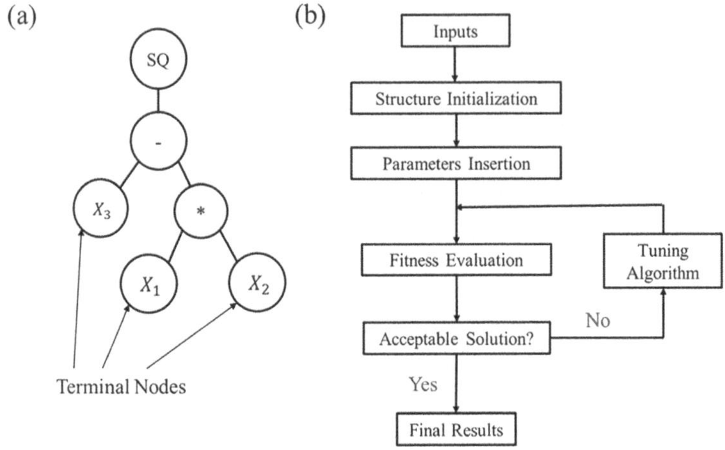

Fig. 2.26 Genetic programming representation and workflow (**a**) tree structure for the expression $(x_1 \times x_2 - x_3)^2$, (**b**) a brief flowchart of genetic programming process. (Reprinted from Huang et al. [46], copyright 2025, with permission from Elsevier)

entire design space defined in Table 2.9, providing robust approximations of the computationally intensive finite element simulations.

Through systematic implementation of the genetic programming framework and comprehensive finite element analysis of 60 design points, a prediction model for the dynamic strength of RULCC is developed, as expressed in Eq. (2.41) below:

$$\mathrm{DIS} = 1.43f_1 - 0.68f_4 - 1.82V_r - 0.32\dot{\varepsilon} + \sqrt{|1.43f_7 - \dot{\varepsilon}|} \tag{2.41}$$

where $f_1 = \sqrt{(69 - V_\mathrm{r})\dot{\varepsilon}}$, $f_2 = \sqrt{\frac{\dot{\varepsilon}}{1+V_\mathrm{r}} + V_\mathrm{f}}$, $f_3 = \dot{\varepsilon}\cdot V_\mathrm{f} - \sqrt{V_\mathrm{r}}$, $f_4 = \frac{f_3}{f_2}$, $f_5 = \sqrt{V_\mathrm{r}}\left(12 - \sqrt{\dot{\varepsilon}}\right)$, $f_6 = \left(\dot{\varepsilon} - \sqrt{\dot{\varepsilon}}\right)\cdot\frac{f_5\cdot\dot{\varepsilon}}{f_5+V_\mathrm{r}}$, $f_7 = \sqrt{|70(\dot{\varepsilon} + V_\mathrm{r}) - f_6|}$.

In the above equations, DIS denotes dynamic impact strength of RULCC, V_f, V_r and $\dot{\varepsilon}$ are fiber ratio, rubber ratio, and dynamic strain rate. Figure 2.27 presents the comparisons between the dynamic strength of the RULCC predicted by the simple meta-model in Eq. (2.41) and those from the experimental tests and the FE simulations.

Figure 2.28 provides comprehensive validation of the derived meta-model through multi-faceted comparisons with both training data and independent results. The analysis reveals: (1) excellent agreement between meta-model predictions and the 60 DoE-selected FE simulations used in model development, demonstrating successful approximation of the underlying numerical model (vid. Fig. 2.28a); (2) strong correspondence with independent physical test results not included in the training set, confirming the model's ability to generalize beyond its development data (vid. Fig. 2.28b); and (3) consistent accuracy when compared against additional FE simulations, verifying reliable performance across the full spectrum of reinforcement ratios and strain rates (Fig. 2.28c).

Figure 2.28 illustrates the dynamic impact strength (DIS) distribution at 200/s strain rate across different fiber and rubber reinforcement ratios. The contour

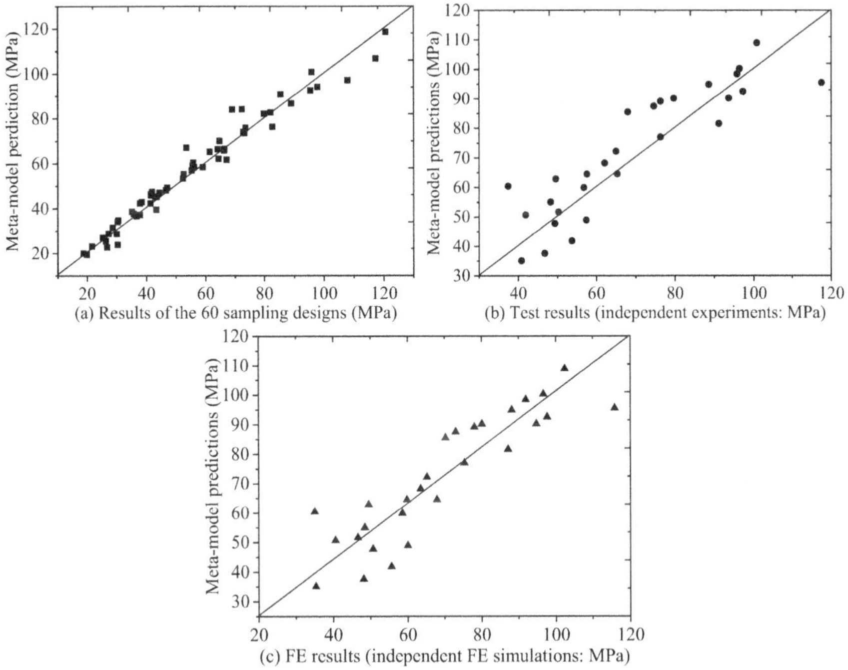

Fig. 2.27 Comparisons between the metamodeling predictions and (**a**) experimental and FE results used in the data training process, (**b**) independent experimental results, and (**c**) Independent finite element results. (Reprinted from Huang et al. [46], copyright 2025, with permission from Elsevier)

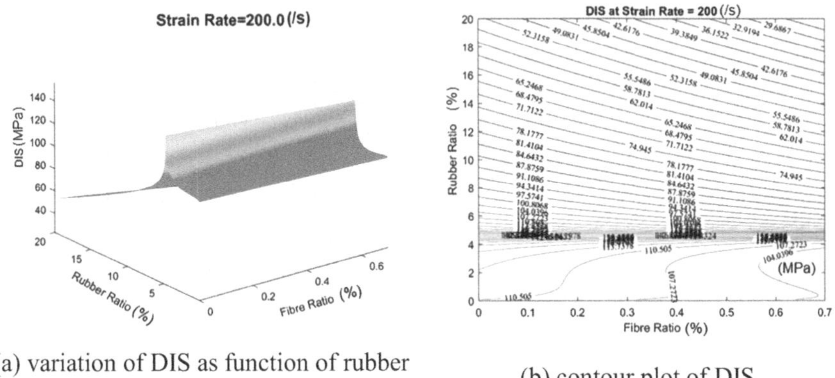

(a) variation of DIS as function of rubber and fiber ratios

(b) contour plot of DIS

Fig. 2.28 Dynamic impact strength of RULCC at an impact strain rate of 200/s. (**a**) variation of DIS as function of rubber and fiber ratios. (**b**) Contour plot of DIS. (Reprinted from Huang et al. [46], copyright 2025, with permission from Elsevier)

analysis reveals a critical reinforcement threshold, showing that: (1) peak DIS occurs at intermediate rubber content (4–5% by volume); (2) beyond this optimal range, increasing rubber content diminishes strength unless compensated by proportional fiber addition; and (3) below the threshold, simultaneous increases in both reinforcements are needed to maintain target strength levels. This nonlinear behavior demonstrates the complex interaction between rubber and fiber components in determining impact resistance. The identified optimal rubber ratio provides valuable guidance for material design, suggesting that maximum dynamic performance requires careful balancing of both reinforcement types rather than simple maximization of either component. The contour patterns further reveal how fiber incorporation can mitigate the strength reduction associated with higher rubber contents, enabling tailored material formulations for specific impact resistance requirements.

References

1. ASTM, Standard test method for compressive strength of cylindrical concrete specimens, ASTM C39/C39M-18, ASTM International, West Conshohocken, 2017.
2. JSCE, Recommendations for design and construction of high-performance fiber reinforced cement composites with multiple fine cracks, Japan Society of Civil Engineers, Tokyo, 2008.
3. J. Hopkinson, On the rupture of iron wire by a blow, in: B. Hopkinson (Ed.), Original papers by the late J. Hopkinson, Scientific papers, Cambridge University Press, 1901, pp. 316-320.
4. Y. Wu, J. Wang, P.J.M. Monteiro, M.H. Zhang, Development of ultralightweight cement composites with low thermal conductivity and high specific strength for energy efficient buildings, Constr. Build. Mater. 87 (2015) 100-112.
5. T. Gupta, R.K. Sharma, S. Chaudhary, Impact resistance of concrete containing waste rubber fiber and silica fume, Int. J. Impact Eng. 83 (2015) 76-87.
6. M. Nili, V. Afroughsabet, Combined effect of silica fume and steel fibers on the impact resistance and mechanical properties of concrete, Int. J. Impact Eng. 37(8) (2010) 879-886.
7. G.X. Chen, Impact resistance research on rubberized concrete, Master's thesis, (in Chinese), 2011.
8. F. Liu, G.X. Chen, L.J. Li, Y.C. Guo, Study of impact performance of rubber reinforced concrete, Constr. Build. Mater. 36 (2012) 604-616.
9. W.H. Fan, F. Liu, F. Yuan, L.J. Li, J. Lu, Experimental study on the effect of strain rates on the dynamic flexural properties of rubber concrete, Constr. Build. Mater. 224 (2019) 408-419.
10. CEB-FIP, Model code 2010 – first complete draft, Vol. 1, Comité Euro-International du Béton, Lausanne, 2010.
11. T. Ngo, P. Mendis, T. Krauthammer, Behavior of ultrahigh-strength prestressed concrete panels subjected to blast loading, J. Struct. Eng. 133(11) (2007) 1582-1590.
12. E. Kharun, M. Abadel, A.M. Abadel, Impact resistance of rubberized self-compacting concrete, Water Sci. 29(1) (2019) 45-53.
13. A. Hameed, U. Youssf, O. ElGawady, Mechanical and dynamic properties of high strength concrete with well graded coarse and fine tire rubber, Constr. Build. Mater. 246 (2020) 118534.
14. Z. Huang, L. Sui, F. Wang, S. Du, Y. Zhou, J. Ye, Dynamic compressive behavior of a novel ultra-lightweight cement composite incorporated with rubber powder, Compos. Struct. 244 (2020) 112300.
15. Y.F. Wang, J. Wang, Z. Wu, X. Chen, Effect of compression casting method on the compressive strength, elastic modulus and microstructure of rubber concrete, J. Clean. Prod. 264 (2020) 121572.

16. M. Hao, K. Wang, Experimental impact analysis on ultra-high performance concrete for achieving stress equilibrium and constant strain rate in split Hopkinson pressure bar using pulse shaping technique, Constr. Build. Mater. 144 (2017) 747-757.
17. A.B. Shi, R. Nemat-Nasser, M.J. Asaro, Experimental and numerical study on choosing proper pulse shapers for testing concrete specimens by split Hopkinson pressure bar apparatus, Constr. Build. Mater. 125 (2016) 326-336.
18. R.M.K.E. Goodman, The mechanical behavior of anisotropic sedimentary rocks, J. Eng. Ind. 89 (4) (1967) 765-771.
19. J.A. Iturrioz, Rock mechanics for gas storage in bedded salt caverns, Int. J. Rock Mech. Min. Sci. 34(3-4) (1997) 1420-1435.
20. E. Denneman, E.P. Kearsley, A.T. Visser, Splitting tensile test for fibre reinforced concrete, Mater. Struct. 44(8) (2011) 1441-1449.
21. S.P. Timoshenko, J.N. Goodier, Theory of elasticity, 3rd ed., McGraw-Hill, New York, 1970.
22. W.H. Fan, F. Liu, F. Yuan, L.J. Li, J. Lu, Experimental study on dynamic split tensile properties of rubber concrete, Constr. Build. Mater. 165 (2018) 675-687.
23. F. Liu, W.H. Fan, Z. Xu, G.R. Tian, L.J. Li, Static and impact behaviour of recycled aggregate concretes under daily temperature variations, J. Clean. Prod. 191 (2018) 283-296.
24. C.A. Ross, J.W. Thompson, S.T. Kurtz, Effects of strain rate on concrete strength, ACI Mater. J. 92(1) (1995) 37-47.
25. J. Weerheijm, J.C.A.M. van Doormaal, Tensile failure of concrete at high loading rates: New test data on strength and fracture energy from instrumented spalling tests, Int. J. Impact Eng. 34 (3) (2007) 609-626.
26. J.R. Klepaczko, A. Brara, An experimental method for dynamic tensile testing of concrete by spalling, Int. J. Impact Eng. 25(4) (2001) 387-409.
27. J. Hallquist, LS-DYNA keyword user's manual, Version 971, Livermore Software Technology Corporation, Livermore, 2007.
28. S.L. Xu, C. Chen, Q.H. Li, X. Zhang, Numerical simulation on dynamic compressive behavior of ultra-high toughness cementitious composites, Eng. Mech. 36(8) (2019) 1-8 (in Chinese).
29. X. Kong, Q. Fang, H. Wu, Y. Peng, Numerical predictions of cratering and scabbing in concrete slabs subjected to projectile impact using a modified version of HJC material model, Int. J. Impact Eng. 95 (2016) 61-71.
30. M. Polanco-Loria, O. Hassanein, T. Børvik, T. Berstad, Numerical predictions of ballistic limits for concrete slabs using a modified version of the HJC concrete model, Int. J. Impact Eng. 35 (2008) 290-303.
31. H. Yang, Study on dynamic mechanical behavior and constitutive model of fiber reinforced cementitious composites, Master's thesis, South China University of Technology, (in Chinese), 2016.
32. J. Chen, Dynamic constitutive model of steel fiber reinforced concrete and its finite element, Doctoral dissertation, Southwest Jiaotong University, (in Chinese), 2010.
33. Z.B. Yu, R.Y. He, Y.C. Liu, L. Li, K. Zhang, Y.L. Zhang, Steel fiber-reinforced concrete under impact loading dynamic constitutive equation, Constr. Build. Mater. 190 (2018) 1049-1055.
34. ASTM, Standard specification for wire cloth and sieves for testing purposes, ASTM E11-01, ASTM, 2001.
35. G.R. Johnson, W.H. Cook, A constitutive model and data for metals subjected to large strains, high strain rates and high temperatures, Eng. Fract. Mech. 21 (1983) 541-548.
36. T.J. Holmquist, G.R. Johnson, A computational constitutive model for glass subjected to large strains, high strain rates and high pressures, J. Appl. Mech. 78 (2011) 011016.
37. H. Xu, H. Wen, A computational constitutive model for concrete subjected to dynamic loadings, Int. J. Impact Eng. 91 (2016) 116-125.
38. X. Li, K.S. Chia, M.H. Zhang, Development of lightweight concrete with high resistance to water and chloride-ion penetration, Cem. Concr. Compos. 32 (2010) 757-766.
39. Y. Qin, H. Qin, Study on the inhomogeneity of spatial distribution of porosity in hardened cement paste, J. Build. Mater. (1994) (in Chinese).

40. Y. Liu, Research on HJC dynamic constitutive model for concrete, Hefei Polytechnic University, (in Chinese), 2009.
41. European Concrete Platform, Eurocode 2: Design of concrete structures – part 1-2: General rules – structural fire design, Brussels, 2004.
42. Z. Huang, F. Wang, Y. Zhou, L. Sui, P. Kang, J.Y.R. Liew, A novel, multifunctional, floatable, lightweight cement composite: Development and properties, Materials 11 (2018) 2043.
43. ASTM, Standard test method for static modulus of elasticity and Poisson's ratio of concrete in compression, ASTM C469/C469M, ASTM International, West Conshohocken, 2014.
44. F. Wang, Study on dynamic properties of ultra lightweight cement composites modified by rubber powder, Master's thesis, Shenzhen University, (in Chinese), 2020.
45. Z. Huang, W. Pan, S. Du, Y. Zhou, J. Ye, Dynamic splitting behavior of ultra-lightweight cement composite with rubber and polyethylene fiber, Constr. Build. Mater. 399 (2023) 132577.
46. Z. Huang, W. Deng, S. Du, Z. Gu, W. Long, J. Ye, Effect of rubber particles and fibers on the dynamic compressive behavior of novel ultra-lightweight cement composites: Numerical simulations and metamodeling, Compos. Struct. 258 (2021) 113210.

Chapter 3
Structural Behaviour of Steel–Concrete Sandwich Composites Infilled with Ultra-Lightweight Cement Composite

3.1 Introduction

The Arctic continental shelf is recognized globally as holding substantial reserves of untapped oil, natural gas, and unconventional hydrocarbons, including gas hydrates. This interest has been further stimulated by the increasing worldwide demand for oil, which is expected to rise by 37% by 2030 [1], as well as findings from the United States Geological Survey (USGS) indicating that 30% of the planet's yet-to-be-discovered natural gas and 13% of its undiscovered oil are located in the Arctic offshore region. Nevertheless, developing these resources poses significant challenges due to the region's extensive ice-covered marine environment. Offshore structures in this area must be designed to endure the dynamic and sometimes static forces imposed by moving ice, which can include massive features like pressure ridges reaching heights of over 30 m. These ice-induced loads, which can exceed 100 MN, often surpass the forces caused by ocean waves on platforms in ice-free waters. As a result, there is a critical need for robust structural designs with high load-bearing capacity and exceptional ductility to ensure safety and stability in such harsh conditions.

Leveraging the enhanced structural performance of steel-concrete composites, the steel-concrete-steel (SCS) sandwich configuration offers a highly advantageous solution for both civil and offshore engineering applications. Recognized for its excellent cost-to-strength efficiency [2, 3], this composite system—when filled with lightweight concrete—delivers substantial structural and economic advantages compared to traditional reinforced concrete alternatives. The system exhibits superior flexural rigidity and exceptional energy dissipation characteristics, making it particularly suitable for resisting extreme environmental conditions and accidental loading scenarios. The outer steel plates function as permanent formwork during concrete placement, significantly improving construction productivity while reducing on-site labor requirements and project timelines. The inherent impermeability of the

Z. Huang, Y. Zhou, *Innovative Lightweight High-performance Steel-concrete Sandwich Composite for Marine Structures*,
https://doi.org/10.1007/978-981-95-4681-7_3

external steel layers decreases the surface area needing costly corrosion protection measures while simplifying inspection and maintenance procedures. As a result, SCS sandwich composites have found successful implementation in demanding applications including marine vessel hulls, ice-resistant structures, tunnel linings, and protective walls in nuclear facilities where exceptional load resistance is paramount.

This chapter primarily examines the structural performance of sandwich composite elements under various loading conditions, including axial compression behavior of walls, flexural and shear responses of flat and curved beams and panels, and punching shear capacity of shell structures. Given the complex load transfer mechanisms and potential failure modes inherent in thick steel-concrete-steel (SCS) sandwich sections, a combined approach of experimental investigation and analytical modeling proves essential. Research typically involves quasi-static testing of SCS wall, beam, and shell specimens to establish baseline performance data. These experimental findings, supplemented by results from existing literature, provide the foundation for developing refined analytical models that elucidate both load distribution patterns and failure mechanisms, enabling accurate prediction of ultimate resistance capacities. Following experimental validation, numerical simulations using finite element software ABAQUS are implemented, incorporating material nonlinearity and accounting for interactions between concrete, steel plates, and shear connectors. The validated computational model subsequently serves as a tool for comprehensive parametric analysis, enabling systematic investigation of key design variables influencing the behavior of SCS sandwich composites incorporating ULCC fill material.

3.2 Materials and Specimens

3.2.1 Ultra-Lightweight Cement Composite

The lightweight steel-concrete-steel (SCS) sandwich configuration employs ultra-lightweight cement composite (ULCC) as its core material, encapsulated between inner and outer steel face plates. The specific mixture proportions for the ULCC formulation are detailed in Table 3.1, with comprehensive mechanical characteristics provided in Table 3.2. The ULCC material implemented in this investigation exhibits a density of 1361 kg/m^3 and achieves a 28-day compressive cylinder strength of 56.1 MPa.

Table 3.1 Mix proportion of ULCC (kg/m^3)

Water	Cement	Silica fume	Cenosphere	SRA	SP	Fiber
245.6	705.2	61.3	318.6	19.1	5.0	6.5

Table 3.2 Mechanical properties of ULCC

Density (kg/m^3)	Cylinder strength (MPa)	Peak strain (10^{-3})	Elastic modulus (GPa)	Poisson ratio	Splitting tensile strength (MPa)
1361 (5)	56.10 (1.47)	4.45 (0.22)	14.30 (0.04)	0.247 (0.04)	5.4 (0.58)

Table 3.3 Mechanical properties of steel plate and headed shear stud for beam and panel

Curved panels	E_s (GPa)	f_y (MPa)	f_u (MPa)	Poisson ratio, ν_s
Mild steel plate	200	304	460	0.3
Headed shear stud	225	524	584	0.3

Table 3.4 Mechanical properties of steel plate and J-hook connectors for wall

Sandwich wall	Component	E_s (MPa)	f_y (MPa)	f_u (MPa)	ν_s
Mild steel plate	6 mm	202.0	309.4	426.8	0.3
	8 mm	179.6	393.9	522.9	
	12 mm	211.5	374.6	507.0	
J-hook bar	HPB 235φ13	189.9	356.8	521.7	

3.2.2 Steel Plate and Shear Connectors

For the fabrication of beam and panel specimens, mild steel plate S275 was employed for the sandwich face plates. The material properties of the mild steel, including Young's modulus E_s, Poisson's ratio ν_s, yield strength f_y, and ultimate strength f_u, were obtained through tensile testing of steel coupon samples. To achieve composite action, conventional Nelson-headed shear studs were utilized as shear connectors at the steel and concrete interface. Detailed mechanical properties of the mild steel plate and shear stud are listed in Table 3.3.

Conventional reinforcing bars were bent to form the J-hook shear connectors, which were then welded onto the steel plates. The mechanical properties, including the yield strength, ultimate strength, Poisson ratio, and Young's modulus, of the steel face plate and J-hook connectors are listed in Table 3.4.

3.2.3 Design of Steel–Concrete Sandwich Wall, Beam, and Shell Elements

The experimental program involved the fabrication and testing of a comprehensive series of structural elements, comprising one flat beam, nine curved beams, two shell segments, and fifteen wall specimens with varying design parameters, as systematically cataloged in Tables 3.5 and 3.6. The principal variables investigated for the beam and shell specimens included the rise-to-span ratio (r/L), the steel contribution ratio (defined as A_{sf}/A_{cf}), the spacing of shear connectors (s), and the thickness of the

Table 3.5 Dimension and material properties of beam and panel specimens

Specimen	t_s (mm)	h_c (mm)	L (mm)	W (mm)	L/h_c ratio	r/L	s/t_s	$d \times h_s$ @s (mm)	f_{ck} (MPa)
Steel–concrete–steel composite beam									
FSB-01	4	100	1450	300	14.5	0(flat)	25	13 × 75@110	58.3
CSB-02	4	180	1179	300	6.55	0.21	25	13 × 150@110	58.3
CSB-03	4	100	1250	300	12.5	0.21	25	13 × 75@110	41
CSB-04	8	100	1250	300	12.5	0.21	12.5	13 × 75@110	58..3
CSB-05	12	100	1250	300	12.5	0.21	8.33	13 × 75@110	57.4
CSB-06	4	80	1250	300	15.6	0.21	25	13 × 50@110	57.4
CSB-07	4	100	1250	300	12.5	0.21	50	13 × 75@110	57.4
CSB-08	4	100	1250	300	12.5	0.21	25	13 × 75@110	57.4
CSB-09	4	100	1250	300	12.5	0.5	25	13 × 75@110	57.4
CSB-10	4	100	1250	300	12.5	0.21	25	13 × 75@110	57.4
Steel–concrete–steel composite shell									
SCSS-1	4	80	1250	1250	15.6	0.1	27.5	13 × 50@110	56.1
SCSS-2	4	80	1250	1250	15.6	0.21	27.5	13 × 50@110	56.1

Note: "FSB" refers to the flat sandwich beam, "CSB" denotes the curved double skin beam specimen, t_s represents the thickness of the steel face plates, h_c is the height of the concrete core, L signifies the clear span of the sandwich beam, W indicates the overall width of the section, s is the spacing between the headed shear studs, d, h_s refer to the diameter and height of the studs, respectively, f_{ck} defines the characteristic compressive strength of the concrete cylinder

concrete core (h_c). The corresponding cross-sectional geometries for these specimens are illustrated in Fig. 3.1a, b. To ensure uniform and integral load transfer during testing, two end plates were affixed to each specimen. Furthermore, to maintain stringent quality control over the workability of the ultra-lightweight cement composite (ULCC) and the precision of its placement, each specimen was cast individually using small, carefully monitored batches of the material.

The test parameters of composite sandwich walls include the concrete compressive strength (f_{ck}), concrete core thickness (h_c), steel contribution ratio ($A_s f_y / A_c f_c$), connector spacing (s), and overall slenderness (λ). In Table 3.6, SCSW-R1 and SCSW-R2 are reference specimens. SCSW-H1 and SCSW-H2 are configured as slender walls with slenderness ratios of $\lambda = 0.54$ and 0.98, achieved by altering the height of the specimen (H). For SCSW-S1 and SCSW-S2, a sparser connector arrangement is employed to examine the effect of connector spacing (partial composite design). SCSW-C1 and SCSW-C2 are designed with thinner and thicker concrete cores, respectively. Headed shear stud connectors are utilized for SCSW-HS, differing from the reference specimens with J-hook connectors. A series of specimens, designated SCSW-C30, SCSW-C45, SCSW-T8, and SCSW-T12, were fabricated to systematically evaluate the effects of concrete compressive strength and the steel contribution factor on structural performance. To further examine the influence of confinement, specimens SCSW-P1 and SCSW-P2 were constructed

Table 3.6 Dimension and material properties of sandwich wall specimens

Specimen	H (mm)	b (mm)	h_c (mm)	t_s (mm)	S (mm)	Connector	w/t	s/t_s	λ	f_{ck} (MPa)	f_y (MPa)
SCSW-R1	400	590	120	6.0	100	JH	4.47	16.67	0.29	54.3	309.4
SCSW-R2	400	590	120	6.0	100	JH	4.47	16.67	0.29	54.3	309.4
SCSW-H1	1000	590	120	5.0	100	JH	4.54	20.00	0.51	60.9	309.4
SCSW-H2	2500	590	120	6.0	100	JH	4.47	16.67	0.92	58.9	309.4
SCSW-S1	500	590	120	6.0	201	JH	4.44	33.50	0.32	54.6	309.4
SCSW-S2	700	590	120	6.0	300	JH	4.44	50.00	0.38	55.0	309.4
SCSW-C1	400	590	90	6.0	100	JH	5.78	16.67	0.37	60.0	309.4
SCSW-HS	400	590	120	6.0	100	HSS	4.47	16.67	0.30	61.1	309.4
SCSW-C30	400	590	120	6.0	100	JH	4.47	16.67	0.27	37.8	309.4
SCSW-C45	400	590	120	6.0	100	JH	4.47	16.67	0.29	54.4	309.4
SCSW-T8	400	590	120	8.0	100	JH	4.34	12.50	0.30	57.1	393.9
SCSW-T12	400	590	120	12.0	102	JH	4.10	8.50	0.25	52.0	375.0
SCSW-P1	400	590	120	6.0	103	JH	4.47	17.17	0.30	55.3	309.4
SCSW-P2	400	590	120	6.0	100	JH	4.47	16.67	0.29	53.8	309.4
SCSW-C2	400	590	135	6.0	100	JH	4.01	16.67	0.26	52.6	309.4

H denotes the total height of the specimen, h_c signifies the thickness of the concrete core, t_s indicates the thickness of the steel face plate, and b represents the width of the wall section, s defines the spacing between shear connectors, while the ratio s/t_s describes the plate slenderness. λ corresponds to the member slenderness ratio. f_{ck} represents the compressive strength of a standard $\varnothing 100 \times 200$ mm concrete cylinder, and f_y indicates the yield strength of the steel. HSS and JH refer to headed shear stud and J-hook type connectors, respectively with closed-edge and open-edge plates, respectively. It is important to note that in these configurations, the side plates function primarily as confinement elements and do not contribute to the overall axial load capacity. The typical configuration of a sandwich shell is depicted in Fig. 3.1, while Fig. 3.2 presents images of the sandwich wall specimens captured during the fabrication stage and the subsequent compression test setup.

Fig. 3.1 Test specimens (**a**) before casting and (**b**) after casting. (Reprinted from Huang et al. (2015), copyright 2025, with permission from Elsevier)

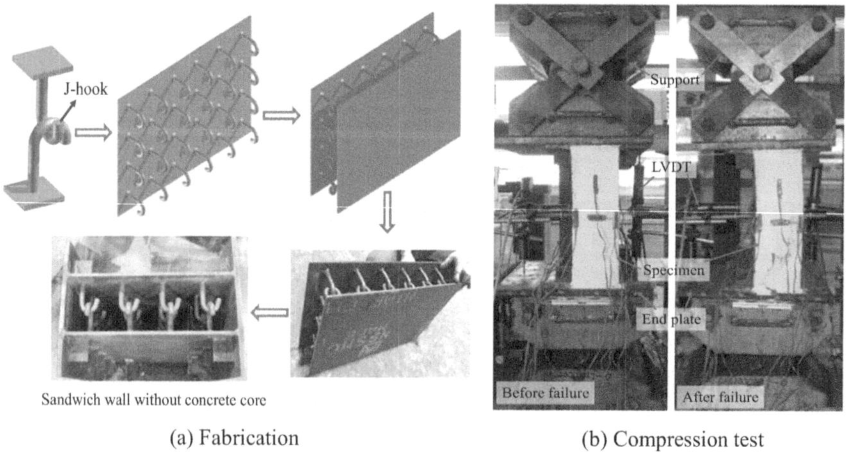

Fig. 3.2 Sandwich wall specimens. (**a**) Fabrication. (**b**) Compression test. (Reprinted from Huang and Liew [4], copyright 2025, with permission from Elsevier)

3.3 Axial Load Behavior of Steel–Concrete Sandwich Wall

3.3.1 Failure and Ultimate Resistance of Sandwich Wall Under Pure Compression

Figure 3.3 shows the principal failure modes identified during compressive test of the specimens, with four distinct categories being observed. The initial failure type, illustrated in Fig. 3.3a, was characterized by either splitting or crushing of the concrete core, which subsequently triggered localized buckling of the steel face

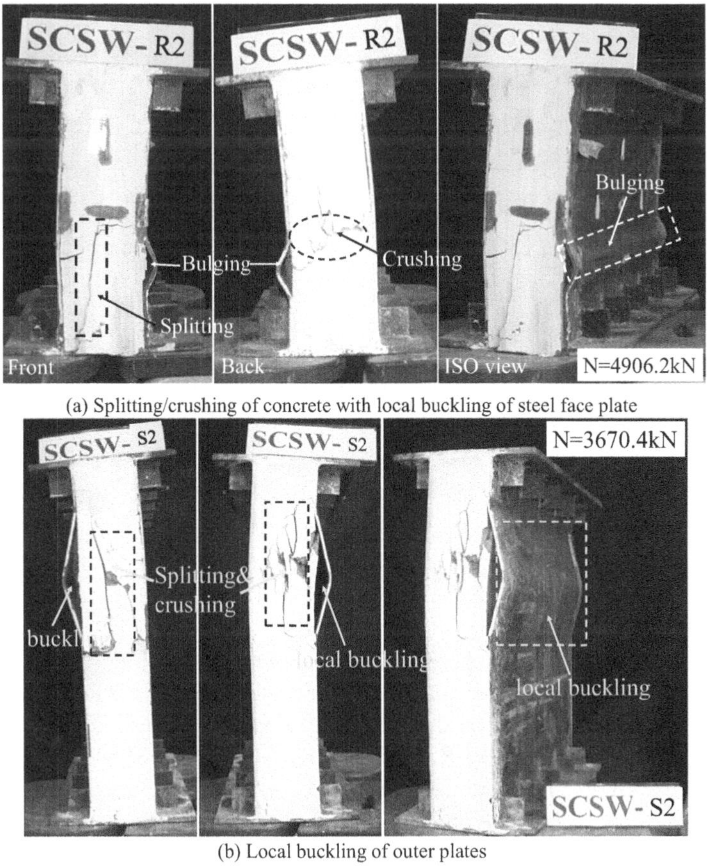

Fig. 3.3 Failure of SCS sandwich wall. (**a**) Splitting/crushing of concrete with local buckling of steel face plate. (**b**) Local buckling of outer plates. (**c**) Tensile fracture of J-hook connector. (**d**) Global buckling of slender wall. (Reprinted from Huang and Liew [4], copyright 2025, with permission from Elsevier)

(c) Tensile fracture of J-hook connector

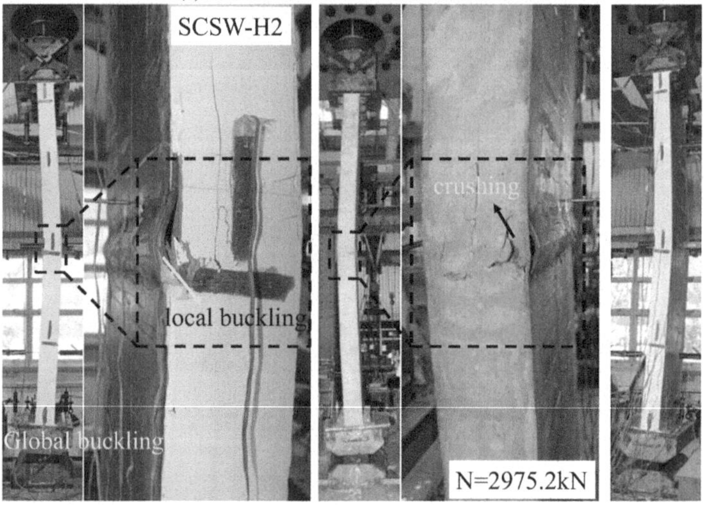

(d) Global buckling of slender wall

Fig. 3.3 (continued)

plates. This failure initiated severe concrete splitting and crushing cracks, with the outer steel plate yielding. Subsequently, the steel plates tended to buckle outward as the sectional stiffness degraded due to the developed cracks in the concrete core. Specimens SCSW-R1, SCSW-R2, SCSW-H1, SCSW-C1, SCSW-C30, SCSW-C50, and SCSW-C2 all exhibited this mode of failure, achieving a maximum compressive resistance averaging 88% of the cross-sectional resistance calculated by the Eurocode 4 approach for encased composite sections.

The second failure type, illustrated in Fig. 3.3b, manifested as a bond failure accompanied by severe, abrupt buckling of the external steel plates. This mode resulted from an inadequate shear and tensile bond at the steel-concrete interface, attributable to insufficient shear connector design, which caused significant interfacial slip. Specimens SCSW-S1 and SCSW-S2 exhibited this brittle failure, characterized by an audible loud pop at the point of collapse, with their maximum load-bearing capacity averaging only 75% of the cross-sectional resistance predicted by the Eurocode 4 method for encased composite sections.

The third failure mode, shown in Fig. 3.3c, involved tensile fracture of the shear connectors. This was observed in specimens SCSW-T8 and SCSW-T12, which featured thicker steel plates. The failure initiated with necking of the J-hook connector shanks, culminating in a complete tensile rupture. The reduced connector spacing-to-plate thickness ratio (s/t_s) in these specimens successfully delayed local plate buckling. Consequently, these specimens performed exceptionally well, achieving an average maximum resistance of 99% of the Eurocode 4 predicted capacity.

The fourth and final failure mode was global buckling of a slender wall specimen, depicted in Fig. 3.3d. Specimen SCSW-H2 failed in this manner, with concrete crushing observed on the concave side and steel yielding on the convex side at the mid-length region due to overall member bending. This instability led to a significant reduction in capacity, with the specimen reaching only 52% of its theoretical cross-sectional resistance, underscoring the critical need to incorporate slenderness reduction factors in the design calculations for such members.

3.3.2 *Failure and Ultimate Resistance of Sandwich Wall Under Combined Loads*

3.3.2.1 Failure Modes and Ultimate Resistance

The behavior of sandwich walls under eccentric loading (SCSW-e20, e40, e70, and e105) was characterized by the formation of visible cracks on the exposed ULCC core as the applied load approached its maximum, accompanied by audible cracking sounds. The primary failure mechanism involved the crushing and splitting of the concrete, which typically occurred after the tensile steel faceplate had yielded. As shown in Fig. 3.4, the failure mode under combined compression and bending was similar to that under pure axial compression, with concrete failure initiating before any local buckling of the steel plates. Contrary to some failure sequences, local buckling was not a precursor but a consequence of the failure, manifesting only after the degradation of the ULCC core's stiffness. This observation supports the finding by Sohel et al. [5] that appropriate shear connector spacing can effectively suppress premature local buckling. The absence of loud reports indicated that the J-hook connectors remained intact. A clear trend emerged where larger load eccentricities resulted in reduced axial capacity, increased specimen curvature, and higher bending moments, though composite action between the steel and concrete was maintained until failure without any signs of separation. For the beam specimen SCSW-B tested in four-point bending, failure commenced with yielding of the bottom steel plate, followed by the development of extensive flexural cracking. The experimental results, including the maximum axial loads (N) and the corresponding bending moments (M) calculated as N multiplied by the sum of eccentricity (e) and measured

(a) e=20 mm

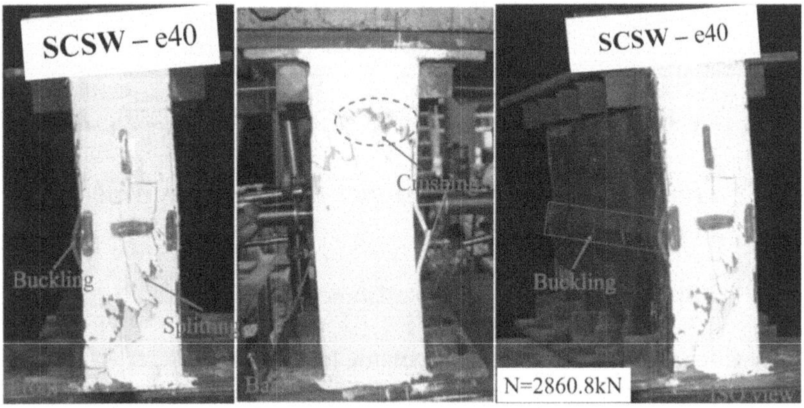

(b) e=40 mm

(c) e=70 mm

Fig. 3.4 Failure modes of sandwich wall under combined loads. (**a**) e = 20 mm. (**b**) e = 40 mm. (**c**) e = 70 mm. (**d**) e = 105 mm. (**f**) bending test. (Reprinted from Huang and Liew [4], copyright 2025, with permission from Elsevier)

(d) e=105 mm

(f) bending test

Fig. 3.4 (continued)

lateral displacement, are consolidated in Table 3.7, confirming the inverse relationship between axial load capacity and eccentricity.

3.3.2.2 Load Deflection Curve Behaviour

Figure 3.5 presents the load versus deflection responses for the tested specimens. The family of curves, which correspond to different load eccentricities, demonstrates a consistent pattern: both the failure load and the initial elastic stiffness of the specimens decrease as the eccentricity increases. Conversely, the structural ductility exhibits an inverse relationship, showing a tendency to improve under conditions of greater eccentricity.

Table 3.7 Failure loads and failure mechanism of sandwich walls subjected to combined axial and bending loads

Specimen	Load conditions	Maximum axial load N (kN)	Specimen M (kN.m)	Load conditions
SCSW-R2	P-C, e = 0 mm	4906.2	0	Concrete failure and local buckling
SCSW-e20	C-L, e = 20 mm	3976.8	98.4	Concrete failure and local buckling
SCSW-e40	C-L, e = 40 mm	2860.6	128.8	Concrete failure and local buckling
SCSW-e70	C-L, e = 70 mm	2034.1	154.5	Concrete failure and local buckling
SCSW-e105	C-L, e = 105 mm	1570.4	180.2	Concrete failure and local buckling
SCSW-e105'	C-L, e = 105 mm	1621.7	186.2	Concrete failure and local buckling
SCSW-B	Pure bending	0	117.8	Flexural failure

P-C pure compression, *C-L* combined loads

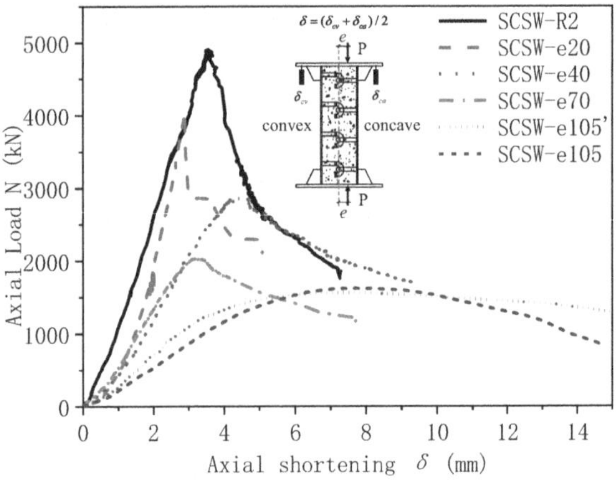

Fig. 3.5 Load-shortening curves for sandwich wall subjected to combined axial and bending loads. (Reprinted from Huang and Liew [4], copyright 2025, with permission from Elsevier)

3.4 Flexural-Shear and Punching-Shear of Curved Sandwich Panels

3.4.1 Failure Mechanism of Sandwich Beam

3.4.1.1 Failure Modes and Ultimate Strength

A summary of the experimental results for the beam and shell specimens, detailing their ultimate failure loads and corresponding failure mechanisms, is provided in Table 3.8. The testing of curved beams revealed three distinct modes of failure: flexural failure, shear-compression failure, and shear-tension failure, which are illustrated in Fig. 3.6a–c, respectively.

Table 3.8 Failure load and failure mechanism of SCS sandwich beams

Specimen	Steel contribution ratio ($A_s f_y / A_c f_c$)	Plate slenderness (s/t_s)	Stiffness (N/mm)	Loading type	Failure load (kN)	Failure mode
FSB-01	0.41	27.5	14,122	Central patch load	133	Flexural
CSB-02	0.23	27.5	294,974		1101	Shear-compression
CSB-03	0.41	27.5	47,645		556	
CSB-04	0.81	13.8	51,218		452	Shear-tension
CSB-05	1.22	9.2	57,022		800	
CSB-06	0.51	27.5	29,691		239	
CSB-07	0.41	55.0	31,793		310	
CSB-08	0.41	27.5	76,181		458	
CSB-09	0.41	27.5	18,983	Central point load	218	
CSB-10	0.41	27.5	41,797	Asymmetric patch load	416	

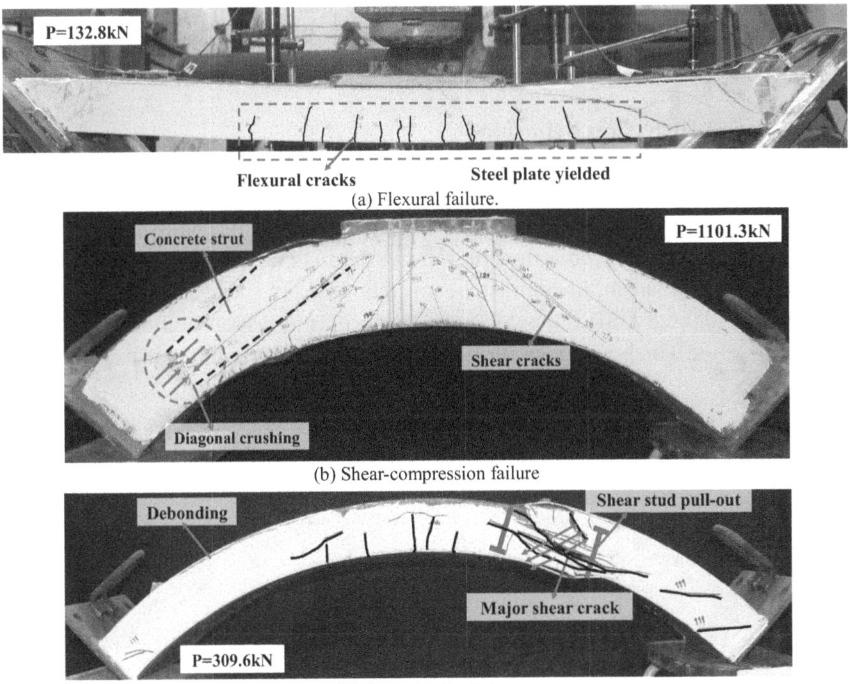

(a) Flexural failure.

(b) Shear-compression failure

(c) Shear-tension failure

Fig. 3.6 Failure modes of sandwich beams. (**a**) Flexural failure. (**b**) Shear-compression failure. (**c**) Shear-tension failure. (Reprinted from Huang et al. (2015), copyright 2025, with permission from Elsevier)

The structural behavior of the specimens varied significantly depending on the failure mode. Beams that failed in flexure demonstrated a clear elastic phase followed by a ductile post-yield response. This failure mechanism initiated with yielding of the tensile steel faceplate, leading to substantial beam deflection and the development of wide cracks in the concrete core during the final loading stages, a phenomenon corroborated by both strain gauge data and visual evidence in Fig. 3.6a. Following the yield of the bottom plate, cracks progressed towards the compression zone, with the neutral axis shifting upwards until it approached the lower surface of the top steel plate ($x_c \approx 0$), coinciding with the attainment of the section's full moment capacity.

In contrast, the shear-compression failure mode, illustrated in Fig. 3.6b, was characterized by initial debonding cracks near the quarter-span at approximately 20% of the ultimate load, accompanied by vertical flexural cracks at mid-span. As the load increased to about 80% of its maximum value, a critical diagonal shear crack formed. Subsequent loading caused this crack to propagate, culminating in the crushing of the diagonal concrete strut, which dictated the ultimate strength of the member.

A third mode, shear-tension failure, is shown in Fig. 3.6c. This failure was marked by the development of a major inclined crack without concrete crushing; instead, the ultimate capacity was governed by the tensile failure of the shear connectors, which were observed to pull out, indicating that strength was limited by the combined shear resistance of the concrete and the tensile capacity of the connectors.

3.4.1.2 Load Deflection Behaviour

Figure 3.7 illustrates the load-deflection responses for all beam specimens, revealing distinct behavioral patterns corresponding to their failure mechanisms. Specimen DSB-01, which experienced a flexural failure, demonstrated a ductile curve with a prolonged plastic plateau. In contrast, beams that failed in shear-compression mode, such as DSB-02 and DSB-03, exhibited a linear elastic response until the onset of debonding, after which a sudden, brittle failure occurred due to the crushing of the

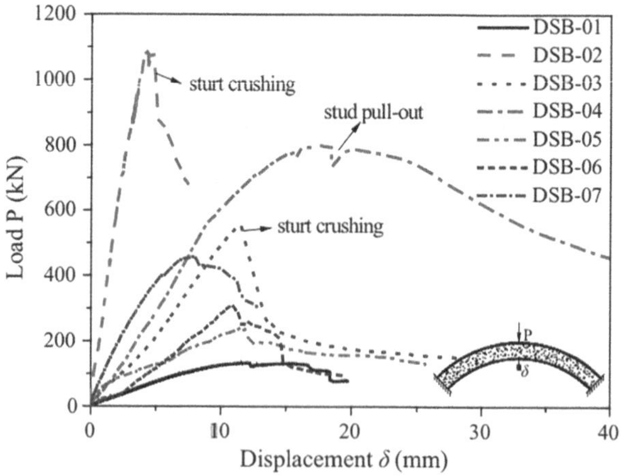

Fig. 3.7 Load–deflection curves of sandwich beams. (Reprinted from Huang et al. (2015), copyright 2025, with permission from Elsevier)

diagonal concrete strut, resulting in an abrupt drop in the load-deflection curve without yielding of the steel face plates.

Specimens that failed in shear-tension mode, including DSB-04 through DSB-07, displayed a more ductile post-peak behavior compared to the shear-compression failures. Notably, DSB-04, characterized by a high steel contribution ratio ($A_s f_y / A_c f_{ck} > 0.8$), showed a softer unloading response. An audible failure of a shear stud was observed in this specimen, causing a temporary load drop. However, the load subsequently increased again as the remaining intact studs engaged, allowing the beam to maintain a degree of composite action and continue carrying additional load.

3.4.2 Failure Mechanism of Sandwich Shell

3.4.2.1 Failure Modes and Ultimate Strength

The structural response of the shell specimen is evident from its load-deflection curve presented in Fig. 3.8. The initial phase shows a linear increase in load with mid-span deflection up to a first peak, after which unloading occurs. This corresponds to a localized punching-shear failure within the concrete core around the load application area, as shown in Fig. 3.9. Following this initial failure, the load increased once more to a second peak under substantial deflection, though this secondary peak did not surpass the initial maximum. The ultimate failure involved a punching fracture of the outer steel plate. Post-failure examination revealed an indentation on the outer plate and a corresponding convex deformation on the inner plate, illustrated in Fig. 3.9b, c. Fracturing of the inner plate along two rows of shear connectors was also observed, a consequence of excessive downward displacement following the loss of concrete support. In summary, the first peak resistance of the shell was controlled by the punching shear capacity of the concrete core, while the second peak was limited by the fracture strength of the weaker of the two steel face plates.

Upon reaching the first peak load, a punched concrete frustum was formed in the middle arch strip. In this case, the middle arch strip lost its load-carrying capacity,

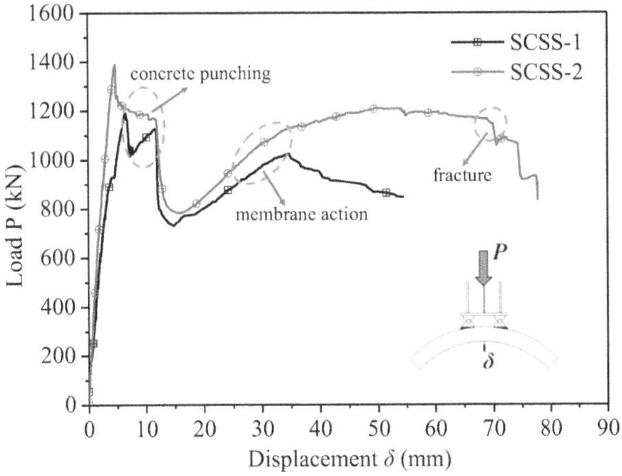

Fig. 3.8 Load deflection curves of sandwich shells. (Reprinted from Huang et al. (2015), copyright 2025, with permission from Elsevier)

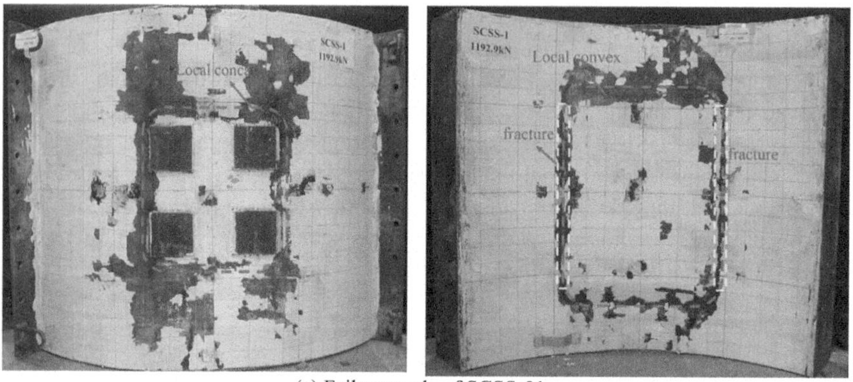

(a) Failure mode of SCSS-01.

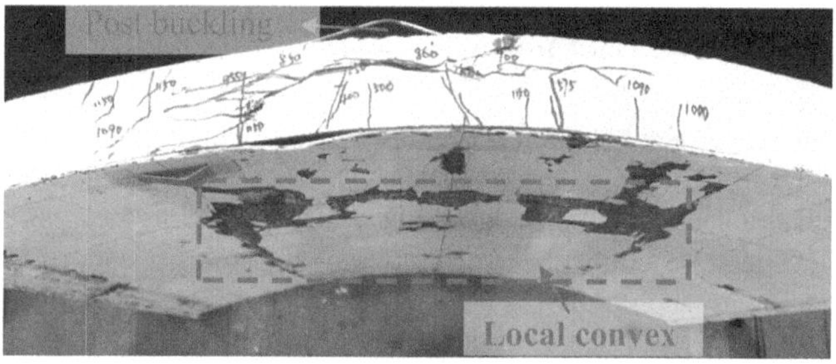

(b) Post buckling and local convex of plates.

(c) Punching shear fracture of plate.

Fig. 3.9 Failure of sandwich shells. (**a**) Failure mode of SCSS-01. (**b**) Post buckling and local convex of plates. (**c**) Punching shear fracture of plate. (Reprinted from Huang et al. (2015), copyright 2015, with permission from Elsevier)

causing the load to drop. However, benefiting from the tension membrane effect of the outer steel plate, the load may be transferred to the neighboring width strip and arch strip through the outer steel plate. Due to the fixed boundary conditions at both edges of the shell, the applied load was ultimately transferred to the supports via the undamaged arch-shaped strip of the composite section. This load redistribution mechanism is reflected in the load-deflection curve, which exhibits a characteristic strength hardening behavior, as emphasized in Fig. 3.8. Consequently, the ultimate load-bearing capacity of the shell is governed by the weaker of two potential failure mechanisms: the punching shear strength of the outer steel plate or the bending moment capacity of the intact arch strip within the shell section.

3.4.2.2 Load Deflection Behaviour

Based on the experimental data, the characteristic load-deflection response of sandwich shells subjected to patch loading is illustrated in Fig. 3.10. Both tested shells exhibited a comparable behavioral pattern, characterized by load-deflection curves displaying two distinct peak loads. The initial response involved a linear ascent to the first peak strength (Stage OA), during which minor tensile cracking occurred in the concrete core. This was followed by an unloading phase (Stage AB), signifying a punching shear failure within the concrete that created a conical fragment. The non-uniform stress distribution around the loaded perimeter, attributable to the large loading area, occasionally resulted in an oscillatory load pattern (Stage AA') and delayed the concrete's complete punching failure. Subsequently, the load increased again (Stage BC) due to the development of tensile membrane action in the outer steel plates. Ultimate failure ensued at a deflection of approximately 30–50 mm, triggered by either a punching fracture of the steel face plate accompanied by local buckling of the outer and inner plates, or by concrete spalling at the mid-span. A summary of the two peak loads, their corresponding deflections, and the final failure modes for each shell is provided in Table 3.9.

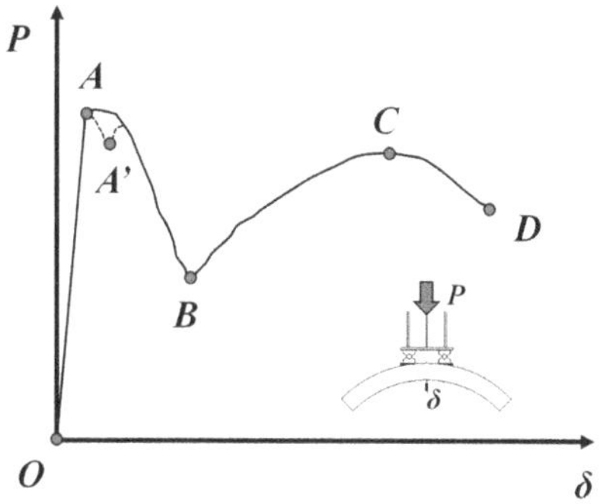

Fig. 3.10 Typical load-deflection curve of SCS shell under patch loading. (Reprinted from Huang et al. (2015), copyright 2025, with permission from Elsevier)

Table 3.9 Summary of failure mechanisms, ultimate load and the associated mid-span deflection for sandwich shell

Specimen	P_1 (kN)	δ_1 (mm)	Failure mode	P_2 (mm)	δ_2 (mm)
SCSS-01	1192.9	6.5	Punching shear of concrete core	1025.0	34.7
SCSS-02	1390.9	4.8	Punching shear of concrete core	1210.0	50.1

P_1 first peak load, P_2 second peak load

3.5 Finite Element Analysis of Curved Steel–Concrete Sandwich Panels

3.5.1 Material Model and Composite Interaction

The Concrete Damaged Plasticity (CDP) model is widely employed in finite element simulations of steel-concrete composite structures, with its application frequently validated against experimental results as documented in literature [6–10]. Following this established methodology, the current study also utilizes the CDP model. This constitutive model integrates non-associated multi-hardening plasticity with isotropic damage elasticity to represent the irreversible material degradation that occurs during fracture processes. It is particularly suitable for analyzing concrete behavior under monotonic, cyclic, and dynamic loading conditions at low confinement levels, effectively capturing key failure mechanisms such as compressive crushing and tensile cracking within both ABAQUS/Standard and Explicit solvers. The definition of the model requires comprehensive uniaxial stress-strain relationships for both tension and compression to characterize the material's elastic, plastic, and damage responses, as illustrated in Fig. 3.11. Key parameters include E_0 (the initial undamaged elastic modulus), ε_t^{el} and ε_t^{pl} (tensile elastic and plastic strains), ε_c^{el} and ε_c^{pl} (compressive elastic and plastic strains), d_t and d_c (damage variables ranging from 0 to 1 for tension and compression), σ_{t0} (tensile yield stress), σ_{c0} (compressive yield stress), and σ_{cu} (ultimate compressive stress). The corresponding mathematical formulations describing the uniaxial stress-strain behavior under both tensile and compressive loading are provided subsequently:

$$\sigma_c = (1 - d_c)E_0\left(\varepsilon_c - \widetilde{\varepsilon}_c^{\,pl}\right) \tag{3.1}$$

$$\sigma_t = (1 - d_t)E_0\left(\varepsilon_t - \widetilde{\varepsilon}_t^{\,pl}\right) \tag{3.2}$$

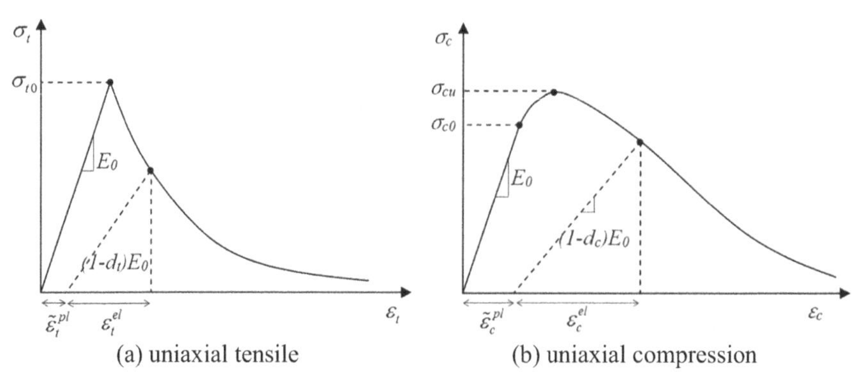

Fig. 3.11 Stress-strain curves of CDP model. (**a**) uniaxial tensile. (**b**) uniaxial compression

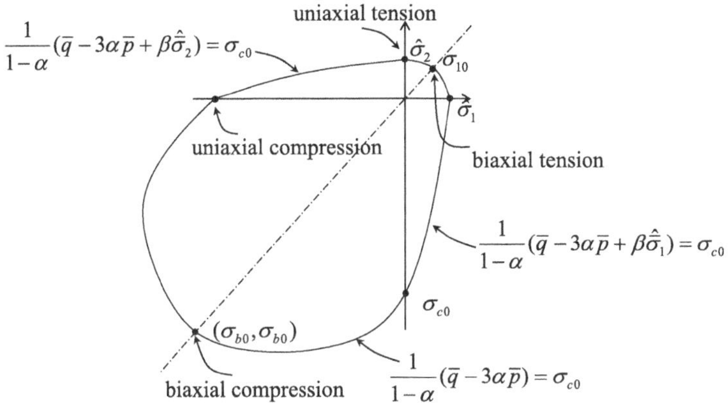

Fig. 3.12 Yield surface in plane stress

The yield function, which defines the failure or damage envelope within the effective stress space, is formulated based on the model initially introduced by Lubliner et al. [11] and subsequently refined by the modifications suggested by Lee and Fenves [12]. The progression of this yield surface is controlled by the development of the hardening variables, denoted as $\tilde{\varepsilon}_t^{pl}$ and $\tilde{\varepsilon}_c^{pl}$, as expressed in the following equation:

$$F=\frac{1}{1-\alpha}\left(\bar{q}-3\alpha\bar{p}+\beta\left(\tilde{\varepsilon}^{pl}\right)<\hat{\bar{\sigma}}_{\max}>-\gamma<-\hat{\bar{\sigma}}_{\max}>\right)-\bar{\sigma}_c\left(\tilde{\varepsilon}_c^{pl}\right)=0 \qquad (3.3)$$

Figure 3.12 shows the yield surface, where, α and γ are dimensionless material constants, $\alpha=(\sigma_{b0}/\sigma_{c0})/[2(\sigma_{b0}/\sigma_{c0})-1]$; $0<\alpha<0.5$; σ_0/σ_0 denotes the ratio of initial equi-bi-axial compressive yield stress to initial uniaxial compressive yield stress (default value 1.16); $\beta=(1-\alpha)\bar{\sigma}_c\left(\varepsilon_c^{pl}\right)/\bar{\sigma}_t\left(\varepsilon_t^{pl}\right)-(1+\alpha)$, where, $\bar{\sigma}_c\left(\varepsilon_c^{pl}\right)$ and $\bar{\sigma}_t\left(\varepsilon_t^{pl}\right)$ represent the effective compressive and tensile cohesion stress; $\hat{\bar{\sigma}}_{\max}$ is the maximum principal effective stress; $\gamma=3(1-K_c)/(2K_c-1)$, $K_c=\sqrt{J_{2TM}}/\sqrt{J_{2CM}}\in(0.5\sim1.0]$ represents the ratio of the second stress invariant (J_2) on the tensile meridian to that on the compressive meridian at initial yield for any given value of the pressure invariant p such that the maximum principal stress is negative ($\hat{\bar{\sigma}}_{\max}<0$) and the value is 2/3;$\bar{p}=-\frac{1}{3}tr(\bar{\boldsymbol{\sigma}})$ is the effective hydrostatic pressure; $\bar{q}=\sqrt{\frac{3}{2}\left(\bar{\mathbf{S}}:\bar{\mathbf{S}}\right)}$ is the Mises equivalent effective stress, in which $\bar{\mathbf{S}}$ is the effective stress deviator.

Plastic flow is governed by a flow potential function $G(\bar{\boldsymbol{\sigma}})$ based on non-associated flow rule $d\varepsilon^{pl}=d\lambda\partial G(\bar{\sigma})/\partial\bar{\sigma}$. Drucker-Prager hyperbolic function is used in the model as given below:

$$G=\sqrt{(\varsigma\sigma_{t0}\tan\psi)^2+\bar{q}^2}-\bar{p}\tan\psi \qquad (3.4)$$

where, ψ denotes dilation angle measured in p-q plane at high confining pressure; ς is a parameter referred to as the eccentricity that defines the rate at which the function approaches the asymptote (the flow potential tends to a straight line as ς tends to zero).

Since the ULCC is a novel kind of lightweight cement composite with limited mathematical model, this study calibrates the parameters of CDP model by the test data. The stress-strain characteristics of concrete, as determined experimentally under both tensile and compressive loading conditions, served as the fundamental input parameters for the damaged plasticity model. The critical damage parameters, d_c and d_t, which are integral to defining the concrete's compression hardening and tension stiffening behavior, can be accurately calibrated using data from uniaxial compression tests and four-point bending tests. Through verification studies conducted within the ABAQUS framework [13], Wang and Chen [14] established that a first-order exponential decay function effectively represents the concrete's damage evolution. Their research demonstrated a definitive relationship between the normalized compressive damage variable, $D_{c,norm}$, and the normalized compressive inelastic strain $\tilde{\varepsilon}^{in}_{c,norm}$, which is mathematically expressed by the following equation:

$$D_{c,norm} = A_0 e^{-\tilde{\varepsilon}^{in}_{c,norm}/t_0} + B_0 \tag{3.5}$$

where, $D_{c,norm} = d_c$, $\tilde{\varepsilon}^{in}_{c,norm} = \tilde{\varepsilon}^{in}_c/\varepsilon_{cm}$, $\tilde{\varepsilon}^{in}_c = \varepsilon_c - \sigma_0/E_0$, $A_0 = 1/(e^{-1/t0-1})$, $B_0 = -1/(e^{-1/t0-1})$.

The single unknown parameter, t_0, can be calibrated by optimizing the correlation between the simulated uniaxial stress-strain curves and the corresponding experimental data. In a similar methodology, the relationship between the normalized tensile damage variable, $D_{t,norm}$, and the normalized crack width, $w_{t,norm}$, can also be characterized using a first-order exponential decay function, expressed as follows:

$$D_{t,norm} = A_1 e^{-w_{t,norm}/t_1} + B_1 \tag{3.6}$$

where, $D_{t,norm} = d_t$, $w_{t,norm} = w/w_u$, $A_1 = 1/(e^{-1/t1-1})$, $B_1 = -1/(e^{-1/t1-1})$. The value of the remaining unknown parameter, t_1, can be ascertained by aligning the simulated load-displacement curves with the experimental data obtained from four-point bending tests. Utilizing this calibration methodology, the parameters for damage in compression and tension, t_0 and t_1, were determined to have values of 0.1 and 0.05, respectively. Figure 3.13 presents the parameters of the CDP model for ULCC based on Eqs. (3.5) and (3.6).

In the finite element simulations, the concrete material parameters—specifically the dilation angle, the ratio of biaxial to uniaxial compressive strength, and the eccentricity—were assigned values of 26.5, 1.16, and 0.1, respectively. These values were selected in accordance with parameters established and validated in previous research [9, 15]. To enhance convergence rate in the concrete, a viscosity parameter $\kappa = 0.0001$ was employed, following recommendations from Barth and Wu [16] and ABAQUS documentation [13].

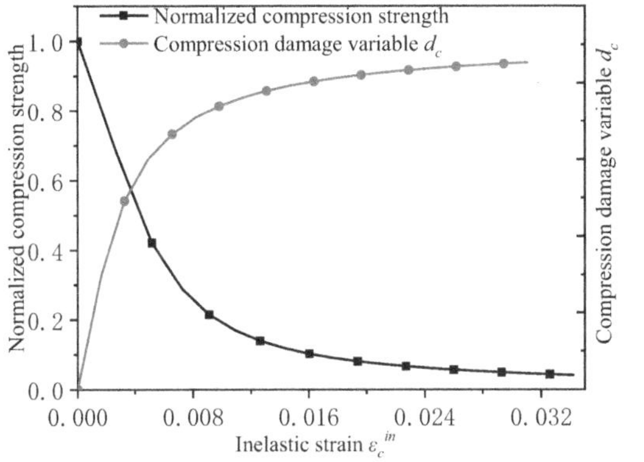

(a) Correlation between normalized compressive strength, damage variable v.s. inelastic strain

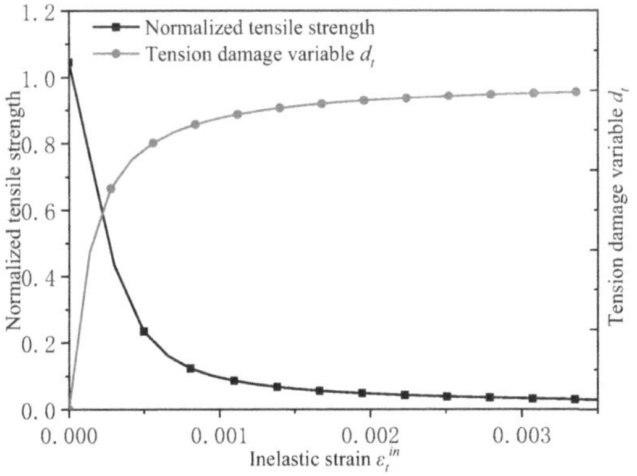

(b) Correlation between normalized tensile strength, damage variable v.s. inelastic strain

Fig. 3.13 Concrete damaged plasticity model for ULCC. (**a**) Correlation between normalized compressive strength, damage variable v.s. inelastic strain. (**b**) Correlation between normalized tensile strength, damage variable v.s. inelastic strain. (Reprinted from Huang and Liew [4], copyright 2025, with permission from Elsevier)

However, preliminary analysis of SCS sandwich structure showed that the solution diverged when the concrete tension resistance was reduced to zero, and the element stiffness was correspondingly reduced to zero in the direction normal to the crack. It was, therefore, the uniaxial tensile strength of concrete assumed to be linear until the tensile strength of concrete was reached, which was taken as $0.1 f_{ck}$. Beyond this failure stress, the tensile softening response was characterized by fracture energy G_F. Equation (3.7) was used to determine G_F [17], which is given below:

$$G_F = \left(0.0469 d_{max}^2 - 0.5 d_{max} + 26\right)\left(\frac{f_c}{10}\right)^{0.7} N/m \tag{3.7}$$

where, f_c is the concrete compressive strength of cylinder in MPa, d_{max} is the maximum coarse aggregate size (in mm). d_{max} was taken as 20 mm for analysis in this work.

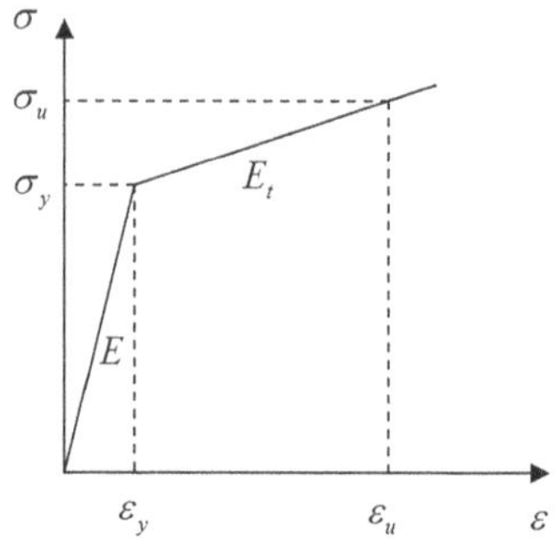

Fig. 3.14 Bilinear stress-strain model for steel under tension. (Reprinted from Huang and Liew [4], copyright 2025, with permission from Elsevier)

In the finite element analysis conducted for this study, the constitutive behavior of steel was represented using an elastic-plastic model incorporating the von Mises yield criterion and an isotropic hardening law within ABAQUS. The material response is characterized by a typical bilinear stress-strain relationship, as shown in Fig. 3.14. This relationship demonstrates linear elastic behavior up to the yield stress (σ_y), followed by a linear hardening regime that continues until the material reaches its ultimate stress (σ_u).

3.5.2 *Steel-to-Concrete Interaction*

In the finite element modeling, the steel plates, shear connectors, and concrete core were all discretized using C3D8R elements, which are 8-node linear bricks with reduced integration and hourglass control. The interaction between the loading plate and the top steel plate, as well as the interface between the steel plates and the concrete core, was simulated using a surface-to-surface contact algorithm with a penalty enforcement method. This approach requires designating a master surface, typically belonging to the stiffer body or the component with a coarser mesh, and a slave surface. Accordingly, the steel surfaces were assigned as master surfaces in contact with the concrete (slave), while the top surface of the specimen was defined as the slave surface interacting with the loading plate (master). The contact algorithm manages penetration by introducing a fictitious spring force when a slave node penetrates the master surface. The model incorporated a defined friction coefficient for tangential behavior, while the normal direction utilized a hard contact pressure-overclosure relationship.

The friction coefficient is a critical parameter, and a sensitivity study was conducted to determine its reasonable value, ranging from 0.2 to 0.6. Based on the mesh sensitivity study, a friction coefficient value of 0.5 was selected for subsequent FE simulation in this work.

3.5.3 *Modelling of Shear Connectors*

Shear connectors serve as a common method to enhance composite action between concrete and steel in composite structures, playing a crucial role in load transfer, resisting longitudinal slip, and providing transverse shear resistance. There are several methods available to model shear connectors: (i) Beam element: This method uses beam elements to represent shear connectors. However, it tends to neglect concrete bearing capacity, failing to capture the interaction between concrete and shear connectors adequately. (ii) Embedment in concrete element: This approach involves embedding shear connectors in concrete elements, with both parts sharing nodes according to the strain compatibility rule. While straightforward, it overlooks slip between concrete and connectors, potentially overestimating the structural stiffness. (iii) Surface-to-surface contact technique: This method utilizes surface-to-surface contact between connector and concrete surfaces. Though seemingly reasonable, it can face challenges related to convergence or time-consuming computations, particularly with numerous contact pairs and high material and geometric nonlinearity. In this work, a simplified modeling approach utilizing nonlinear connector elements is introduced to simulate the behavior of shear connectors, incorporating tensile test data to accurately represent concrete bearing and tensile response under partial interfacial contact conditions. This method models both headed shear studs and J-hook connectors as straight round bars connected by nonlinear elements, significantly reducing the number of contact pairs required—particularly for complex geometries—and thereby improving computational efficiency. As illustrated in Fig. 3.15a, b, the proposed connector element is implemented in the finite element model for both connector types. It is important to emphasize that the tensile load-displacement relationships must be specific to each connector type; for example, headed studs require data from tests on studs embedded in concrete, while J-hook connectors need corresponding data for their specific geometry. The tensile test configurations and resulting load-displacement curves for the headed stud and J-hook connector used in this study are provided in Figs. 3.16 and 3.17, respectively.

3.5.4 *Verification*

3.5.4.1 Steel–Concrete Sandwich Wall

A representative FE model of the sandwich wall under compressive loading is depicted in Fig. 3.18. To simulate symmetry, all nodes on the symmetric plane were restrained from displacing in the out-of-plane direction. The mechanical interaction between the J-hook connectors and the concrete core was modeled using the proposed simplified connector element, which integrated the actual tensile-elongation behavior of the J-hook connectors as determined from experimental data.

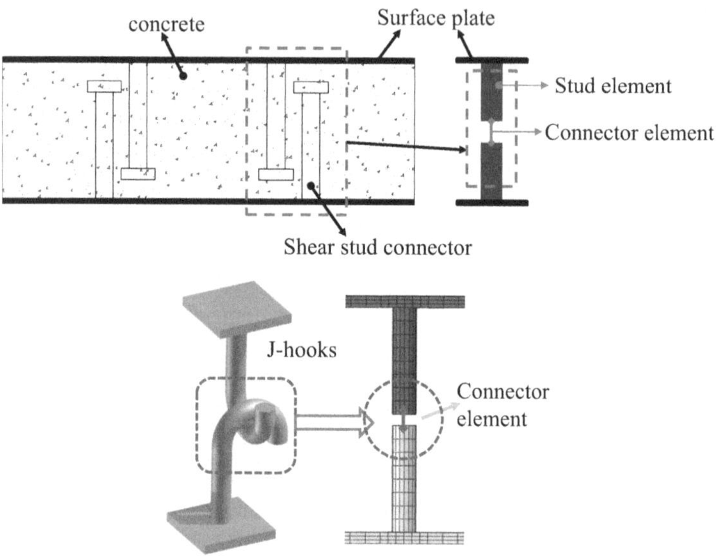

Fig. 3.15 Simplified model for a pair of headed shear studs and J-hook connectors. (Reprinted from Huang and Liew [4], copyright 2025, with permission from Elsevier)

Furthermore, the FE model incorporated initial geometric imperfections for the long wall specimens, based on precise measurements taken during testing. As illustrated in Fig. 3.19, the maximum lateral deviation was quantified using a digital spirit level, with its magnitude calculated via Eq. (3.8). To account for this imperfection in the nonlinear analysis, the model was initialized with an out-of-straightness profile corresponding to the shape of the first buckling mode, scaled to the measured maximum amplitude. This approach ensured that the influence of initial geometric imperfections was accurately represented in the simulation.

$$\delta_i = \frac{\alpha_t l_1 + \alpha_b l_2}{2} \tag{3.8}$$

A comprehensive comparison between numerical simulations and experimental data is presented in Table 3.10 and accompanying figures. The numerical model demonstrates high accuracy, with an average ratio of experimental maximum load to predicted load of 1.05 and a standard deviation of 0.05. Figure 3.20a–e reveal that the finite element analysis successfully replicates the primary failure modes observed in physical tests. Specifically, the model accurately captures concrete crushing and local buckling of steel face plates, with damage values approaching unity in the concrete core indicating material failure. While the precise location of steel plate buckling shows some variation due to unavoidable initial imperfections such as residual stresses, aggregate distribution, and early-age concrete effects, the characteristic buckling waveform between adjacent shear connectors is effectively captured, as visible in Fig. 3.20f.

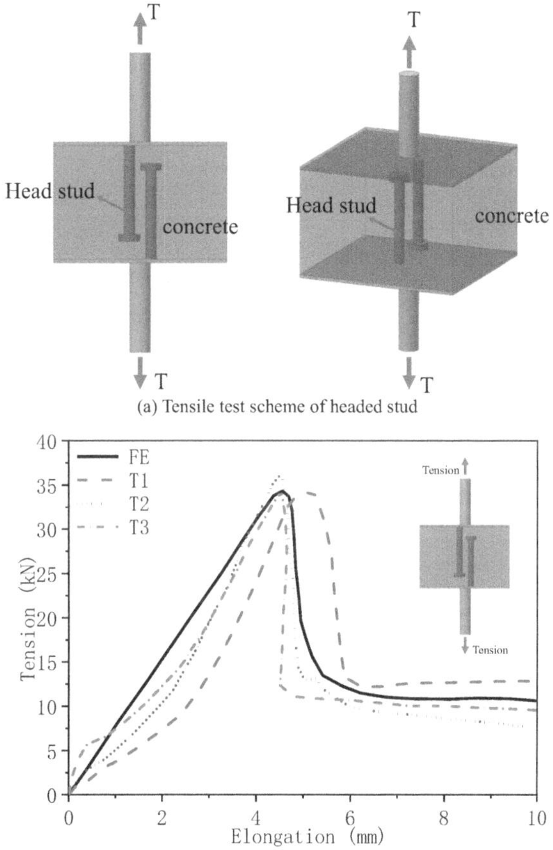

(a) Tensile test scheme of headed stud

(b) Pull-out test results [9]

Fig. 3.16 Tensile-elongation of headed stud in concrete. (**a**) Tensile test scheme of headed stud. (**b**) Pull-out test results [10]. (Reprinted from Huang and Liew (2015), copyright 2025, with permission from Elsevier)

The load-deflection relationships derived from FE analysis show strong correlation with experimental measurements, as evidenced in Fig. 3.21. Furthermore, the model provides reasonable estimations of strain history at critical monitoring locations (S1 and SG7) for selected wall specimens, as demonstrated in Fig. 3.22a–d. For members under eccentric compression and four-point bending conditions, Fig. 3.23 confirms that the simulated load-deformation paths closely match the experimental curves. Table 3.10 compares the maximum loads obtained from experiments and FE analysis for SCS sandwich walls, showing good agreement between the two methods, with a mean N_{test}/N_{FE} ratio of 1.05 and a standard deviation of 0.05. Similarly, quantitative assessment shows average test-to-prediction ratios of 0.967 for compressive load and 1.012 for bending moment, with standard deviations of 0.064 and 0.113 respectively. These consistent validation results across multiple loading conditions confirm that the developed finite element model reliably predicts the structural behavior of steel-concrete-steel sandwich walls under combined loading scenarios (see Table 3.11).

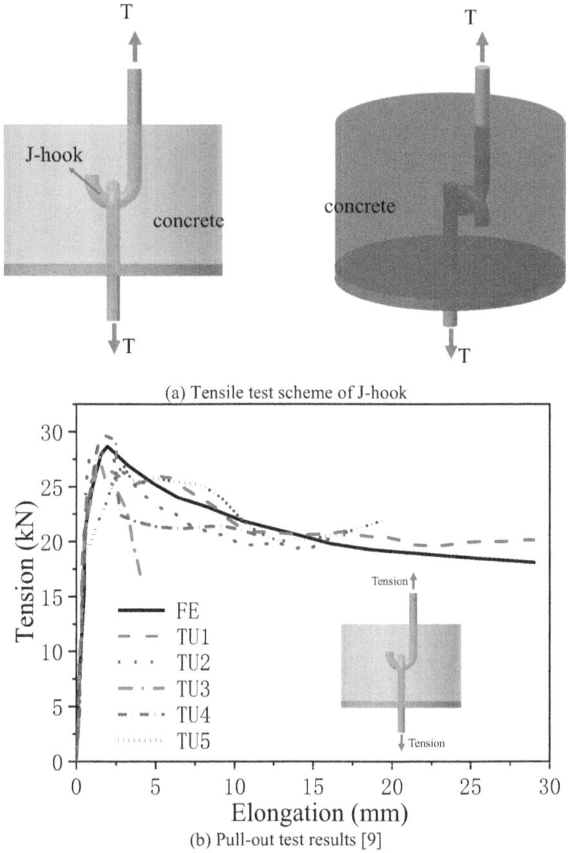

Fig. 3.17 Tensile-elongation of J-hook connector embedded in concrete. (**a**) Tensile test scheme of J-hook. (**b**) Pull-out test results [10]. (Reprinted from Huang and Liew [4], copyright 2025, with permission from Elsevier)

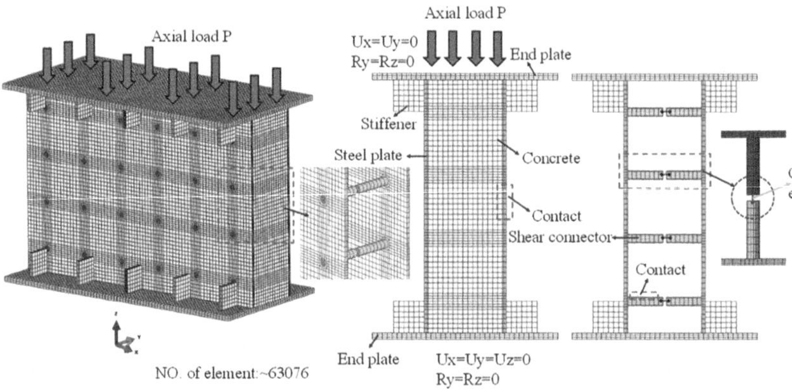

Fig. 3.18 Typical FE model for SCS sandwich wall. (Reprinted from Huang and Liew [4], copyright 2025, with permission from Elsevier)

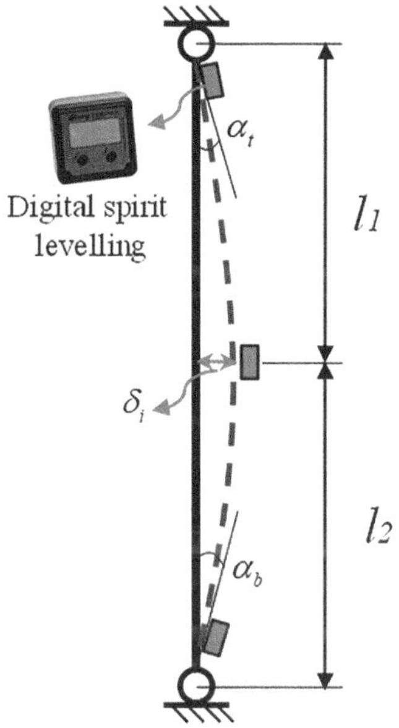

Fig. 3.19 Imperfection measuring of slender SCS wall

Table 3.10 Comparison of maximum load capacities obtained from FE and test results for sandwich walls

Specimen	Maximum load		
	N_{test} (kN)	N_{FE} (kN)	N_{test}/N_{FE}
SCSW-R1	4191	4295.8	0.98
SCSW-R2	4906	4727.8	1.04
SCSW-H1	4788	4405.1	1.09
SCSW-H2	2975	2826.0	1.05
SCSW-S1	4656	4394.3	1.06
SCSW-S2	3670	3376.7	1.09
SCSW-C1	4248	4013.8	1.06
SCSW-HS	5317	5083.3	1.05
SCSW-C30	3916	3626.9	1.08
SCSW-C45	4689	4614.1	1.02
SCSW-T8	6889	5939.2	1.16
SCSW-T12	8418	8460.7	0.99
SCSW-P1	5120	5117.1	1.00
SCSW-P2	4933	4636.1	1.06
SCSW-C2	5467	5554.0	0.98
Mean value			1.05
Std. Dev.			0.05

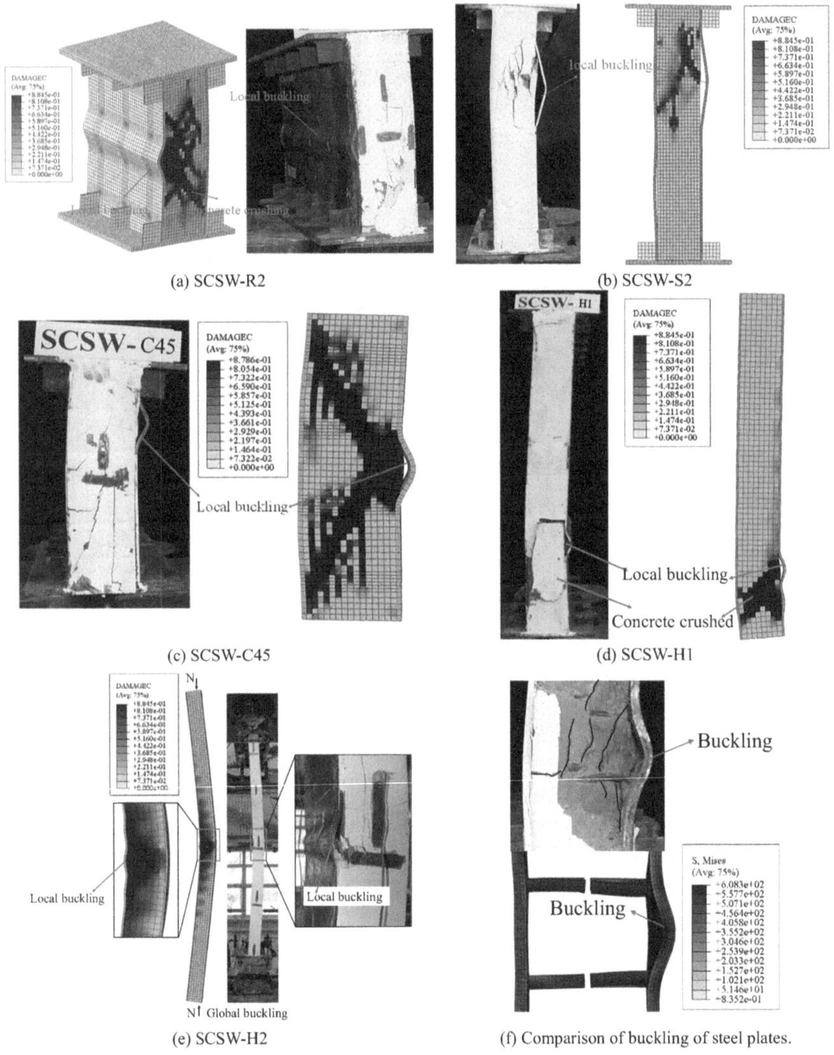

Fig. 3.20 Comparison of failure modes obtained from test and FE results for sandwich walls. (**a**) SCSW-R2. (**b**) SCSW-S2. (**c**) SCSW-C45. (**d**) SCSW-H1. (**e**) SCSW-H2. (**f**) Comparison of buckling of steel plates. (Reprinted from Huang and Liew [4], copyright 2025, with permission from Elsevier)

3.5.4.2 Steel-Concrete Sandwich Beam

Figure 3.24 depicts a typical quarter-symmetric FE model of the curved SCS sandwich beam with headed shear studs. In general, the FE models yield reasonable estimations of the load-deflection behavior for most test specimens, including maximum load, shear crack locations, damaged concrete, and failure modes.

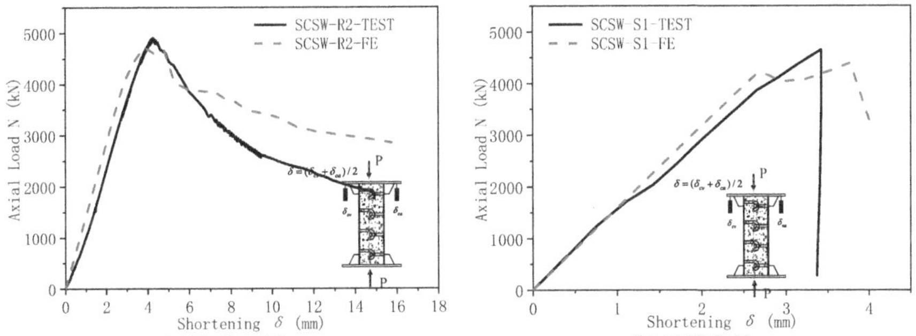

(a) SCSW-R2 (b) SCSW-S1

Fig. 3.21 Comparison of load-defection curves obtained from test and FE results for sandwich walls. (**a**) SCSW-R2. (**b**) SCSW-S1. (Reprinted from Huang and Liew [4], copyright 2025, with permission from Elsevier)

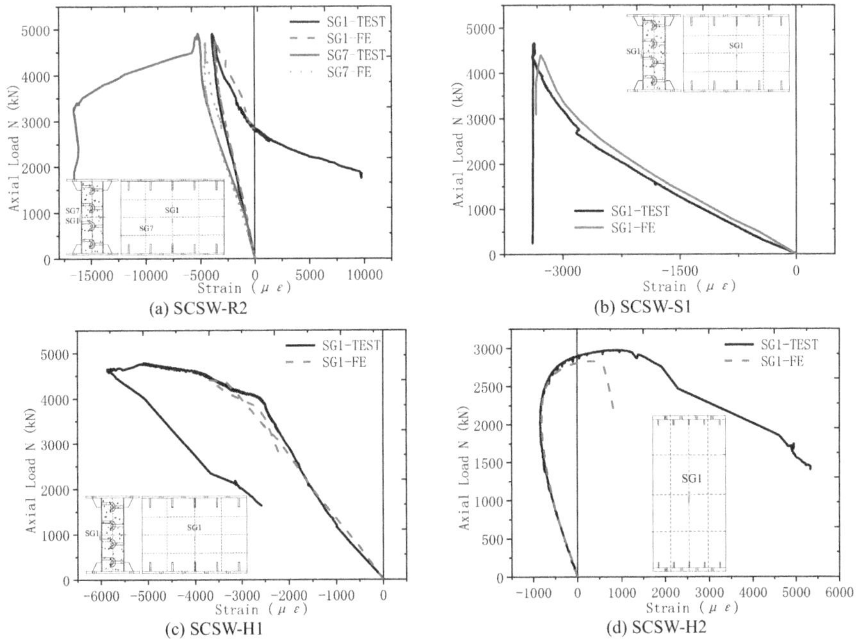

Fig. 3.22 Comparison of load-strain curves obtained from FE and test results for sandwich walls. (**a**) SCSW-R2. (**b**) SCSW-S1. (**c**) SCSW-H1. (**d**) SCSW-H2

Table 3.12 summarizes the numerical simulation results along with a comparison to the test results. The FE model demonstrates strong predictive capability for the beam specimens, with an average ratio of experimental to simulated maximum load of 1.03 and a standard deviation of 0.06. As evidenced by the comparative load-deflection curves in Fig. 3.25a–j, the numerical simulations capture the structural response with considerable accuracy, consistently predicting the ultimate load within

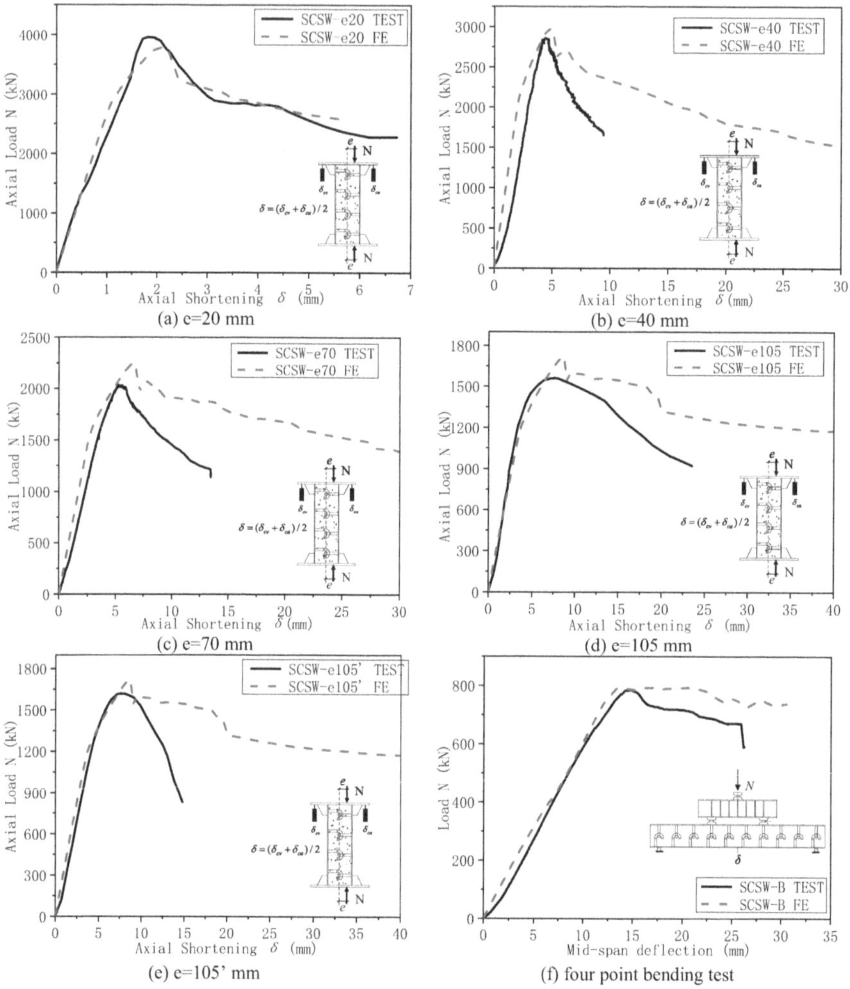

Fig. 3.23 Comparison of load-shortening curves obtained from test and FE for sandwich walls. (**a**) e = 20 mm. (**b**) e = 40 mm. (**c**) e = 70 mm. (**d**) e = 105 mm. (**e**) e = 105′ mm. (**f**) four point bending test. (Reprinted from Huang and Liew [4], copyright 2025, with permission from Elsevier)

a 10% margin of error, which confirms the model's effectiveness in simulating shear behavior. Further validation is provided in Fig. 3.26a–f, which juxtapose the observed failure modes with the predicted concrete damage distribution, where a damage scale from 0 (intact) to 1 (fully damaged) is employed. The analysis successfully replicates critical failure mechanisms, such as the initiation of shear cracks and concrete damage, as exemplified by specimen CSB-06 in Fig. 3.27. In most specimens, the concrete damage value approaches unity at the quarter-span region, indicating material crushing or cracking. Additionally, the model accurately

Table 3.11 Comparison between test and FE results for sandwich walls

	Test results		FE results		Test/FE	
Specimen	N_{test}	M_{test}	N_{FE}	M_{FE}	N_{test}/N_{FE}	M_{test}/M_{FE}
SCSW-R2	4906.2	0	4727.8	0	1.038	–
SCSW-e20	3976.8	98.4	3785.4	81.6	1.051	1.206
SCSW-e40	2860.6	128.8	2976.1	129.0	0.961	0.998
SCSW-e70	2034.1	154.5	2251.7	167.6	0.903	0.922
SCSW-e105	1570.4	180.2	1729.0	189.7	0.908	0.950
SCSW-e105′	1621.7	186.2	1729.0	189.7	0.908	0.982
SCSW-B	0	117.8	0	112.7	–	1.045
Mean value					0.967	1.012
Std. Dev.					0.064	0.113
SCSW-e90(FE)			1938.6	184.1		
SCSW-e125(FE)			1523.8	199.5		
SCSW-e150(FE)			1274.6	202.2		

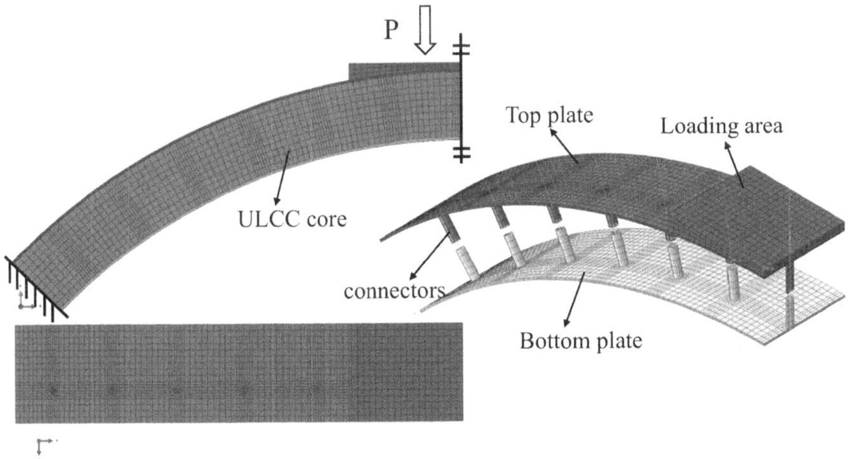

Fig. 3.24 Quarter-symmetric FE model for curved SCS beam. (Reprinted from Huang and Liew (2015), copyright 2025, with permission from Elsevier)

predicts global deformation patterns, including post-buckling behavior of the compression steel plate, as visible in Fig. 3.26d.

3.5.4.3 Steel-Concrete Sandwich Shell

The validity of the finite element (FE) model is confirmed through comparisons with experimental data, including load-deflection behavior, ultimate load resistance, and failure mechanisms. Utilizing a quarter-model as shown in Fig. 3.28 that exploits the double symmetry of the double skin composite shells, the analysis shows a

Table 3.12 Comparison of maximum loads obtained from FE and tested results for sandwich beam

Specimen	Maximum load		
	P_{test} (kN)	P_{FE} (kN)	P_{FE} / P_{test}
FSB-01	132.8	124.7	1.06
CSB-02	1101.3	1216.1	0.91
CSB-03	555.8	494.0	1.13
CSB-04	452.4	435.3	1.04
CSB-05	800.4	802.3	1.00
CSB-06	238.6	228.9	1.04
CSB-07	309.6	291.2	1.06
CSB-08	458.2	468.0	0.98
CSB-09	218.3	216.4	1.01
CSB-10	416.4	402.3	1.03
Mean			1.03
Std. dev.			0.06

reasonable correlation with test results. For example, the FE-predicted midspan deflection under applied load aligns well with experimental findings, as shown in Fig. 3.29 and compared to tests conducted by Shukry and Goode [18]. A more detailed evaluation for specific specimens further demonstrates this agreement, as the numerical analysis accurately captures critical failure modes observed in testing, such as concrete punching and the post-local buckling of steel face plates, as shown in Fig. 3.30. The load-deflection response from the model successfully reproduces a two-peak pattern, characterized by a drop caused by concrete punching shear followed by a hardening phase from the tensile membrane action of the steel plates, as shown in Fig. 3.31. Although the overall match is good, a minor discrepancy emerges after the initial peak, likely due to the complexity of modeling the concrete's behavior at the point of crushing and beyond. Similarly, the model provides a satisfactory prediction of strain histories at designated locations, with the general shape of the load-strain curves matching test measurements, as shown in Fig. 3.32. Any minor variations here may be ascribed to potential experimental factors like strain gauge misalignment, as well as numerical instabilities following material failure in the simulation.

3.6 Analytical Predictions of Axial Load Resistance

3.6.1 Compressive Resistance of Sandwich Wall

The J-hook connectors significantly enhance the structural performance of SCS sandwich walls by effectively mitigating outward buckling of the face plates. This improvement is achieved through optimal spacing of J-hook connectors, which confines potential buckling to occur only between adjacent rows of connectors, as illustrated in Fig. 3.33. Under compressive loading, paired J-hooks develop tensile forces that restrain the steel face plates from outward dilation. This restraining action

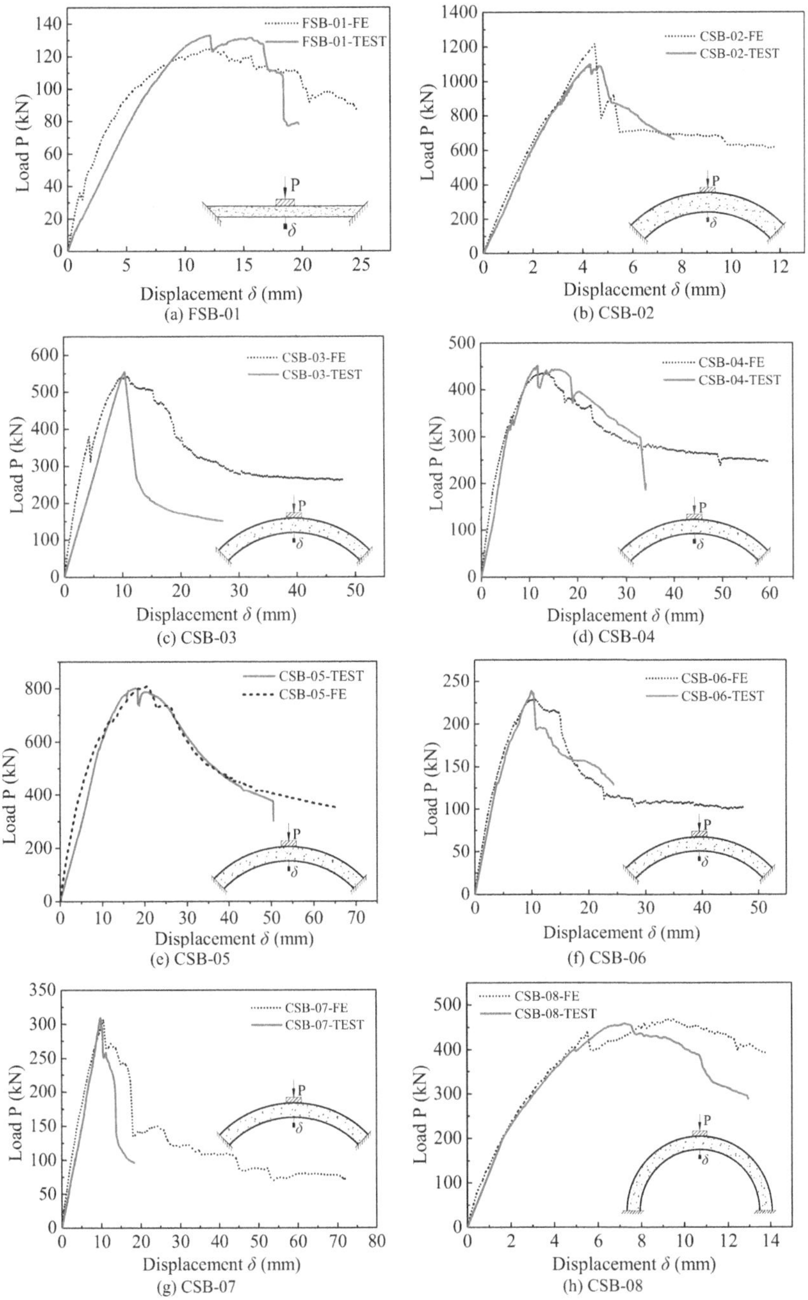

Fig. 3.25 Comparison of load-deflection curves obtained from FE and test results for sandwich beams. (**a**) FSB-01. (**b**) CSB-02. (**c**) CSB-03. (**d**) CSB-04. (**e**) CSB-05. (**f**) CSB-06. (**g**) CSB-07. (**h**) CSB-08. (**i**) CSB-09. (**j**) CSB-10. (Reprinted from Huang and Liew (2015), copyright 2025, with permission from Elsevier)

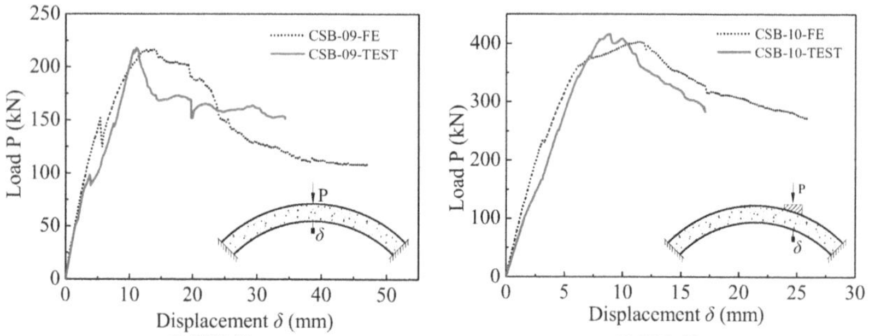

(i) CSB-09 (j) CSB-10

Fig. 3.25 (continued)

simultaneously places the concrete core in a beneficial biaxial compressive state, resulting in increased concrete compressive strength due to the confinement effect.

Conversely, systems lacking stiffeners or with sparsely distributed connectors demonstrate significantly reduced performance. In such cases, J-hook connectors fail to provide adequate shear and tensile bond strength at the steel-concrete interface, resulting in minimal composite action. This deficiency leads to independent load-bearing behavior of the steel plates and concrete core, with premature outward buckling of steel plates occurring at early loading stages [19, 20]. Experimental results in Table 3.13 clearly demonstrate this phenomenon: specimens SCSW-S1 and SCSW-S2, featuring wider connector spacing and consequently higher effective width-to-thickness ratios of the steel plates, failed through local buckling of the face plates.

The buckling mechanism initiates a cascade of failure events: once the steel plates buckle and unload, the concrete core becomes solely responsible for resisting the applied load, ultimately leading to concrete failure. This failure progression manifests as a sudden load drop in the displacement curve. Measurements revealed substantial buckling wave heights of 200 and 300 mm for SCSW-S1 and SCSW-S2 respectively, corresponding directly to the shear connector spacing. These specimens exhibited particularly poor performance characteristics, including low ductility indices and brittle post-peak behavior, as evidenced by both load-shortening curves and visual observations during testing.

For practical design, the compressive resistance of the SCS sandwich wall may be conservatively estimated by superimposing the individual resistances of its components:

$$N_{\mathrm{ul}} = \phi_{\mathrm{a}} A_{\mathrm{s}} f_{\mathrm{y}} + \phi_{\mathrm{c}} A_{\mathrm{c}} f_{\mathrm{ck}} \tag{3.9}$$

where ϕ_{a}, ϕ_{c} are the reduction factor for steel and concrete respectively; A_{s}, A_{c} are cross sectional area of steel plates and concrete; f_{ck}, f_{y} are compressive strength of concrete and yield strength of steel.

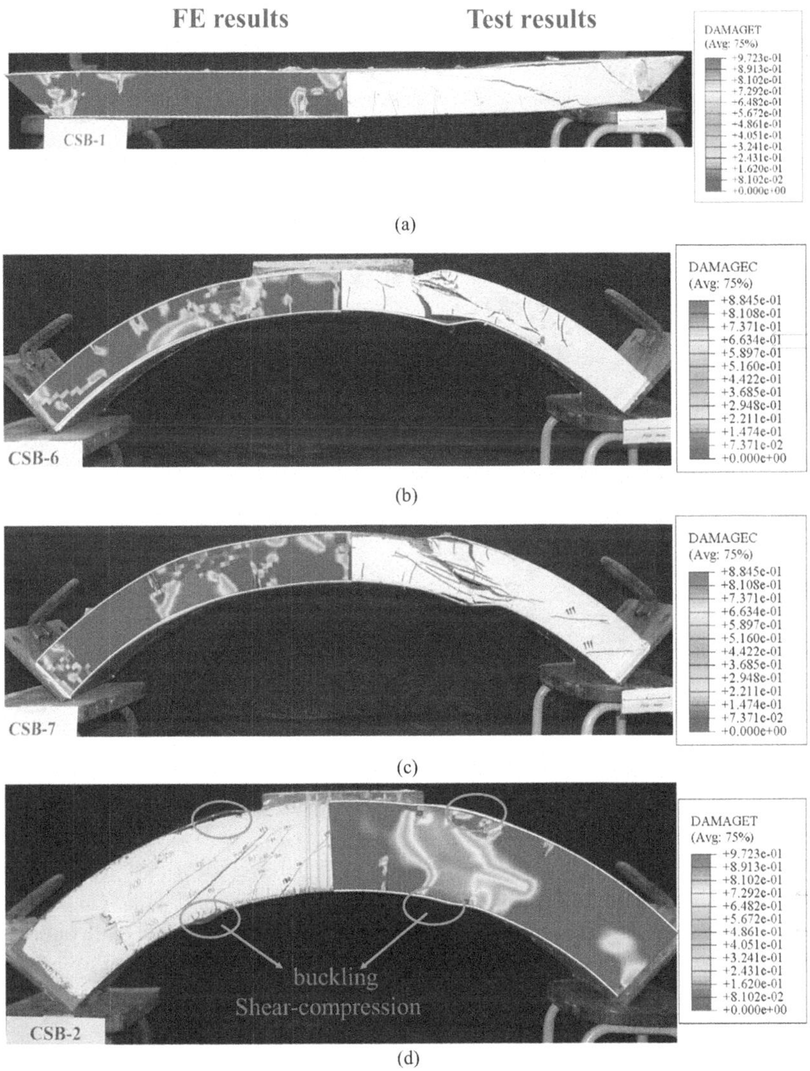

Fig. 3.26 Comparison of failure modes between FE and test results for sandwich beams. (**a**) CSB-1. (**b**) CSB-6. (**c**) CSB-7. (**d**) CSB-2. (**e**) CSB-5. (**f**) CSB-8. (Reprinted from Huang and Liew (2015), copyright 2025, with permission from Elsevier)

All steel components intersecting the concrete failure plane may be considered to provide confinement. The resulting confinement stress can be expressed as:

$$\sigma_c = \frac{n_J P_J}{\mathrm{WH}} \tag{3.10}$$

where, n_J is the number of J-hook connectors, which is calculated by $n_J = HW/s_H\, s_W$; H, W, s_H, s_W are the height of the wall, the width of the wall, the connector spacing in

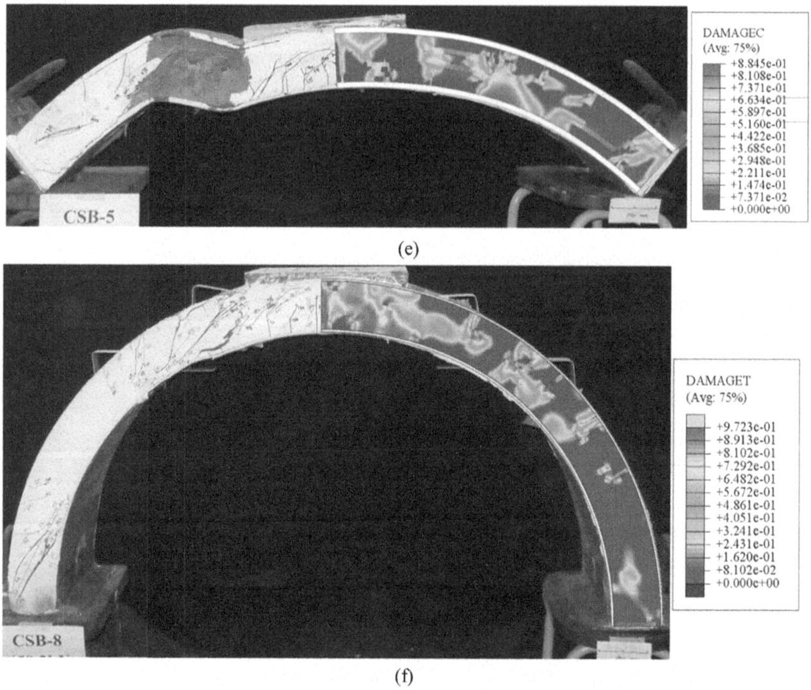

Fig. 3.26 (continued)

height direction, and the connector spacing in width direction. P_J represents the tensile resistance of J-hook connector.

Experimental and analytical studies (Ref. [21]) have identified four primary failure mechanisms for J-hook connectors under tensile loading: Concrete breakout failure P_{cb}, J-hook pullout failure P_{pl}, connector fracture P_u, and steel plate punching shear P_{ps}. Therefore, the tensile resistance of J-hook connector is given by:

$$P_J = \min\begin{cases} P_{cb} = 0.33\sqrt{f_{ck}}A_N \\ P_{pl} = 0.9\varphi f_{ck} e_h d + 0.116 d^2 f_y \\ P_u = \phi A_{se} f_{ut} \\ P_{ps} = A_v f_u / \sqrt{3} \end{cases} \tag{3.11}$$

Introducing the non-dimensional form to Eq. (3.12) results in:

$$\frac{\sigma_c}{f_{ck}} = \frac{100 n_J P_J}{\mathrm{WH} f_{ck}} \tag{3.12}$$

To develop the regression model, additional test data from the literature [22–24] on double-skin lightweight composite walls under compression were incorporated, as summarized in Table 3.14. A linear regression analysis of this data yielded the

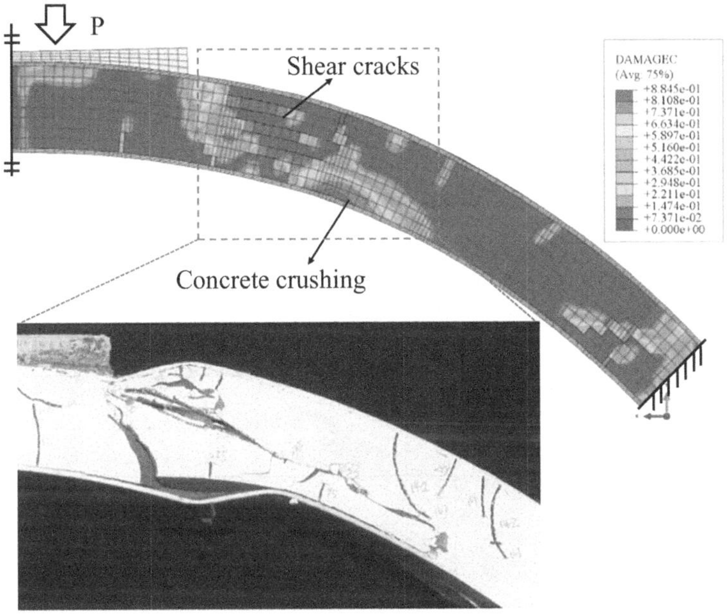

Fig. 3.27 Crack pattern of CSB-06. (Reprinted from Huang and Liew (2015), copyright 2025, with permission from Elsevier)

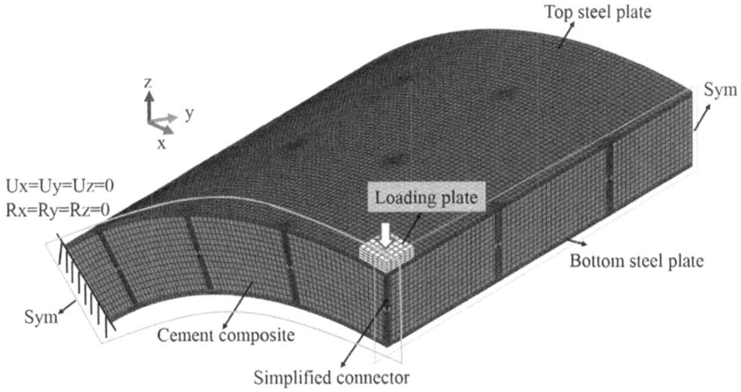

Fig. 3.28 Quarter FE model of sandwich shell subject to patch load. (Reprinted from Huang and Liew (2015), copyright 2025, with permission from Elsevier)

relationship between concrete failure stress and confinement stress for SCS sandwich walls with connectors, which is presented in Fig. 3.34 and expressed by the following equation:

$$\phi_c = \frac{\sigma_f}{f_{ck}} = 0.2995 + 0.0489\left(\frac{100 n_J P_J}{WHf_{ck}}\right) \tag{3.13}$$

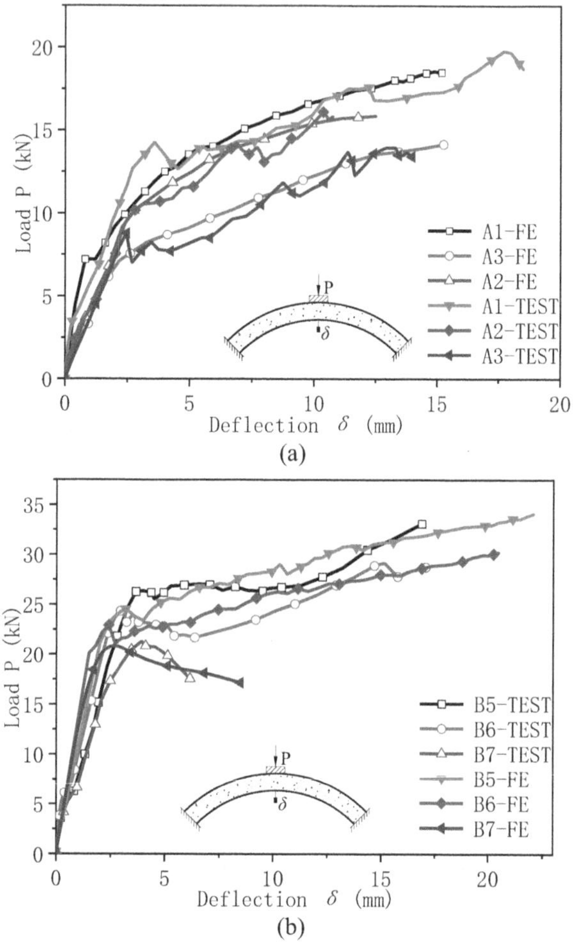

Fig. 3.29 Comparison of load-defection curves obtained from FE and test results in Shukry's test [18]. (**a**) A-series sandwich beams (A1-A3). (**b**) B-series sandwich beams (B5-B7). (Reprinted from Huang and Liew (2015), copyright 2025, with permission from Elsevier)

Experimental studies reveal that even with confinement effects, the lightweight concrete in sandwich walls often fails to achieve its theoretical cross-sectional resistance ($f'_{c,\,eff}$). This phenomenon has been consistently demonstrated by Prabha et al. [23] and Mydin and Wang [24], and can be primarily attributed to: (1) higher width-to-thickness ratios in SCS walls compared to compact composite columns; (2) lower fracture toughness of lightweight concrete; and (3) premature cracking under combined stresses. A strength reduction factor is necessary to adjust the nominal compressive strength of the lightweight concrete core in a sandwich wall.

The cross-section resistance of SCS sandwich composite wall can also be calculated by the following equation that replaces the first term on the right-hand side of the below equation, as follow:

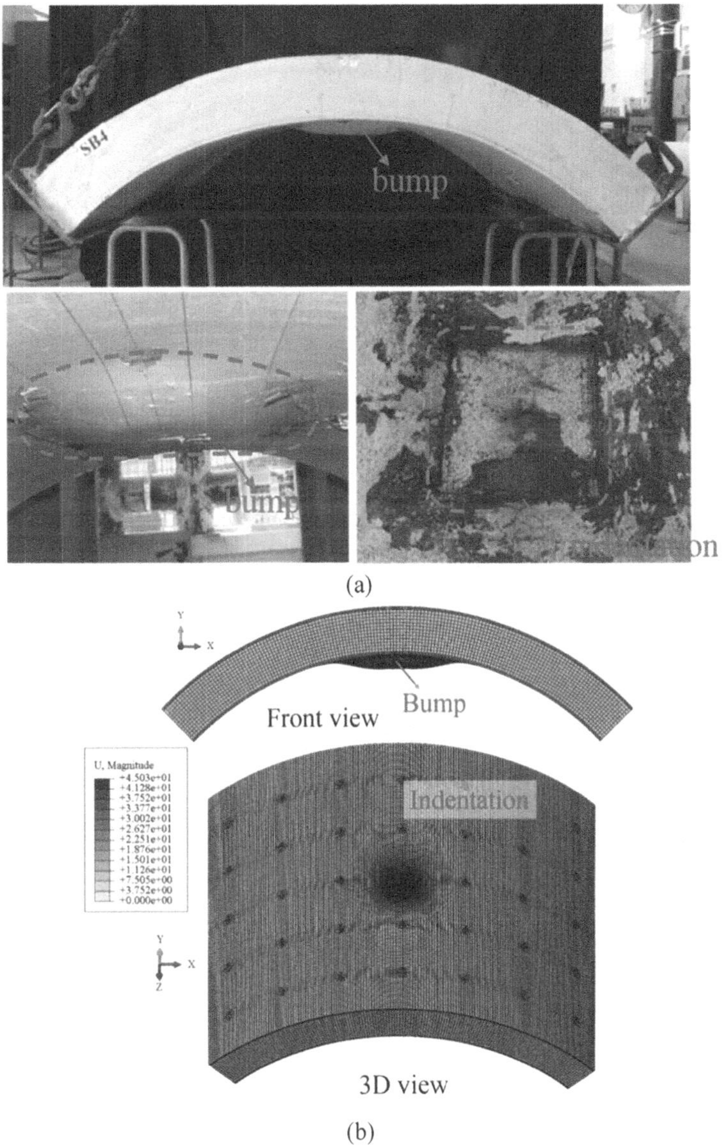

Fig. 3.30 Comparison of failure mode obtained from test and FE results for sandwich shells. (**a**) Failure mode of SCS double skin shell SB4 in. (**b**) Failure mode of SCS double skin shell SB4 (FE). (Reprinted from Huang and Liew (2015), copyright 2025, with permission from Elsevier)

$$N_{\text{ull}} = b_{\text{eff}} t f_{\text{y}} + \phi_{\text{c}} A_{\text{c}} f_{\text{ck}} \tag{3.14}$$

An analytical method utilizing the effective width concept was proposed by Liang and Uy [25] to characterize the post-local buckling response of steel plates in concrete-infilled steel box columns. The proposed model is presented below:

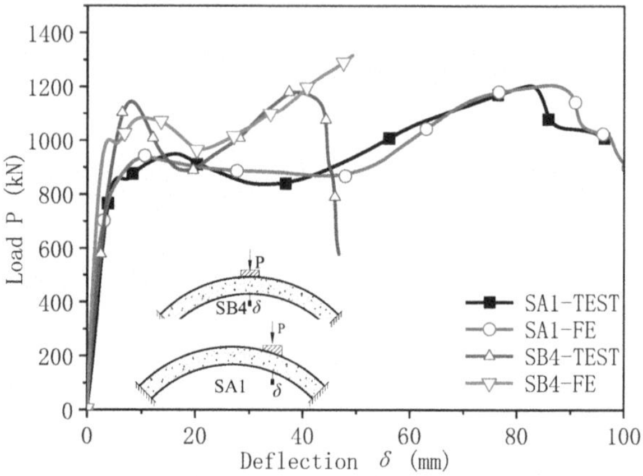

Fig. 3.31 Comparison of load–deflection curves obtained from test and FE results for sandwich shells. (Reprinted from Huang and Liew (2015), copyright 2025, with permission from Elsevier)

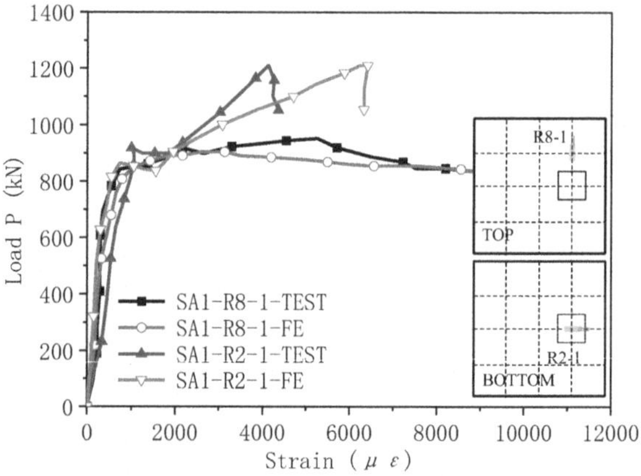

Fig. 3.32 Comparison of load–strain curves obtained from test and FE results for sandwich shells. (Reprinted from Huang and Liew (2015), copyright 2025, with permission from Elsevier)

$$\frac{b_{\text{eff}}}{b} = 0.675\left(\frac{\sigma_{\text{cr}}}{f_{\text{y}}}\right)^{1/3} \quad \text{for } \sigma_{\text{cr}} \le f_{\text{y}} \tag{3.15}$$

$$\frac{b_{\text{eff}}}{b} = 0.915\left(\frac{\sigma_{\text{cr}}}{\sigma_{\text{cr}} + f_{\text{y}}}\right)^{1/3} \quad \text{for } \sigma_{\text{cr}} > f_{\text{y}} \tag{3.16}$$

The proposed formulae demonstrate reasonable accuracy in predicting the compressive resistance of sandwich walls, with average test-to-prediction ratios of 1.09 and 1.08 accompanied by low coefficients of variance (0.14 and 0.13, respectively), while Fig. 3.35 scatter plot confirms their reliable predictive capability through close agreement between experimental results and analytical predictions.

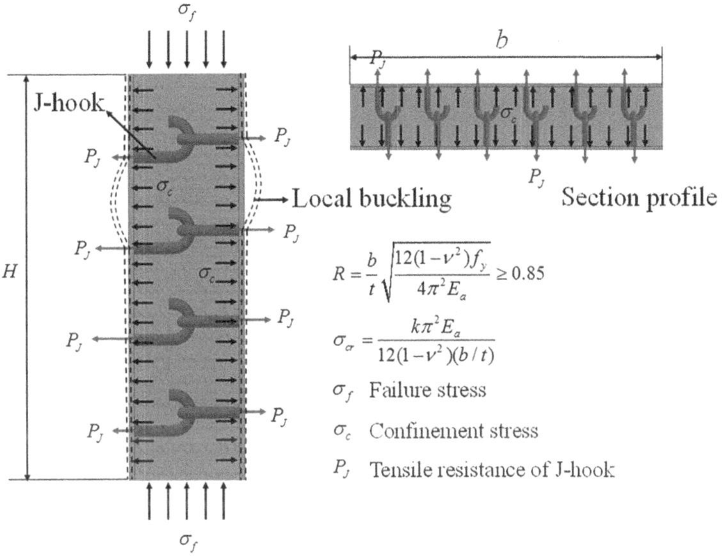

Fig. 3.33 Buckling mode of outer plates restrained by J-hook connector. (Reprinted from Huang and Liew [4], copyright 2025, with permission from Elsevier)

Table 3.13 Failure loads, failure modes and ductility index of SCS sandwich walls

Specimen	N_{test} (kN)	$0.85A_cf_{ck} + A_sf_y$ (kN)	SI	δ_{test} (mm)	$\delta_{0.85}$ (mm)	μ	Failure mode
SCSW-R1	4191	5403	0.776	3.93	5.85	1.489	CS + LB
SCSW-R2	4906	5457	0.899	3.56	4.03	1.132	CS + CC + LB
SCSW-H1	4788	5466	0.876	4.85	6.02	1.241	CC + LB
SCSW-H2	2975	5735	0.519	8.91	9.46	1.062	GB
SCSW-S1	4656	5516	0.844	4.11	4.11	1.000	LB
SCSW-S2	3670	5530	0.664	3.73	4.09	1.097	LB
SCSW-C1	4248	4832	0.879	3.97	4.2	1.058	CC + LB
SCSW-HS	5317	5855	0.908	4.65	6.05	1.301	CS + LB
SCSW-C30	3916	4445	0.881	3.44	3.84	1.116	CS + LB
SCSW-C45	4689	5400	0.868	4.08	4.16	1.020	CS + CC + LB
SCSW-T8	6889	7099	0.970	4.47	5.57	1.246	CTF + LB
SCSW-T12	8418	8411	1.001	4.99	8.13	1.629	CTF + LB
SCSW-P1	5120	5954	0.860	3.81	4.77	1.252	CC + LB
SCSW-P2	4933	5873	0.840	4.27	5.42	1.269	CC + LB
SCSW-C2	5467	5767	0.948	4.73	5.41	1.144	CS + CC + LB
Mean			0.849			1.204	
Std.Dev			0.121			0.174	

CS concrete splitting, *CC* concrete crushing, *LB* local buckling of steel, *CTF* connector tensile fracture, *GB* Global buckling; Strength index: $SI = N_{test}/(0.85A_cf_{ck} + A_sf_y)$; Ductility Index $\mu = \delta_{0.85}/\delta_{test}$

Table 3.14 Test data on lightweight sandwich walls from published literature

Specimen	P_c (kN)	σ_f (MPa)	σ_f/f_{ck}	n	A_h (mm^2)	A_{sc} (mm^2)	σ_c (MPa)	σ_c/f_{ck}	References
1	389	5.66	0.719	30	0	1508	66.1	8.40	
2	314	4.57	0.581	20	0	1005	44.1	5.60	
3	378	5.50	0.699	32	0	1609	70.5	8.97	
4	325	4.73	0.601	6	1566	1085	47.6	6.05	[23]
5	270	3.93	0.500	0	1566	783	34.3	4.36	
PP1	149	2.17	0.276	0	0	0	0	0	
PP2	191	2.78	0.353	0	0	0	0	0	
NSE1[a]	127	3.37	0.571	6	0	170	29.7	5.03	[24]
7	1868	9.29	0.300	0	5400	2700	46.0	1.48	[22]

[a] P_c denotes the load on concrete only; n = number of connecting bolts; f_y = 280 MPa and E_a = 200 GPa for all tests. Prabha et al. [23]: f_{ck} = 7.87 MPa, W = 685 mm, t_c = 120 mm, d_{stu} = 8 mm, H = 870 mm; $A_h = 2tH$; $A_{sc} = (0.5A_h + nA_{stu})$; Mydin and Wang [24]: f_{ck} = 5.9 MPa, W = 400 mm, t_c = 100 mm, d_{stu} = 6 *mm*, H = 400 mm; Wright [22]: f_{ck} = 31 MPa, W = 914 mm, t_c = 220 mm, d_{stu} = -nil-; H = 1800 mm

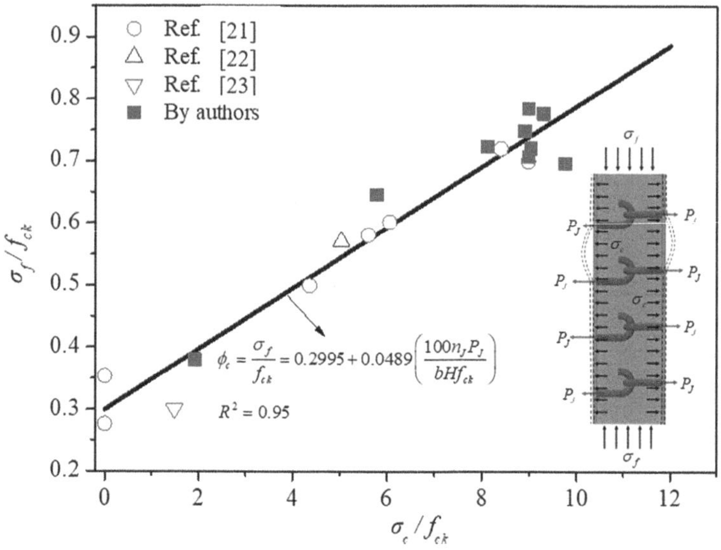

Fig. 3.34 Regression of failure stress ratio (σ_f/f_{ck}) with confinement stress ratio (σ_c/f_{ck}). (Reprinted from Huang and Liew [4], copyright 2025, with permission from Elsevier)

3.6.2 *Compression-Bending Interaction Resistance of Sandwich Walls*

The N-M interaction diagram for the sandwich section is presented in Fig. 3.36. Point A on the curve indicates the modified axial resistance under pure compression,

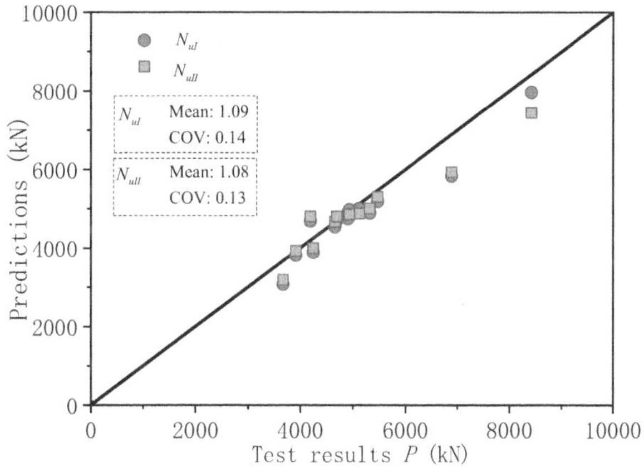

Fig. 3.35 Comparison of test results and predictions by proposed methods. (Reprinted from Huang and Liew [4], copyright 2025, with permission from Elsevier)

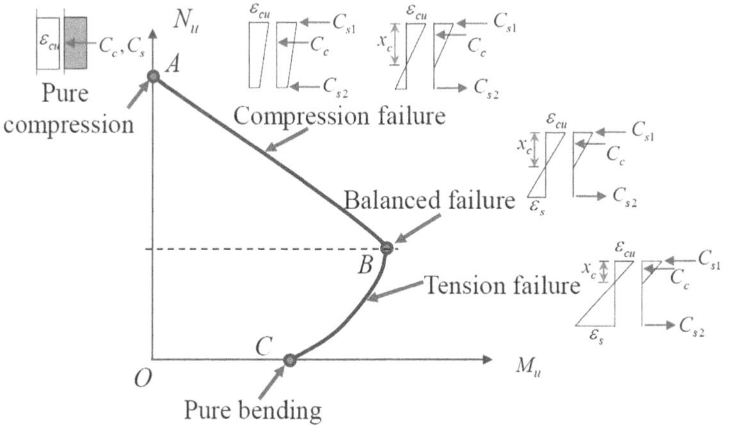

Fig. 3.36 N-M interaction diagram of sandwich section. (Reprinted from Huang and Liew [4], copyright 2025, with permission from Elsevier)

while Point B corresponds to the balanced failure condition. The segment of the curve between Points A and B represents the region of compression failure, which occurs under small eccentricities. Point C represents the status of pure bending while the range from point B to point C would be in tension failure (large eccentricity). Figure 3.37 depicts the force equilibrium status of compression- and tension-controlled section for each branch, respectively.

The generation of an N-M interaction diagram requires evaluating multiple combinations of axial load (N) and bending moment (M), beyond the known points of pure compression and pure bending. The diagram is constructed for a specific section by iteratively selecting different neutral axis depths (x_c). The process begins

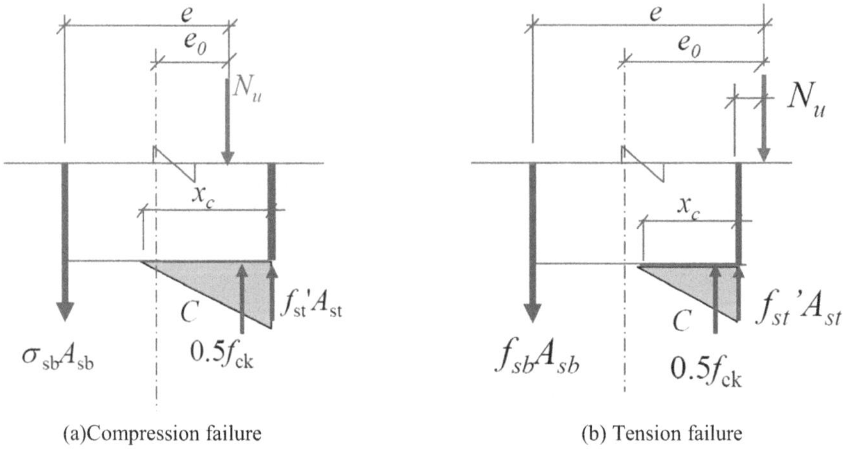

Fig. 3.37 Equilibrium state in compression- and tension-dominated sections. (**a**) Compression failure. (**b**) Tension failure. (Reprinted from Huang and Liew [4], copyright 2025, with permission from Elsevier)

with x_c at infinity (representing pure axial load) and progresses to a minimal value (representing pure bending). For the compression failure region (between points A and B), the calculation assumes the strain in the extreme concrete fiber has reached its ultimate value of $\varepsilon_{cu} = 0.0045$. For the tension failure region of the interaction diagram (from point B to C), the analysis assumes the tensile strain in the outermost steel fiber reaches the yield strain, $\varepsilon_y = f_y/E_s$. The distribution of strain and stress across the section is then determined by applying the fundamental principles of plane sections remaining plane, material compatibility, and the constitutive stress-strain relationships of the materials. From this resulting internal stress profile, the corresponding axial compressive force (N) and bending moment (M) at failure can be calculated. The equilibrium equations governing this state are provided below.

1. Pure compression: Point A

   The compression resistance of sandwich wall is obtained by considering the buckling effect of steel face plate:

$$N_A = N_{pl,Rd} = 0.85A_c f_{cd} + \phi_a A_s f_{yd} \tag{3.17}$$

   where, $\phi_a = 0.898R^{-0.771}$ $(R \geq 0.85)$, R represents the width-thickness ratio, defined $R = \frac{W}{t}\sqrt{\frac{12(1-\nu^2)f_y}{4\pi^2 E_a}} \geq 0.85$ [26, 27].

2. Compression failure: Branch AB

$$N_u = \frac{1}{2}f_{ck}bx_c + f'_{st}A_{st} - \sigma_{sb}A_{sb} \tag{3.18}$$

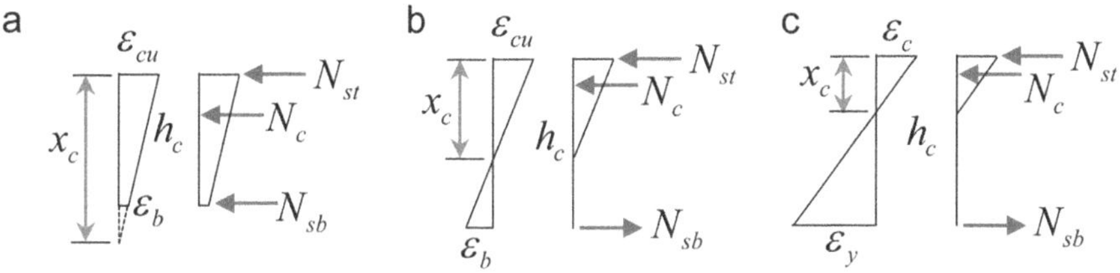

Fig. 3.38 Analysis of strain distribution and force equilibrium in cross-sections. (**a**) Compression-controlled failure. (**b**) Transition stage (initation of steel yielding). (**c**) Tension-controlled failure. (Reprinted from Huang and Liew [4], copyright 2025, with permission from Elsevier)

$$N_{\mathrm{u}}e=\frac{1}{2}f_{\mathrm{ck}}bx_{\mathrm{c}}\left(h_{\mathrm{c}}-\frac{x_{\mathrm{c}}}{3}+\frac{t_{\mathrm{b}}}{2}\right)+f'_{\mathrm{st}}A_{\mathrm{st}}(h_{\mathrm{c}}+t_{\mathrm{b}}) \tag{3.19}$$

where, $e=e_0+\frac{h_c}{2}+\frac{t_b}{2}$, $f'_{st}A_{st}$ represents the compression force carried by steel plate in compression side, $\sigma_{sb}A_{sb}$ represents the compression force or tension force carried by steel plate in tension side, as shown in Fig. 3.37a.

The analysis assumes that the symmetrical steel plates yield simultaneously as the extreme concrete fiber in compression attains its ultimate strain. Referring to Fig. 3.38a, b, the strain (ε_b) and the corresponding stress (σ_{sb}) in the bottom steel plate are determined as follows:

$$\frac{\varepsilon_{\mathrm{cu}}}{\varepsilon_{\mathrm{b}}}=\frac{x_{\mathrm{c}}}{h_{\mathrm{c}}-x_{\mathrm{c}}},\ \frac{\varepsilon_{\mathrm{cu}}}{\varepsilon_{\mathrm{b}}}=\frac{x_{\mathrm{c}}}{x_{\mathrm{c}}-h_{\mathrm{c}}} \tag{3.20}$$

$$\sigma_{\mathrm{sb}}=E_{\mathrm{s}}\varepsilon_{\mathrm{b}}\leq f_{\mathrm{y}} \tag{3.21}$$

3. Tension failure: Branch BC

Similarly, the force equilibrium is expressed based on Fig. 3.37b:

$$N_{\mathrm{u}}=\frac{1}{2}f_{\mathrm{ck}}bx_{\mathrm{c}}+f'_{\mathrm{st}}A_{\mathrm{st}}-f'_{\mathrm{sb}}A_{\mathrm{sb}} \tag{3.22}$$

$$M_{\mathrm{u}}=N_{\mathrm{u}}e=\frac{1}{2}f_{\mathrm{ck}}bx_{\mathrm{c}}\left(h_{\mathrm{c}}-\frac{x_{\mathrm{c}}}{3}+\frac{t_{\mathrm{b}}}{2}\right)+f'_{\mathrm{st}}A_{\mathrm{st}}(h_{\mathrm{c}}+t_{\mathrm{b}}) \tag{3.23}$$

where, $e=e_0+\frac{h_c}{2}+\frac{t_b}{2}$, $\frac{\varepsilon_y}{\varepsilon_c}=\frac{h_c-x_c}{x_c}$, $\sigma_c=E_c\varepsilon_c\leq f_{ck}$ based on Fig. 3.38c.

The section analysis employs an iterative procedure that, for a given eccentricity, continues until force equilibrium is achieved—reaching either concrete crushing strain in compression failure or steel yield strain in tension failure—with curvature increments repeated until section failure occurs via concrete ultimate strain attainment or steel yielding, ultimately generating the N-M interaction diagram for short

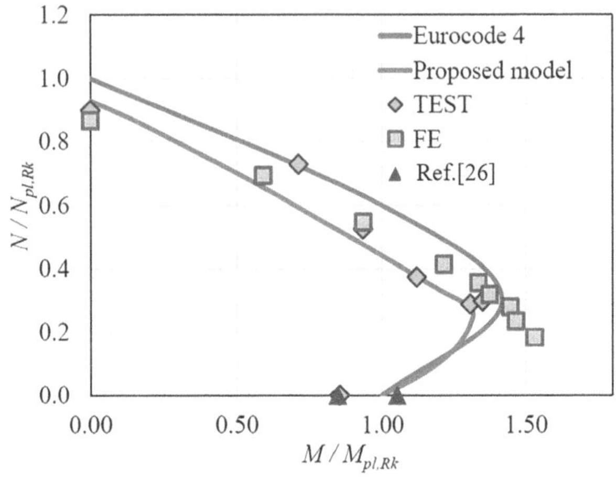

Fig. 3.39 Proposed N-M interaction model against Eurocode 4 prediction, test, and FE results. (Reprinted from Huang and Liew [4], copyright 2025, with permission from Elsevier)

sandwich walls by compiling maximum bending moments at each axial load level; Fig. 3.39 compares the resulting generalized N-M diagrams (normalized for material/dimensional independence) between Eurocode 4 and the developed model, showing test/FE data mostly lying outside the model's conservative predictions except for one pure bending case that marginally underperforms due to observed premature flexural-shear failure—a phenomenon corroborated by Ref. [27] where specimen J6 (flexural-shear failure) aligned with this study's data while J7 (pure flexure) exceeded predictions, collectively validating the model's reliability for combined compression-bending resistance checks of ULCC-filled walls via Eqs. (3.17)–(3.23).

3.7 Analytical Predictions of Shear Resistance

3.7.1 Shear Resistance of Sandwich Beam

In the shear-compression failure mode, the primary load path is a major diagonal concrete strut that forms between the loading point and the support. Failure occurs due to the crushing of this diagonal strut. Consequently, the governing load capacity can be determined as follows:

$$V_c = f_{ce} A_{str} \tag{3.24}$$

where, A_{str} denotes effective area of the diagonal strut calculated as: $A_{str} = a_s \times b_w$, a_s is the effective depth of the diagonal strut while b_w is the beam width.

The total shear resistance for a shear-tension failure is calculated by combining the contributions from the concrete core and the shear connectors, as defined by the following expression:

$$V = V_c + V_s \tag{3.25}$$

V_c and V_s can be obtained in according to clause 11.1 in ACI 318, as given by Eqs. (3.26) and (3.27):

$$V_c = \min\begin{Bmatrix} \left(2 + \frac{4}{\beta}\right)\lambda\sqrt{f'_c}b_0 d \\ \left(\frac{\alpha_s d}{b_0} + 2\right)\lambda\sqrt{f'_c}b_0 d \\ 4\lambda\sqrt{f'_c}b_0 d \end{Bmatrix} \tag{3.26}$$

$$V_s = \frac{A_v f_{yt} d}{s} \tag{3.27}$$

where, A_v represents the total cross-sectional area of the shear reinforcement (such as stirrups or ties) that is present within a specified spacing distance s.

The flexural moment resistance of composite beam is obtained by Eq. (3.28) below:

$$M_{pl} = N\left(h_c + \frac{t_c}{2} + \frac{t_t}{2}\right) = n_p P_R(h_c + t) \tag{3.28}$$

where, $N_t = n_p P_R$ is compression force at top steel face plate; P_R is the shear resistance of shear connectors which can be calculated by Eurocode 4 as given below:

$$P_{stu} = \min\left\{0.8 f_u\left(\frac{\pi d^2}{4}\right)\right\}/\gamma_v, \frac{0.29 d^2 \alpha\sqrt{f_{ck}E_{cm}}}{\gamma_v} \tag{3.29}$$

where, $\alpha = 0.2\left(\frac{h_{sc}}{d} + 1\right)$ for $3 \leq \frac{h_{sc}}{d} \leq 4; \alpha = 1$ for $\frac{h_{sc}}{d} > 4$; the yield strength of shear stud $f_u < 500$ MPa.

3.7.2 *Punching Shear Resistance of Sandwich Shell*

Experimental observations identified a two-stage failure process, characterized by an initial punching shear failure of the concrete core followed by a punching fracture of the steel face plates. These correspond to the first and second peak loads, respectively. The resistance at the first peak is calculated as the sum of the shear contributions from the concrete core (V_c) and the shear connectors (V_s), following a formulation analogous to Eq. (3.27). The concrete contribution V_c is determined using Eq. (3.30) per Eurocode 2 and Majdzadeh et al. [28], while the contribution from the shear connectors V_s is given by Eq. (3.31).

$$V_c = \left[C_c k_c \eta_1 (100 \rho f_{ck})^{1/3} + k_f \tau_{f,FRC}\right] u_1 \overline{h} \tag{3.30}$$

$$V_s = \sum_{i=1}^{n_{cp}} \frac{F_t}{s} \overline{h}_c \sin \alpha_i \tag{3.31}$$

The parameters F_t, a_i, and n_{cp} represent the tensile resistance of a shear connector, its inclination angle, and the number of connectors, respectively [29]. The value of F_t for a stud embedded in concrete is determined as the minimum resistance governed by four potential failure modes: concrete breakout, stud pullout, tensile failure of the stud itself, and punching shear of the steel plate. This is calculated using Eq. (3.32).

$$F_t = \min \begin{cases} P_{cb} = 0.33\sqrt{f_{ck}} A_N \\ P_{pl} = 0.9 \varphi f_{ck} e_h d \\ P_u = \phi A_{se} f_{ut} \\ P_{ps} = A_v f_u / \sqrt{3} \end{cases} \tag{3.32}$$

where, $A_N = \pi h_{ef}^2 \left(1 + \frac{d_h}{h_{ef}}\right)$ represents the projected area of cone surface to the free concrete surface; A factor of $u = 1.4$ is applied for anchors in concrete regions where analysis confirms no cracking; otherwise $\varphi = 1.0$. The distance e_h is defined as the measurement from the stud's inner shaft to its outer tip, which must satisfy the condition $3d_0 \leq 4.5d_0$; d is the diameter of stud;φ is the reduction factor of the steel; A_{se} and f_{ut} are the cross-sectional area and the ultimate strength of headed shear stud; $A_v = \pi dt$ and f_u are the punching area and ultimate strength of steel plate. The second peak load resistance, corresponding to the punching fracture of the steel face plates, is governed by the failure of either the outer or bottom plate initiated at the shear studs, with the smaller of the two values determining the capacity as specified in Eq. (3.33).

$$P = \min\left\{2n_{cp}\pi d_s t f_u/\sqrt{3}, f_u U t/\sqrt{3}\right\} \tag{3.33}$$

This equation incorporates the following parameters: the number of shear studs along the punch-through perimeter (n_{cp}), the stud diameter (d_s), the ultimate tensile strength of the steel (f_u), and the perimeter of the loaded area (U).

A comparison of analytical predictions with experimental results is provided in Table 3.15. For composite beams, the mean test-to-predicted load ratio (P_{test}/V_{ACI318}) is 1.211, demonstrating reasonable agreement. In contrast, for composite shells, the Eurocode 2 method appears to overestimate the first peak load by

Table 3.15 Comparison of analytical predictions and test results for composite beams and shells

Specimen	P_{test} (kN)	V_{EC2} (kN)	P_{test}/V_{EC2}
Steel-concrete sandwich shell			
DSS-01	1192.9[a]	1540.3	0.774
	1025.0[b]	971.1	1.056
DSS-02	1390.9[a]	1594.1	0.873
	1210.0[b]	996.0	1.215
Steel-concrete sandwich beam			
DSB-01	132.8	137.2	0.968
DSB-02	1101.3	957.7	1.150
DSB-03	555.8	408.7	1.360
DSB-04	800.4	540.8	1.480
DSB-05	238.6	256.6	0.930
DSB-06	309.6	227.6	1.360
DSB-07	458.2	372.5	1.230
Mean			1.211
Std. dev.			0.208

Note: Values marked in [a] represents the first peak load while marked in [b] represents the second peak load for composite shell

approximately 17.7%. This discrepancy may stem from an overestimation of the punching control perimeter, which is defined in Eurocode 2 as being located at a distance of $2h_c$ from the loading area. Predictions for the second peak load using Eq. (3.33) show better correlation, with mean test-to-prediction ratios of 1.056 and 1.215, indicating the predictions are slightly conservative.

Declaration of Interests The authors acknowledge that portions of this chapter incorporate material originally published in the Ph.D. thesis of the first author Huang Zhenyu from the National University of Singapore. This reproduction has been authorized through official permission granted by the National University of Singapore via their ScholarBank repository (scholarbank.nus.edu.sg).

References

1. B.B.C. News, World oil demand to rise by 37%, June 20, 2006, Available at: http://news.bbc.co.uk/2/hi/business/5099400.stm
2. P.W. Marshall, A. Palmer, J.Y.R. Liew, T. Wang, M.K.W. Thein, Bond enhancement for sandwich shell ice wall, in: International Conference and Exhibition on Performance of Ships and Structures in Ice (ICETECH 2010), Anchorage, Alaska, USA, 2010.
3. J.Y.R. Liew, T.Y. Wang, Novel steel–concrete–steel sandwich composite plates subject to impact and blast load, Adv. Struct. Eng. 14(4) (2011) 673–687.
4. Z. Huang, J.Y.R. Liew, Structural behaviour of steel–concrete–steel sandwich composite wall subjected to compression and end moment, Thin-Walled Struct. 98 (2016) 592-606.
5. K.M.A. Sohel, J.Y.R. Liew, J.B. Yan, M.H. Zhang, K.S. Chia, Behavior of steel–concrete–steel sandwich structures with lightweight cement composite and novel shear connectors, Compos. Struct. 94(12) (2012) 3500–3509.
6. N.E. Shanmugam, G. Kumar, V. Thevendran, Finite element modelling of double skin composite slabs, Finite Elem. Anal. Des. 38(7) (2002) 579–599.

7. N. Foundoukos, J.C. Chapman, Finite element analysis of steel–concrete–steel sandwich beams, J. Constr. Steel Res. 64 (2008) 947–961.
8. Z. Tao, Refined finite element modelling of concrete-filled steel stub columns, in: The Australasian Structural Engineering Conference (ASEC 2014), Auckland, New Zealand, 2014, Paper No. 54.
9. A. Chen, X. Castello, M. Duan, R.D.T. Filho, S.F. Estefen, Ultimate strength behaviour of sandwich pipes filled with steel fiber reinforced concrete, Ocean Eng. 55 (2012) 125–135.
10. J.B. Yan, Finite element analysis on steel–concrete–steel sandwich beams, Mater. Struct. 48 (2015) 1645-1667.
11. J. Lubliner, J. Oliver, S. Oller, E. Oñate, A plastic-damage model for concrete, Int. J. Solids Struct. 25(3) (1989) 299–326.
12. J. Lee, G.L. Fenves, Plastic-damage model for cyclic loading of concrete structures, J. Eng. Mech. ASCE 124(8) (1998) 892–900.
13. ABAQUS, ABAQUS standard user's manual, version 6.12, Dassault Systèmes Corp., Providence, RI, 2012.
14. J.C. Wang, Y.K. Chen, Applications of ABAQUS in civil engineering, Zhejiang University Press, Hangzhou, China, 2006 (in Chinese).
15. H.T. Nguyen, S.E. Kim, Finite element modeling of push-out tests for large stud shear connectors, J. Constr. Steel Res. 65(10–11) (2009) 1909–1920.
16. K.E. Barth, H. Wu, Efficient nonlinear finite element modeling of slab on steel stringer bridges, Finite Elem. Anal. Des. 42 (2006) 1304–1313.
17. FIP, CEB-FIP model code 1990, Thomas Telford Ltd., London, 1993.
18. M.E.S. Shukry, C.D. Goode, Punching shear strength of composite construction, ACI Struct. J. 87(1) (1990) 12–22.
19. Y. Yang, Y. Wang, F. Fu, Effect of reinforcement stiffeners on square concrete-filled steel tubular columns subjected to axial compressive load, Thin-Walled Struct. 82 (2014) 132–144.
20. Y. Wang, Y. Yang, S. Zhang, Static behaviors of reinforcement-stiffened square concrete-filled steel tubular columns, Thin-Walled Struct. 58(1) (2012) 18–31.
21. Z. Huang, J. Wang, J.Y.R. Liew, P.W. Marshall, Lightweight steel–concrete–steel sandwich composite shell subject to punching shear, Ocean Eng. 102 (2015) 146–161.
22. H. Wright, Axial and bending behavior of composite walls, J. Struct. Eng. ASCE 124(7) (1998) 758–764.
23. P. Prabha, V. Marimuthu, M. Saravanan, G.S. Palani, N. Lakshmanan, R. Senthil, Effect of confinement on steel–concrete composite light-weight load-bearing wall panels under compression, J. Constr. Steel Res. 81 (2013) 11–19.
24. M.A.O. Mydin, Y.C. Wang, Structural performance of lightweight steel–foamed concrete–steel composite walling system under compression, Thin-Walled Struct. 49(1) (2011) 66–76.
25. Q.Q. Liang, B. Uy, Theoretical study on the post-local buckling of steel plates in concrete-filled box columns, Comput. Struct. 75(5) (2000) 479–490.
26. J. Cai, Z.Q. He, Axial load behavior of square CFT stub column with binding bars, J. Constr. Steel Res. 62(5) (2006) 472–483.
27. J.B. Yan, Ultimate strength behaviour of steel–concrete–steel sandwich beams and shells, National University of Singapore, Singapore, 2012, Ph.D. thesis.
28. F. Majdzadeh, S.M. Soleimani, N. Banthia, Shear strength of reinforced concrete beams with a fiber concrete matrix, Can. J. Civ. Eng. 33(6) (2006) 726–734.
29. J.Y.R. Liew, K.M.A. Sohel, Lightweight steel–concrete–steel sandwich system with J-hook connectors, Eng. Struct. 31(5) (2009) 1166–1178.

Chapter 4
Novel Double-Layer Steel–Lightweight High-Ductility Concrete Composite (LHDCC)–Steel Sandwich Panels Under Punching Loads

4.1 Introduction

Traditional single-layer steel-concrete sandwich panels often experience localized punching shear failure under concentrated impact loads, where large areas of the composite panel fail to contribute effectively to load-bearing capacity. Thus, improving material utilization to enhance the panel's load resistance and deformation capacity has become a critical research focus. To address punching shear failure, this study proposes dual-path improvements through material design and structural design to optimize material efficiency and punching shear resistance. From material design aspect, an ultra-lightweight high-ductility cementitious composite (ULHDCC) was developed using cement as the binder, cenospheres as lightweight aggregates, polyethylene (PE) fibers for toughening, and rubber powder for energy absorption resistance. From structural design aspect, different shear connector configurations and multi-layer steel-concrete sandwich structures were employed to optimize load transfer mechanisms and stress distribution.

This chapter employs experimental test and finite element (FE) modeling to investigate the stiffness and ultimate load resistance of multi-layer SCS panels, validating the model against experimental punching tests and then conducting parametric studies on five key variables: layer number, steel contribution ratio, connector type, connector spacing, and loading area. The top steel plate exhibited significant tensile membrane action, improving overall structural integrity. The ductile material and multi-layer arrangement enhanced material utilization efficiency, energy dissipation capacity, and deformation performance. The proposed multi-layer design effectively mitigated punching shear failure and improved ultimate resistance. Based on these analyses, simplified methods for predicting the stiffness (elastic and plastic) and the ultimate resistance of double layered steel-concrete-steel (SCS) sandwich structures.

Z. Huang, Y. Zhou, *Innovative Lightweight High-performance Steel-concrete Sandwich Composite for Marine Structures*,
https://doi.org/10.1007/978-981-95-4681-7_4

4.2 Punching Shear Test Program

The study investigated the punching shear behavior of ten SCS sandwich panels, including two single-layer and eight double-layer configurations, with a lightweight high-ductility cement composite (LHDCC) core. The LHDCC mixture, incorporating 0.7% PE fibers, achieved a compressive cylinder strength of approximately 45 MPa while maintaining a reduced density of 1450 kg/m^3. Three types of shear connectors—J-hooks, headed studs, and hybrid combinations, were employed to enhance composite action. Material characterization tests determined the mechanical properties of the steel plates and connectors, with complete specifications provided in Table 4.1. As illustrated in Fig. 4.1, the test setup involved applying concentrated loads through a square patch at the panel center while supporting the specimens on four edge rollers. A universal testing machine recorded the applied loads, while a Linear Variable Differential Transformer (LVDT) monitored the central displacement of the bottom plate during loading. This configuration allowed for comprehensive evaluation of the panels' structural response under punching shear conditions.

4.2.1 Materials and Specimens

Table 4.2 presents the material properties of the LHDCC, steel plates, and shear connectors used in the study. The lightweight high-ductility cement composite was formulated with 0.7% PE fibers to achieve a target 28-day cube strength of 55 MPa while maintaining a reduced density of 1450 kg/m^3. Material testing conducted concurrently with panel evaluations demonstrated the LHDCC's exceptional tensile strain capacity (3–4%), significantly exceeding conventional concrete and meeting ECC material standards, as shown in Fig. 4.2.

The mechanical properties of steel components were characterized through standardized coupon tests, revealing thickness-dependent yield (f_{py}) and ultimate (f_{pu}) strengths: 4 mm plates exhibited 272/428 MPa, 6 mm plates 268/418 MPa, and 8 mm plates 262/394 MPa. The J-hooks, manufactured from HRB400 steel, demonstrated yield (f_{y_J}) and ultimate (f_{u_J}) strengths of 326 MPa and 444 MPa, respectively, while headed studs showed superior performance with 405 MPa yield (f_{y_H}) and 509 MPa ultimate (f_{u_H}) strengths.

The test matrix (vid. Table 4.3) systematically varies key parameters: shear span-to-depth ratio ($\lambda = 0.5$–1.25), steel contribution ratio ($\rho = 0.08$–0.32, determined via 4–8 mm plate thickness), shear connector configuration (J-hooks [Φ12 mm], studs [Φ13 mm], or hybrid), loading area (100 mm × 100 mm or 150 × 150 mm), and connector spacing (S_x = 100–200 mm, yielding plate slenderness 16.7–33.3). Single-layer specimens (S90-6, S140-4) feature 90 mm/140 mm concrete thicknesses, while double-layer panels maintain 70 mm per layer. The degree of composite action (η, calculated per Sect. 4.5.2.3) differentiates partial ($\eta < 1.0$: D140-6

Table 4.1 Geometric parameters and material properties of SCS panels tested by the authors [1]

								Concrete	Steel plate		J-hook		Headed stud	
Specimen	L_s (mm)	h_{c1} (mm)	h_{c2} (mm)	H_c (mm)	t_p (mm)	a (mm)	S (mm)	f_{ck} (MPa)	f_{yp} (MPa)	f_{up} (MPa)	f_{y_J} (MPa)	f_{u_J} (MPa)	f_{y_H} (MPa)	f_{u_H} (MPa)
S90-6	500	90	/	90	6	100	100	44.3	268	418	326	444	405	509
S140-6	500	140	/	140	6	100	100	43.3	268	418	326	444	405	509
D140-4	500	70	70	140	4	100	100	42.9	272	428	326	444	405	509
D140-6	500	70	70	140	6	100	100	44.2	268	418	326	444	405	509
D140-8	500	70	70	140	8	100	100	44.9	262	394	326	444	405	509
D140-6(J)	500	70	70	140	6	100	100	43.9	268	418	326	444	/	/
D140-6(H)	500	70	70	140	6	100	100	42.6	268	418	/	/	405	509
D140-6(S150)	500	70	70	140	6	100	150	44.3	268	418	326	444	405	509
D140-6(S200)	500	70	70	140	6	100	200	45.0	268	418	326	444	405	509
D140-6(A150)	500	70	70	140	6	150	100	44.6	268	418	326	444	405	509

Notes: L_s is the shear span from the loading patch to the support roller; h_{c1} is the height of upper concrete core; h_{c2} is the height of lower concrete core; H_c is the total height of concrete core; t_p is the thickness of steel plate; a is the width of the loading patch; S is the spacing between the two adjacent connectors; f_{ck} is the compressive cylinder strength of concrete (MPa); f_{yp}, f_{up} are the yield and ultimate strength of steel plate; f_{y_J}, f_{u_J} are the yield and ultimate strength of J-hook; f_{y_H}, f_{u_H} are the yield and ultimate strength of headed stud

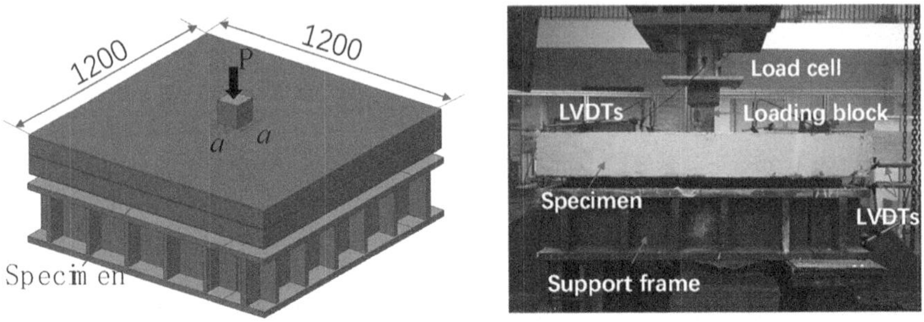

Fig. 4.1 Test set-up and loading scheme. (Reprinted from Huang et al. [1], copyright 2025, with permission from Elsevier)

Table 4.2 Material properties of concrete and steel

Specimen	Concrete	Steel plate		J-hook		Headed stud	
	f_{cu} (MPa)	f_{py} (MPa)	f_{pu} (MPa)	f_{y_J} (MPa)	f_{u_J} (MPa)	f_{y_H} (MPa)	f_{u_H} (MPa)
S90-6	55.4	268	418	326	444	405	509
S140-6	54.1	268	418	326	444	405	509
D140-4	53.6	272	428	326	444	405	509
D140-6	55.3	268	418	326	444	405	509
D140-8	56.1	262	394	326	444	405	509
D140-6(J)	54.9	268	418	326	444	/	/
D140-6(H)	53.3	268	418	/	/	405	509
D140-6(S150)	55.4	268	418	326	444	405	509
D140-6(S200)	56.3	268	418	326	444	405	509
D140-6(A150)	55.8	268	418	326	444	405	509

Notes: f_{cu} is the compressive strength of 100 × 100 mm concrete cube (MPa); f_{py}, f_{pu} are the yield and tensile strength of steel plate; f_{y_J}, f_{u_J} are the yield and tensile strength of J-hook; f_{y_H}, f_{u_H} are the yield and tensile strength of headed stud

(S150), D140-6(S200)) and full composite ($\eta \geq 1.0$) designs, building upon established SCS beam [2, 3] and panel [4–6] configurations to ensure comparable baseline performance.

The double-layer SCS specimens were fabricated under controlled workshop conditions using a systematic assembly process. Each panel incorporated two external steel faceplates and a central steel plate; all interconnected through welded shear connectors applied using a welding gun. The manufacturing sequence involves: (a) attaching hybrid shear connectors to the bottom plate; (b) welding connectors to both surfaces of the middle plate and underside of the top plate; (c) assembling the bottom and middle plates; (d) integrating the top plate; and (e) casting the LHDCC concrete into the completed steel formwork. Laboratory procedures documented the precise formwork preparation and concrete placement, ensuring consistent specimen quality throughout the experimental program as shown in Fig. 4.3a–f.

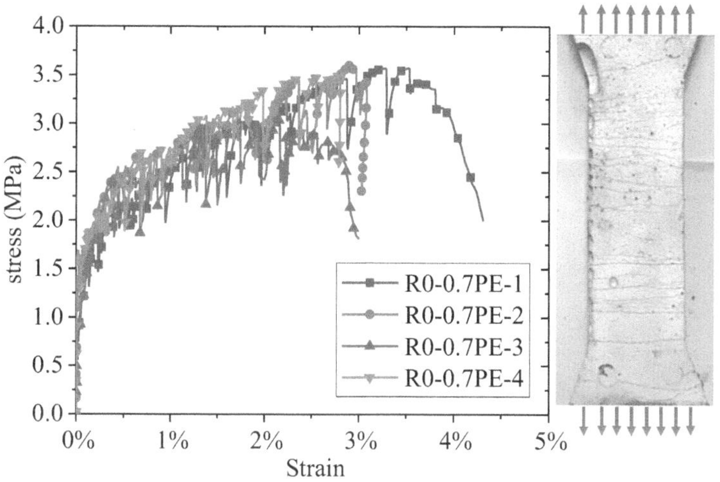

Fig. 4.2 Tensile stress-strain curve and cracking characteristic of LHDCC. (Reprinted from Huang et al. [1], copyright 2025, with permission from Elsevier)

4.2.1.1 Failure Mode

The experimental study examined the failure mechanisms and load-displacement behavior of steel-LHDCC-steel sandwich panels under punching shear. The load-displacement curves consistently displayed two or three distinct peak points, each corresponding to different failure stages. In single-layer panels, the failure mode at the first peak depended on the shear span-to-depth ratio—flexural yielding of the bottom steel plate occurred in specimens with larger ratios, while concrete punching shear dominated in those with smaller ratios. Both types of single-layer panels ultimately experienced punching shear fracture of the top steel plate at the second peak. For double-layer configurations, all specimens initially failed due to concrete punching shear, followed by distinct secondary failure modes depending on composite action. Fully composite panels exhibited top steel plate fracture at the second peak, whereas partially composite ones showed combined plate fracture and interfacial bond-slip. Continued loading beyond the second peak occasionally produced a third peak when the middle steel plate resisted punching, demonstrating the enhanced structural performance of double-layer systems. These panels exhibited superior load capacity, post-peak strength retention, ductility, and energy absorption compared to their single-layer counterparts, as illustrated in Fig. 4.4. The experiment measured the vertical displacement at the central and quarter spans, and the relative slippage between the steel plate and concrete at each load increment. Figure 4.5c shows the arrangement of the LVDTs. The experiment also measured the strain of the bottom and top steel plates, as shown in Fig. 4.5a, b. The loading rate was

Table 4.3 Geometric parameters of sandwich panel specimens

Specimen	h_{c1} (mm)	h_{c2} (mm)	h_c (mm)	$t_t = t_m = t_b$ (mm)	a (mm)	J-hook	Head stud	S_x (mm)	λ	ρ (%)	η
S90-6	90	/	90	6	100	ϕ12@200	ϕ13@200	100	4.41	11.0	1.00
S140-6	140	/	140	6	100	ϕ12@200	ϕ13@200	100	2.96	7.4	1.00
D140-4	70	70	140	4	100	ϕ12@200	ϕ13@200	100	2.96	7.7	1.00
D140-6	70	70	140	6	100	ϕ12@200	ϕ13@200	100	2.85	11.2	1.00
D140-8	70	70	140	8	100	ϕ12@200	ϕ13@200	100	2.74	14.4	1.00
D140-6(J)	70	70	140	6	100	ϕ12@100	/	100	2.85	11.3	1.00
D140-6(H)	70	70	140	6	100	/	ϕ13@100	100	2.85	11.3	1.00
D140-6(S150)	70	70	140	6	100	ϕ12@300	ϕ13@300	150	2.85	11.3	0.65
D140-6(S200)	70	70	140	6	100	ϕ12@400	ϕ13@400	200	2.85	11.3	0.37
D140-6(A150)	70	70	140	6	150	ϕ12@200	ϕ13@200	100	2.69	11.3	1.00

Notes: h_{c1} is the height of upper concrete core; h_{c2} is the height of lower concrete core; h_c is the total height of concrete core; $t_t = t_m = t_b =$ are the thickness of the top, middle and bottom steel plates, respectively; a is the width of the loading patch; the symbols of the J-hook $\phi d_J@S_J$: d_J is the diameter of J-hook, S_J is the J-hook spacing; the symbols of the headed stud $\phi d_H@S_H$; d_H is the diameter of headed stud, S_H is the headed stud spacing; S_x is the spacing between the two adjacent connectors; the shear span-to-depth ratio $\lambda = 0.5(L_s - a)/(t_t + t_m + t_b + h_{c1} + h_{c2})$; the steel contribution ratio $\rho = (t_t + t_m + t_b)/(t_t + t_m + t_b + h_{c1} + h_{c2})$; η is the degree of composite act. Failure mechanism and responses of sandwich panels

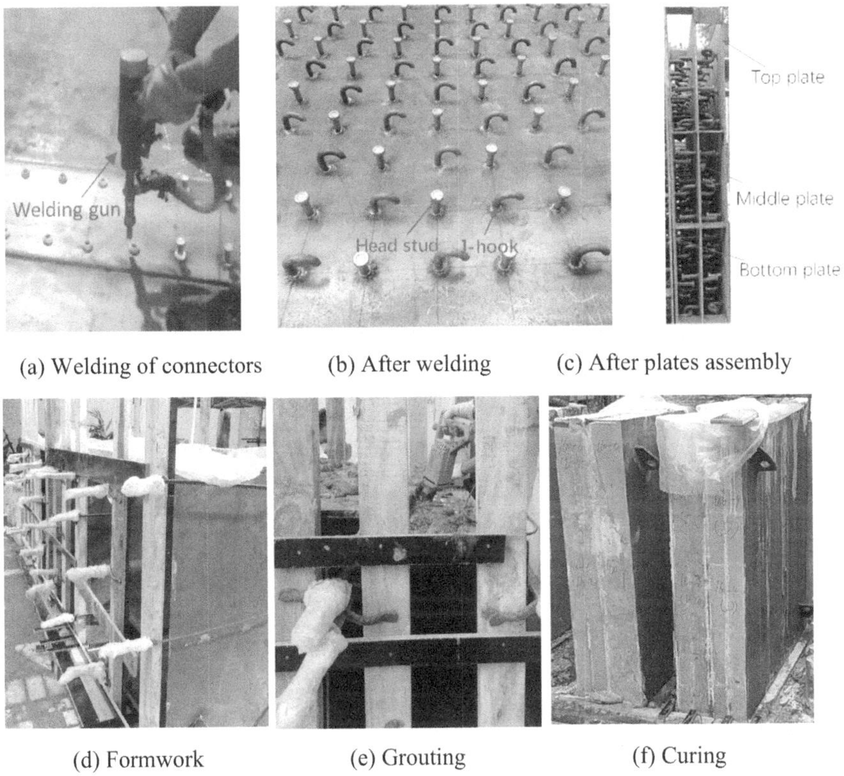

Fig. 4.3 Fabrication of steel formwork and concrete cast in the laboratory. (**a**) Welding of connectors. (**b**) After welding. (**c**) After plates assembly. (**d**) Formwork. (**e**) Grouting. (**f**) Curing. (Reprinted from Huang et al. [1], copyright 2025, with permission from Elsevier)

0.2 mm/min. Before loading to failure, preload procedure on each specimen was conducted 20 cycles to ensure the instrumentation recorded the test data reliably and correctly.

The load transfer mechanism in single-layer panels is fundamentally governed by key parameters including steel contribution ratio (ρ) and shear span-to-depth ratio (λ), which dictate distinct failure progress ions. For specimen S90-6 ($\lambda = 4.41$), flexural behavior dominates initially, producing globally distributed downward deformation until top plate punching shear fracture occurs, as shown in Fig. 4.6. In contrast, S140-6 ($\lambda = 2.96$) exhibits shear-dominated response with localized deformation beneath the loading area, initiating through concrete punching cone formation before progressing to identical top plate failure (see Fig. 4.6b). Post-test examination (see Fig. 4.6c) reveals the complete failure sequence: (i) diagonal cracks propagate through the concrete core, compacting material within the punching cone and shearing connectors across along the crack paths, and (ii) the top plate fails in

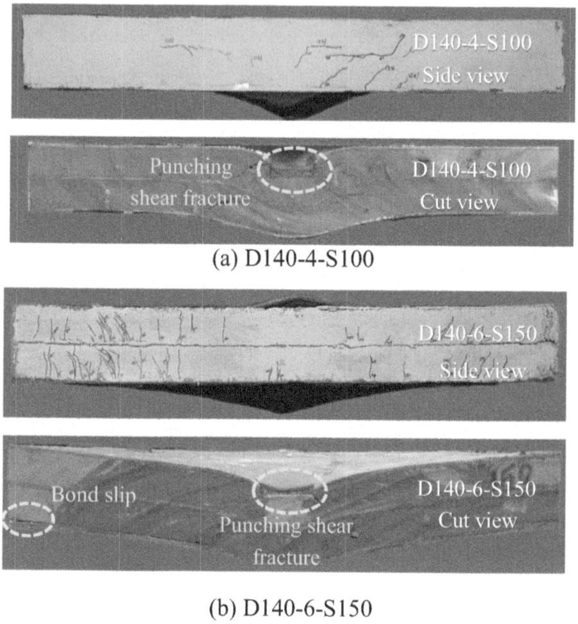

Fig. 4.4 Failure mode. (**a**) D140-4-S100. (**b**) D140-6-S150. (Reprinted from Zhang et al. [27], copyright 2025, with permission from Elsevier)

perimeter shear. These observations confirm that while λ determines initial behavior (flexural- versus shear-dominated), the ultimate failure mode remains consistently governed by top plate punching capacity, with shear connectors actively participating in resistance development throughout the loading history.

The double-layer panels demonstrate two distinct failure progressions governed by composite action levels. Full composite specimens (e.g., D140-4, D140-6,) exhibit localized deformation beneath the loading area, progressing through concrete punching cone formation with clear diagonal cracking, followed by sequential shear fractures of both top and middle plates (see Fig. 4.7). In contrast, partial composite panels (e.g., D140-6(S150) and D140-6(S200)) develop globally distributed flexural deformation due to interfacial bond-slip (see Fig. 4.7a), which reduces composite action by allowing steel-concrete relative displacement [3]. Despite all specimens ultimately experience top plate shear failure (see Fig. 4.7b), their internal damage patterns differ significantly: full composite designs are characterized by pronounced concrete compaction and middle plate fracture along the punching cone, whereas partial composite versions display finer flexural cracks that only yield without fracturing. Notably, even partial composite panels eventually form punching cones despite lacking obvious diagonal cracks, confirming the fundamental role of steel plate continuity in preserving residual capacity after bond-slip occurs.

These general trends are confirmed by the specimen-level results in Fig. 4.8. In the full composite panel D140-4, localized indentation of the middle and bottom plates accompanies the formation of a clear punching cone, with the top and middle

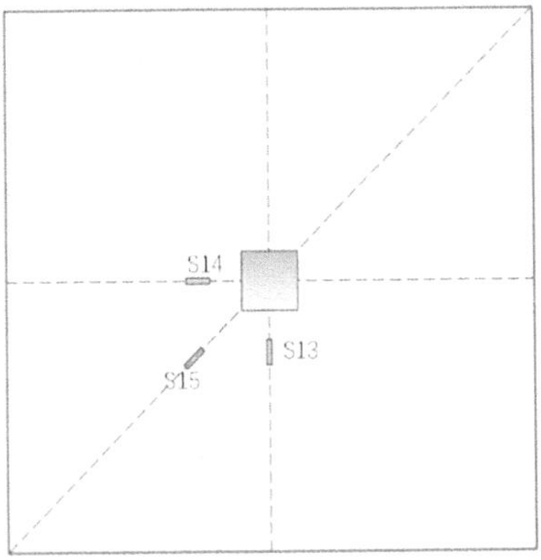

(a) Strain gauge layout on the top plate

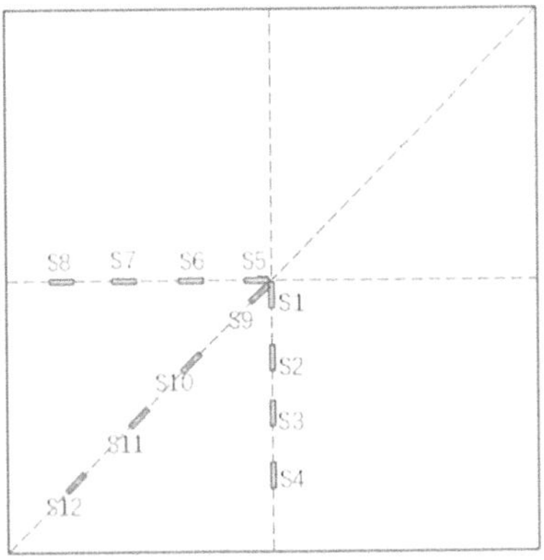

(b) Strain gauge layout on the bottom plate

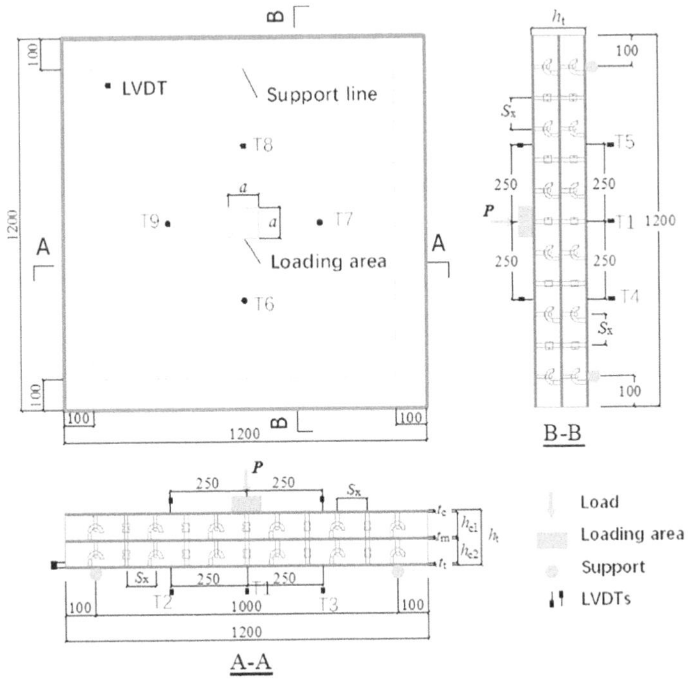

(c) Layout of LVDTs

Fig. 4.5 Test setup and instrumentation. (**a**) Strain gauge layout on the top plate. (**b**) Strain gauge layout on the bottom plate. (**c**) Layout of LVDTs. (Reprinted from Huang et al. [1], copyright 2025, with permission from Elsevier)

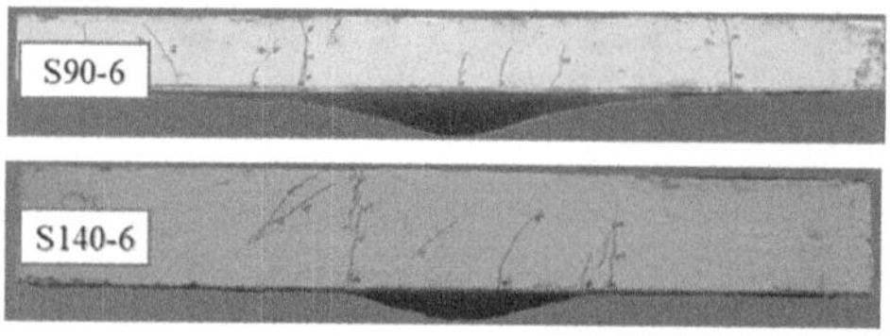

(a) Downward convex deformation by side view

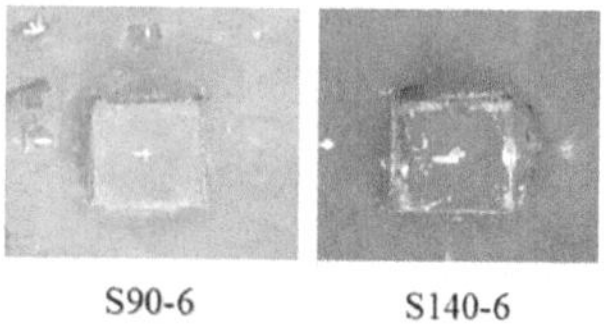

(b) Punching shear fracture of top plate

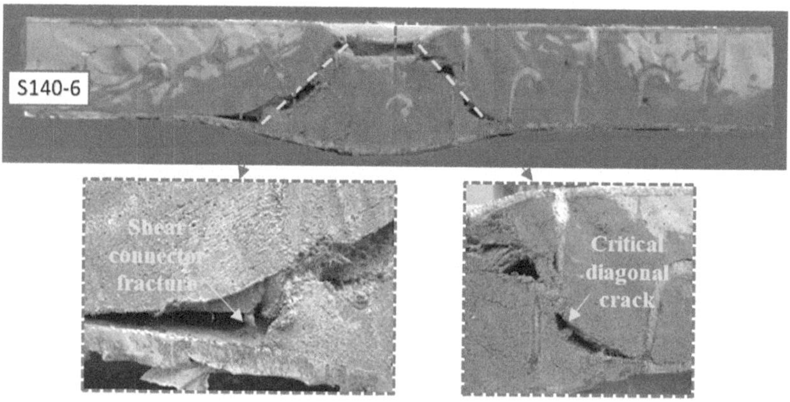

(c) Local punching cone by cut view

Fig. 4.6 Failure shape of single layer panels at the end of test. (**a**) Downward convex deformation by side view. (**b**) Punching shear fracture of top plate. (**c**) Local punching cone by cut view. (Reprinted from Huang et al. [1], copyright 2025, with permission from Elsevier)

plates failing in shear. By contrast, the partial composite panel D140-6(S150) shows pronounced bond-slip that produces global flexural deformation; the top plate is fully sheared, the middle plate only yields, and the concrete develops fine flexural cracks. Nevertheless, a punching cone still forms in this specimen, confirming the consistent role of concrete cone failure across both composite levels.

4.2.1.2 Load-Displacement Curve

The load-displacement responses of single-layer and double-layer sandwich panels are presented in Fig. 4.9a, b, respectively, while Fig. 4.9c, d illustrate their characteristic behavior patterns. These plots clearly reveal the enhanced structural performance of double-layer configurations compared to single-layer panels under punching shear loading.

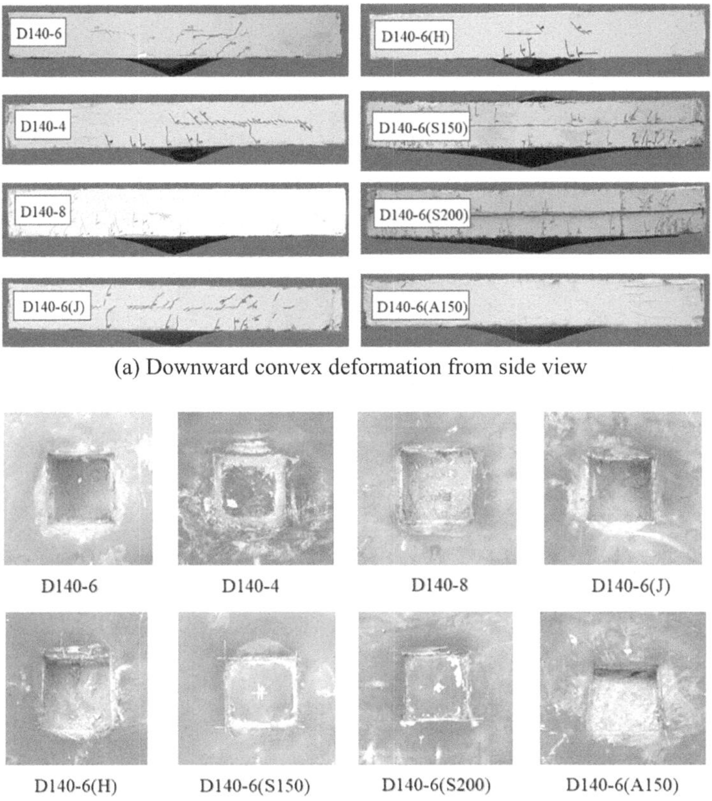

Fig. 4.7 Failure shape of double layer panels at the end of test. (**a**) Downward convex deformation from side view. (**b**) Punching shear fracture of top plate. (Reprinted from Huang et al. [1], copyright 2025, with permission from Elsevier)

The load-displacement responses reveal distinct behavioral patterns between single-layer panels with different shear span-to-depth ratios (λ). The more flexible Type I specimen (S90-6, higher λ) exhibits initial yielding of the bottom steel plate accompanied by through-thickness flexural cracking, reaching its first peak through flexural mechanisms before developing membrane action at large deflections. In contrast, the stiffer Type II panel (S140-6, lower λ) achieves its first peak through concrete punching shear failure, characterized by rapid diagonal crack propagation. Both configurations demonstrate similar secondary load enhancement from steel membrane effects, ultimately failing through sudden punching shear fracture of the top plate—though Type II exhibits greater initial strength (P_1) due to its shear-dominated response, while both specimens converge to comparable ultimate resistances (P_2) governed by identical top plate failure mechanisms. The post-peak behaviors diverge significantly, with Type I maintaining stable load capacity during

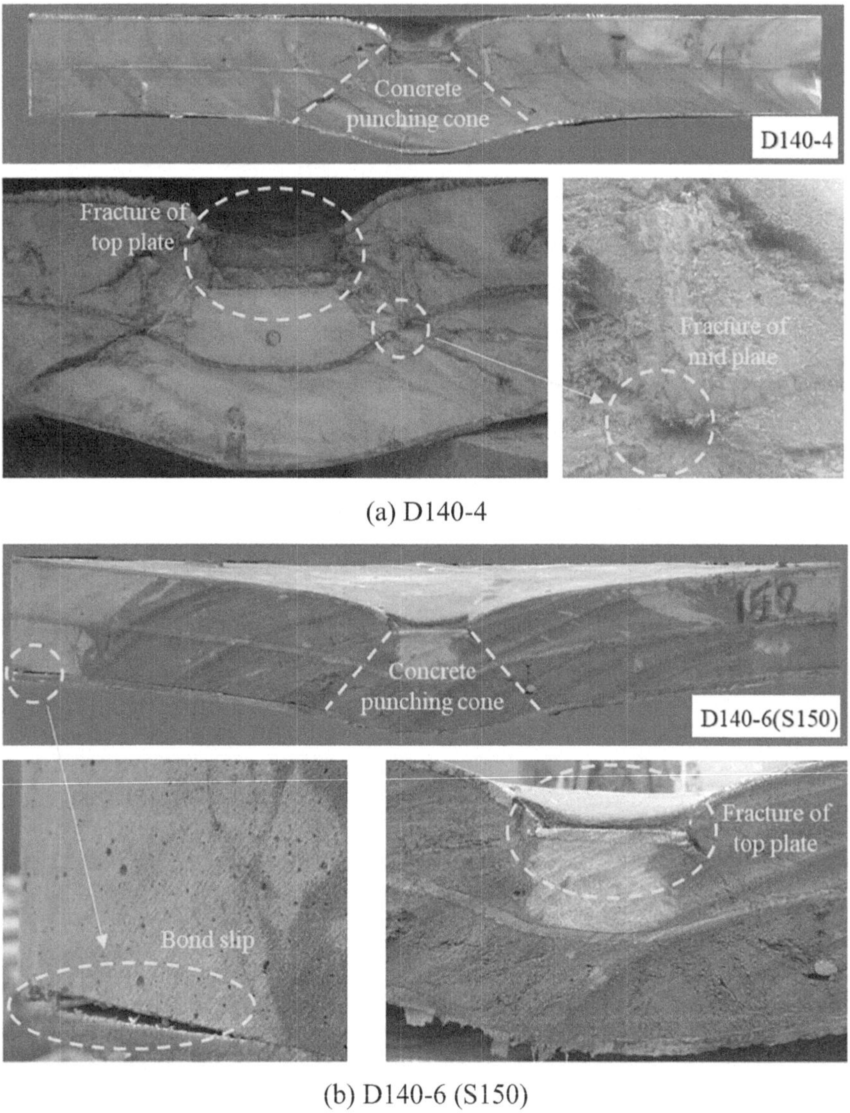

Fig. 4.8 Cut view of double layer panels at the end of test. (**a**) D140-4. (**b**) D140-6 (S150). (Reprinted from Huang et al. [1], copyright 2025, with permission from Elsevier)

its flexural-to-membrane transition, while Type II shows immediate softening after concrete shear failure before membrane action engagement.

The double-layer panels demonstrate distinct load-displacement behaviors based on their composite action levels. Partial composite specimens (Type A) exhibit greater flexibility and ductility than their full composite counterparts (Type B), with their P-δ curves showing two peak resistances compared to three for full

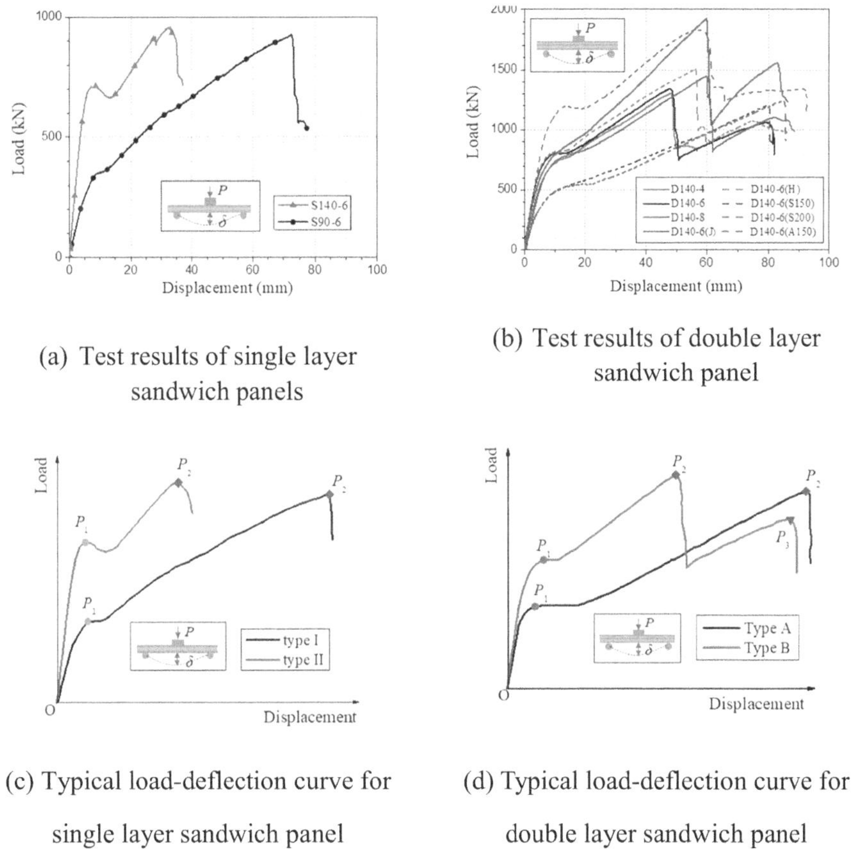

(a) Test results of single layer sandwich panels

(b) Test results of double layer sandwich panel

(c) Typical load-deflection curve for single layer sandwich panel

(d) Typical load-deflection curve for double layer sandwich panel

Fig. 4.9 Load-displacement curves. (**a**) Test results of single layer sandwich panels. (**b**) Test results of double layer sandwich panel. (**c**) Typical load-deflection curve for single layer sandwich panel. (**d**) Typical load-deflection curve for double layer sandwich panel. (Reprinted from Huang et al. [1], copyright 2025, with permission from Elsevier)

composite panels. Both types initially fail through concrete punching shear (P_1), though full composite panels achieve higher initial stiffness and P_1 values due to enhanced connector density. The middle steel plate's presence prevents abrupt post-peak softening observed in single-layer panels, enabling sustained load transfer through membrane action. Partial composite panels experience bond-slip during this phase, reducing stiffness and redistributing loads, while both types reach their ultimate capacity (P_2) through top plate fracture. Full composite panels uniquely develop a third peak (P_3) from middle plate membrane action before its eventual shear failure, typically at lower loads than P_2 when middle plate thickness equals or falls below the top plate's. These behavioral differences highlight how composite action degree fundamentally governs both the deformation capacity and failure progression in multi-layer sandwich systems.

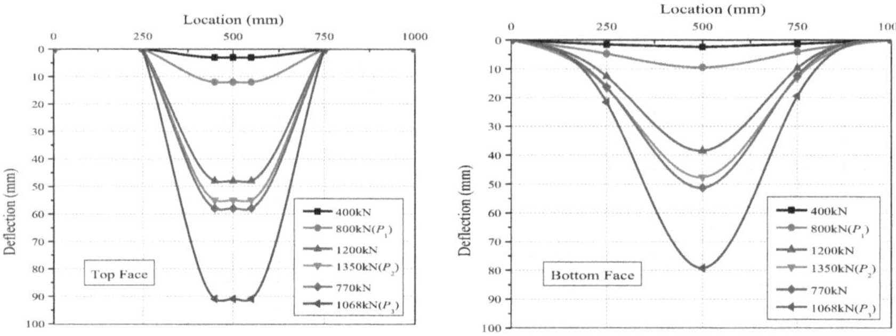

(a) D140-6

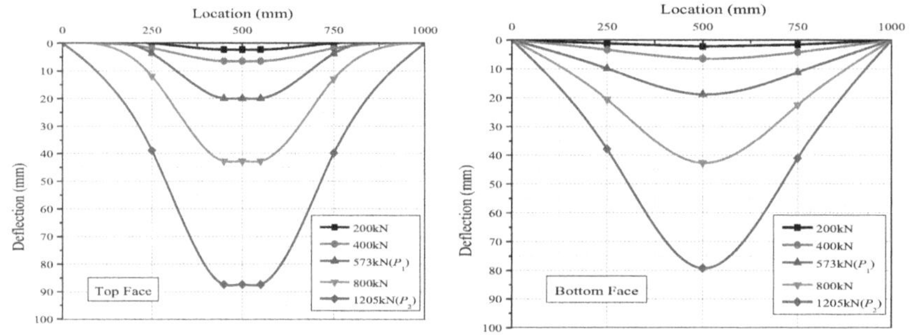

(b) D140-6 (S150)

Fig. 4.10 Vertical deflection shape of sandwich panel. (**a**) D140-6. (**b**) D140-6 (S150). (Reprinted from Huang et al. [1], copyright 2025, with permission from Elsevier)

4.2.1.3 Deflection Shape

The study's investigation of load transfer mechanisms in double-layer panels reveals distinct deformation characteristics through LVDT-measured deflection profiles (see Fig. 4.10). Both full composite (D140-6) and partial composite (D140-6(S150)) specimens exhibit greater top face deflections than bottom face displacements at all load levels, demonstrating the LHDCC core's progressive compaction and energy absorption capacity. The full composite panel shows localized deformation concentrated near the loading patch, while its partial composite counterpart develops global flexural deformation—evidenced by distributed curvature across the span. This behavioral divergence stems from bond-slip effects in partial composite panels, which redistribute deformations by allowing interfacial slippage between steel and concrete components. The characteristic flat midspan regions in top face deflection curves visually confirm localized indentation beneath the loading area, with the deformation patterns directly correlating to the observed failure mode transitions discussed previously. These measurements quantitatively validate how composite action degree fundamentally governs both load redistribution and deformation development in sandwich panel systems.

4.2.1.4 Strain Distribution in Sandwich Panels

Strain measurements (see Fig. 4.11) reveal distinct stress development patterns in the steel plates of double-layer panels. Both specimens show yielding ($\varepsilon_y \approx 1300$ με) in bottom plate elements (S1, S2, S6, S10 for D140-6; S1, S5, S9 for D140-6(S150)) at load level P_1, coinciding with concrete punching cone formation. The partial composite panel (D140-6(S150)) exhibits more extensive yielding beyond the punching cone (S3, S7, S11) during P_1–P_2 loading, demonstrating how bond-slip redistributes stresses across larger plate areas. Top plate strains universally exceed yield limits before reaching P_2 through membrane action, though measurements become unreliable beyond 20,000 με. These strain patterns quantitatively confirm the bond-slip effect's role in transforming load transfer mechanisms; while full composite behavior concentrates plasticity near the loading zone, partial composite action promotes widespread yielding through interfacial slip and stress redistribution.

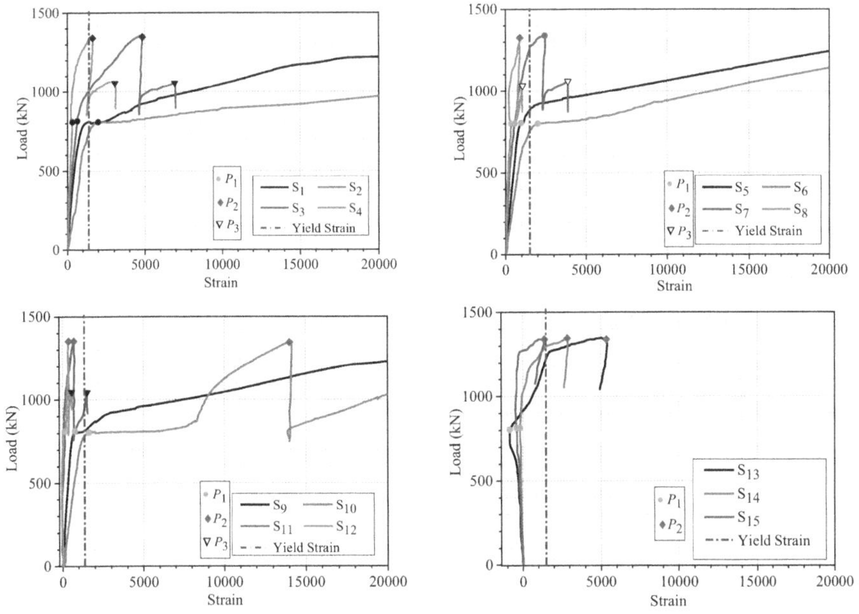

Fig. 4.11 Load versus strain curves. (a) D140-6. (b) D140-6 (S150). (Reprinted from Huang et al. [1], copyright 2025, with permission from Elsevier)

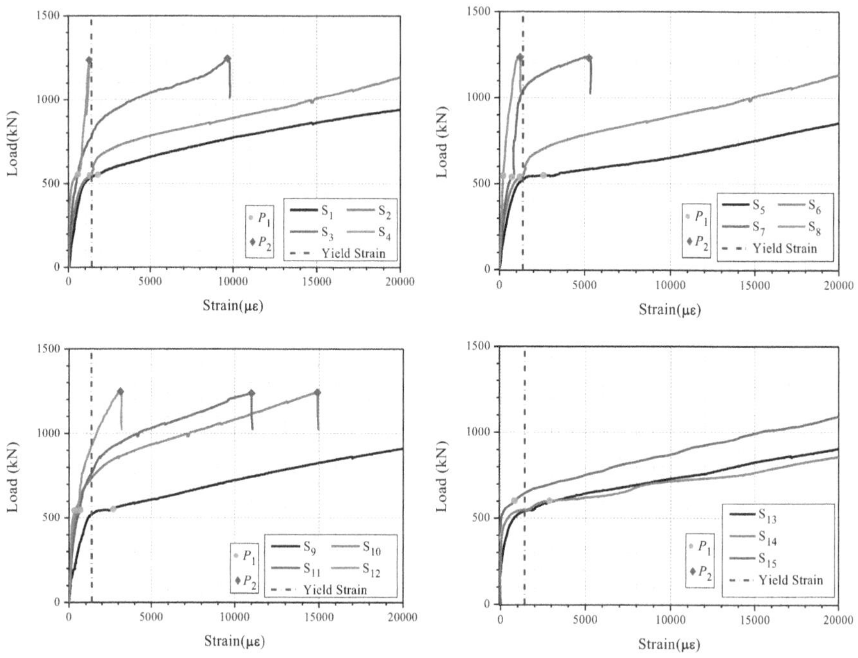

(b) D140-6 (S150)

Fig. 4.11 (continued)

4.3 Numerical Modelling

The study uses ABAQUS to analyze double-layer sandwich panels' structural performance under concentrated loads, examining load transfer mechanism, failure modes, stiffness, ductility, and energy absorption behavior. A systematic modeling approach is developed and validated against experimental data to ensure result reliability.

4.3.1 Material Models of Concrete and Steel

The mechanical properties of the developed LHDCC are characterized through standard compression and tension tests, with the resulting stress-strain curves presented in Fig. 4.12. This engineered cementitious composite contains PII 52.5 R Portland cement, silica fume, GGBFS, and 0.7% PE fibers (mix proportions detailed in Table 4.4). Material testing reveals an elastic modulus of 14 GPa, 45 MPa compressive strength (cylinder), and 1.6 MPa tensile strength. While the PE fibers show minimal impact on compressive behavior, they significantly enhance tensile performance through effective fiber bridging, enabling strain-hardening

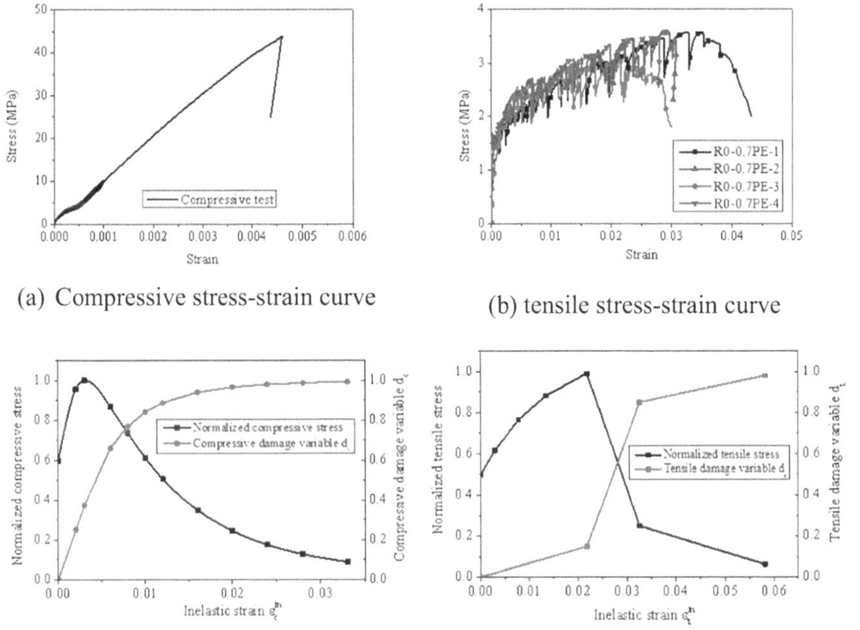

(a) Compressive stress-strain curve

(b) tensile stress-strain curve

(c) Compressive constitutive model in FE

(d) Tensile constitutive model in FE

Fig. 4.12 Material properties of concrete. (**a**) Compressive stress-strain curve. (**b**) Tensile stress-strain curve. (**c**) Compressive constitutive model in FE. (**d**) Tensile constitutive model in FE. (Reprinted from Zhang et al. [27], copyright 2025, with permission from Elsevier)

Table 4.4 Mix proportion of LHDCC (kg/m^3)

Material	Water	OPC	SF	GGBFS	PE fiber	HWRA	SRA
LHDCC	259.0	702.0	78.0	339.9	9.7	7.0	9.0

Note: *OPC* ordinary Portland cement, *SF* silica fume, *GGBFS* ground granulated blast furnace slag, *HWRA* high Water reducing agent, *SRA* shrinkage reducing agent

behavior up to 3–4% tensile strain—a characteristic that qualifies the material as an Engineered Cementitious Composite (ECC) [7].

The accurate modeling of concrete behavior is crucial for finite element analysis of SCS sandwich panels, as the concrete core plays a primary role in load-bearing capacity. Among ABAQUS available concrete models—brittle cracking, smeared cracking, and concrete damage plasticity (CDP)—the CDP model was selected for this study due to its superior capability to simulate inelastic concrete behavior through isotropic compressive/tensile plasticity with damage parameters. This widely-used approach requires comprehensive input data including separate stress-strain relationships for both compression and tension, along with their corresponding damage evolution characteristics, to properly represent concrete's nonlinear response under loading. The model's proven versatility and numerical stability make it particularly suitable for simulating the complex behavior of composite sandwich structures.

Existing constitutive models primarily address normal-weight concrete, with limited options for lightweight and fiber-reinforced lightweight concrete. Liu et al. [8] developed a unified model for both material types using 842 test datasets. Their statistically-based approach combines Weibull distribution for the ascending branch and log-normal distribution for the descending branch of the stress-strain curve, with the complete compressive stress-strain relationship expressed as:

$$\sigma_c = \begin{cases} E_0\varepsilon_c \exp\left[-\dfrac{1}{m}\left(\dfrac{\varepsilon_c}{\varepsilon_{ck}}\right)^m\right] & \varepsilon_c \le \varepsilon_{ck} \\ f_{ck} \exp\left[-0.5\left(\dfrac{\ln\left(\frac{\varepsilon_c}{\varepsilon_{ck}}\right)}{n}\right)^2\right] & \varepsilon_c > \varepsilon_{ck} \end{cases} \tag{4.1}$$

and the compressive damage parameter is given as:

$$d_c = \begin{cases} 1-\exp\left[-\dfrac{1}{m}\left(\dfrac{\varepsilon_c}{\varepsilon_{ck}}\right)^m\right] & \varepsilon_c \le \varepsilon_{ck} \\ 1-f_{ck} \exp\left[-0.5\left(\dfrac{\ln\left(\frac{\varepsilon_c}{\varepsilon_{ck}}\right)}{n}\right)^2\right]/E_0\varepsilon_c & \varepsilon_c > \varepsilon_{ck} \end{cases} \tag{4.2}$$

The compressive behavior is modeled using stress (σ_c), strain (ε_c), elastic modulus (E_0), strength (f_{ck}), and damage parameter (d_c). Shape factors $m = 2.15$ and $n = 0.883$ define the stress-strain curve. Inelastic strain $\varepsilon_c^{in} = \varepsilon_c - \sigma_c/E_0$ accounts for plastic deformation, while normalized stress σ_c/f_{ck} and damage evolution are shown in Fig. 4.12c.

The tensile behavior of LHDCC is characterized using experimental data from coupon tests (see Fig. 4.12b) for the elastic and strain-hardening stages. Since no established models exist for lightweight concrete's tensile softening, this study adopts a simplified bilinear degradation model, following successful applications with normal and high-strength concrete [9]. The complete tensile constitutive relationship, presented in Fig. 4.12d, incorporates normalized stress σ_t/f_{tu}, where f_{tu} is ultimate tensile strength and damage parameters, with inelastic strain (ε_t^{in}) calculated as $\varepsilon_t^{in} = \varepsilon_t - \sigma_t/E_0$. This approach effectively captures the material's post-peak tensile behavior while maintaining computational efficiency.

The CDP model incorporates five key parameters to characterize concrete behavior: dilation angle ($w = 50°$), stress invariant ratio ($K_c = 0.667$), biaxial/uniaxial strength ratio (r_{b0}/r_{c0} = 1.16), eccentricity (0.1), and viscosity parameter (0.0001). These values were carefully selected through calibration with experimental data and validation against established references [9–11], ensuring accurate representation of the material's plastic deformation and failure characteristics under multiaxial stress conditions. For steel, a typical bilinear model based on experimental stress-strain curves is used in simulation.

4.3.2 Element Type, Mesh Scheme, Boundary Condition

The FE model of the double-layer sandwich panel, illustrated in Fig. 4.13, employs a quarter-symmetry approach to reduce computational cost while maintaining accuracy. All components, i.e., steel plates, concrete cores, and shear connectors are discretized using C3D8R elements, with careful attention paid to mesh compatibility at material interfaces to prevent convergence issues. A comprehensive mesh sensitivity study informed the final discretization scheme, featuring a global element size of 10 mm with localized refinement around shear connectors (16 circumferential divisions). The model simplifies certain geometric details of J-hooks and headed studs while preserving their essential mechanical behavior. Boundary conditions enforce symmetry constraints, with fixed roller supports and pressure loading applied through a rigid patch to simulate the experimental loading conditions. This modeling strategy effectively balances computational efficiency with the need to capture complex material interactions and damage progression.

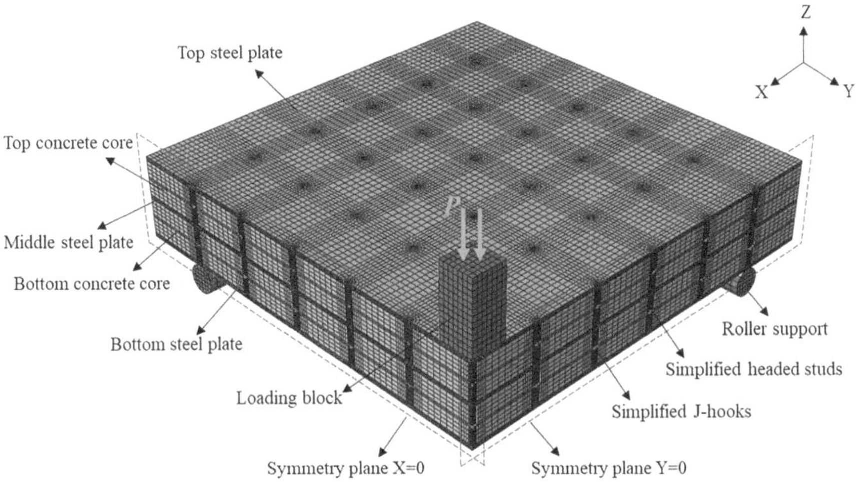

Fig. 4.13 FE model of double-layer sandwich panel. (Reprinted from Zhang et al. [27], copyright 2025, with permission from Elsevier)

4.3.3 Contact Definition

The FE model employs a balanced master-slave contact algorithm for all material interfaces, including concrete-steel plate, concrete-shear connector, loading patch-steel plate, and steel plate-roller support interactions. This approach ensures accurate contact analysis by allowing each surface node to function as both master and slave during iterations. Contact properties incorporate hard normal behavior and frictional tangential response, with coefficients determined through experimental validation. For steel-concrete interfaces, a friction coefficient of 0.7 was adopted based on literature values [12–15] and test data calibration, while steel-steel contacts used a conservative 0.5 coefficient to enhance numerical stability without significantly affecting results. This contact formulation effectively captures the essential interfacial behavior while maintaining computational robustness throughout the nonlinear analysis.

4.3.4 Representation of Shear Connectors

The shear connectors play a critical role in SCS panels by transferring transverse shear forces while maintaining composite action between concrete and steel plates under punching loads. To address the computational challenges of modeling complex J-hook and overlapped stud geometries, this study introduces a simplified approach representing these connectors as steel bars connected by nonlinear spring elements (see Fig. 4.14). The steel bars provide shear and flexural resistance, while the calibrated spring elements capture tensile behavior based on experimental load-displacement curves from dedicated SCS connection tests. This simplified modeling strategy effectively preserves the connectors' mechanical function while significantly improving numerical convergence and computational efficiency.

4.3.5 Validation of FE Model

The FE model intentionally excludes steel damage modeling to avoid convergence issues arising from simultaneous concrete and steel damage simulation, limiting the analysis to elastic-plastic behavior without fracture prediction. As shown in Fig. 4.15, the numerical results demonstrate good agreement with experimental load-displacement curves for both single-layer and double-layer specimens up to the steel plate fracture point. The model accurately captures initial stiffness, first peak resistance (P_1), and plastic hardening behavior for most specimens. However,

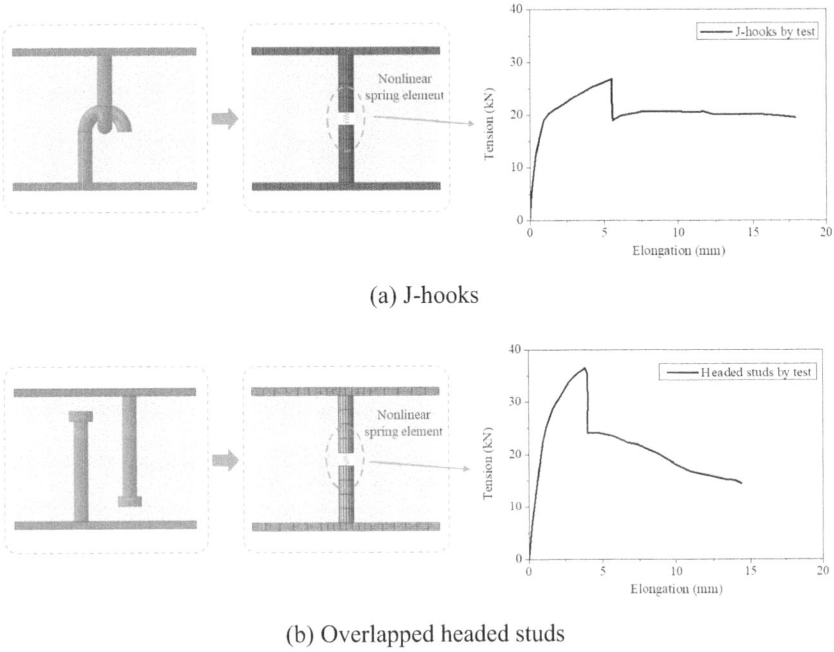

Fig. 4.14 Simplification of J-hooks and overlapped headed studs for FE modeling. (**a**) J-hooks. (**b**) Overlapped headed studs. (Reprinted from Zhang et al. [27], copyright 2025, with permission from Elsevier)

discrepancies observed in cases like D140-4-S100 and D140-6-S150 stem from three primary factors: (1) simplifications in shear connector modeling affecting concrete-connector interaction, particularly in nonlinear regimes; (2) unaccounted residual stresses from welded connectors influencing steel plate behavior; and (3) challenges in precisely simulating concrete cracking and crushing patterns. These limitations highlight the trade-offs between model complexity and computational practicality in simulating composite structural behavior.

Figure 4.16 demonstrates good correlation between experimental and numerical failure modes for panels D140-6-S100 (full composite action) and D140-6-S150 (partial composite action). The FE model accurately captures the distinct failure mechanisms: D140-6-S100 shows concentrated punching cone formation under the load patch with no interfacial slip, while D140-6-S150 exhibits combined global bending and localized indentation due to steel-concrete bond slip that reduces core stress transfer. These comparisons validate the model's ability to simulate both deformation patterns and composite interaction effects, though some differences in localized damage patterns may arise from the simplified treatment of concrete cracking and steel plasticity.

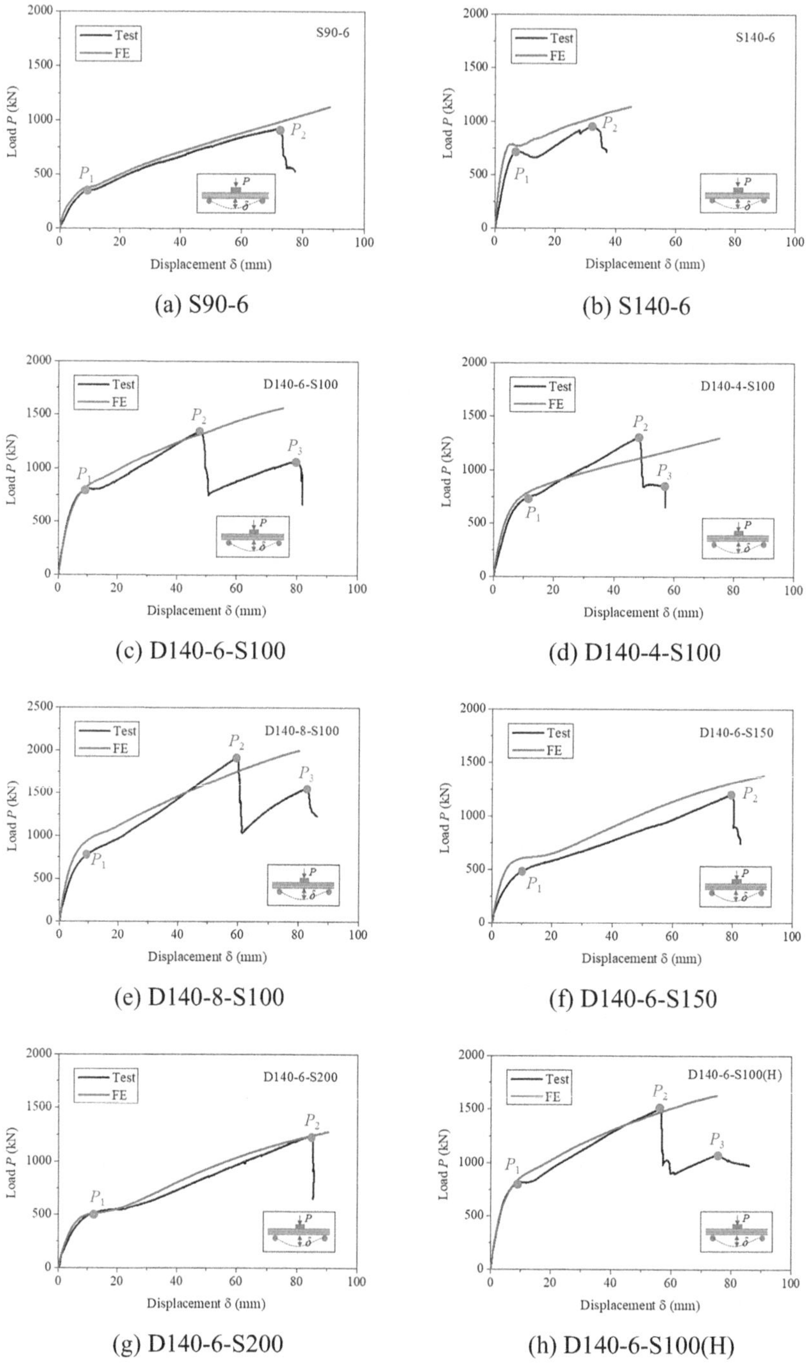

Fig. 4.15 Comparison of load-displacement curves between test and FE. (**a**) S90-6. (**b**) S140-6. (**c**) D140-6-S100. (**d**) D140-4-S100. (**e**) D140-8-S100. (**f**) D140-6-S150. (**g**) D140-6-S200. (**h**) D140-6-S100(H). (**i**) D140-6-S100(J). (**j**) D140-6-S150(A150). (Reprinted from Zhang et al. [27], copyright 2025, with permission from Elsevier)

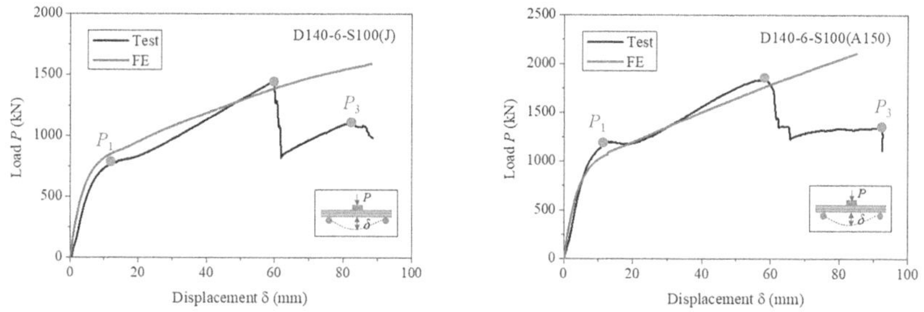

(i) D140-6-S100(J) (j) D140-6-S150(A150)

Fig. 4.15 (continued)

(a) D140-4-S100

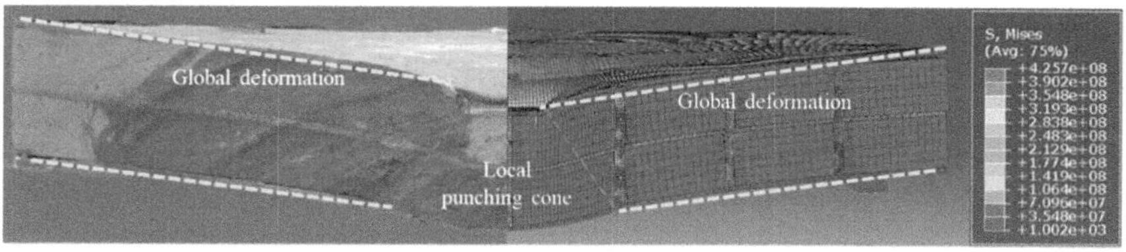

(b) D140-6-S150

Fig. 4.16 Comparison of failure modes between test and FE. (**a**) D140-4-S100. (**b**) D140-6-S150. (Reprinted from Zhang et al. [27], copyright 2025, with permission from Elsevier)

4.4 Effect of Design Parameters on Resistance and Stiffness

4.4.1 Effect of Layer Number

The comparative analysis of S140-6 (single-layer) and D140-4 (double-layer) panels with identical steel contribution ratios (ρ) reveals significant performance differences (see Fig. 4.17a). Both specimens demonstrate similar initial stiffness and first peak resistance, as these characteristics primarily depend on concrete and shear connector properties which were maintained constant. However, the double-layer configuration exhibits substantially greater ultimate capacity due to the additional load-bearing

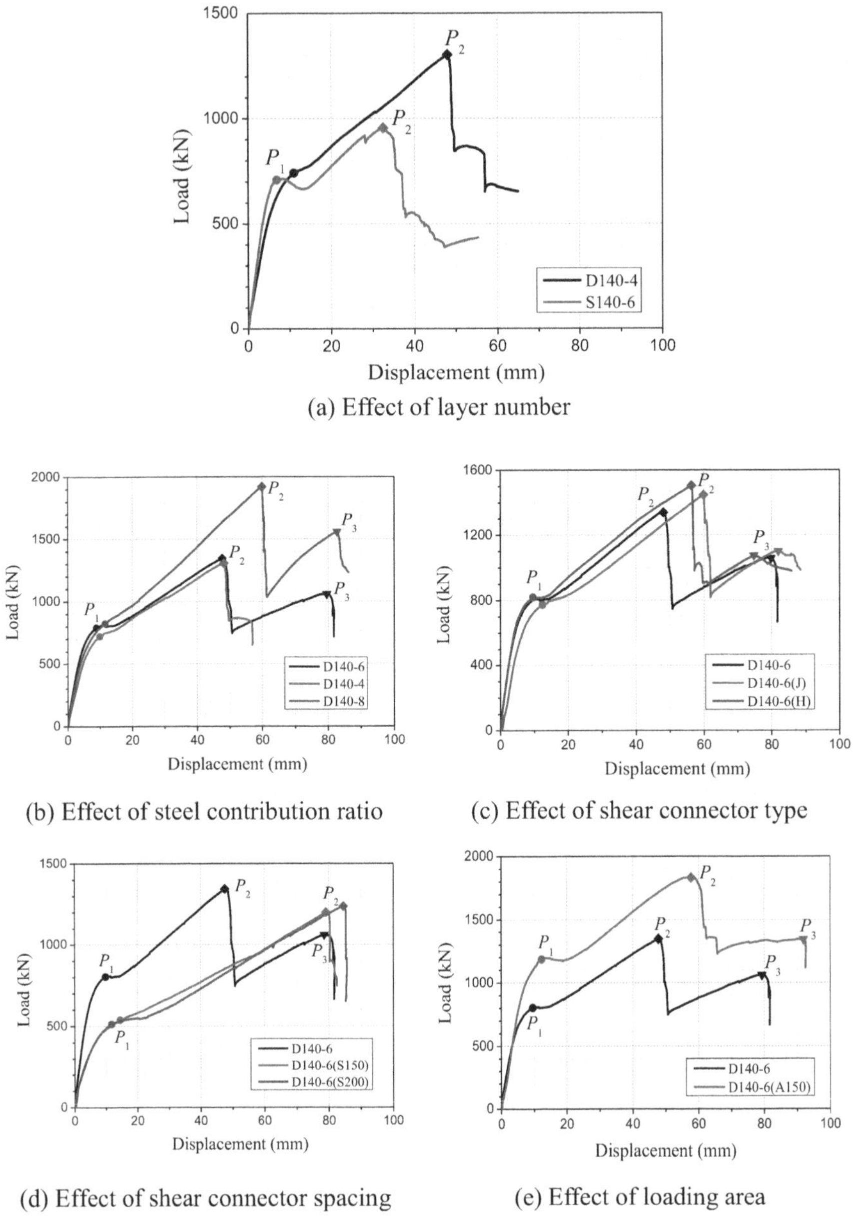

Fig. 4.17 Parametric analysis. (**a**) Effect of layer number. (**b**) Effect of steel contribution ratio. (**c**) Effect of shear connector type. (**d**) Effect of shear connector spacing. (**e**) Effect of loading area. (Reprinted from Huang et al. [1], copyright 2025, with permission from Elsevier)

contribution from its middle steel plate. While both systems develop comparable secondary stiffness through membrane action, the double-layer panel maintains superior post-peak residual strength after top plate fracture, as its lower layers remain structurally active. These results demonstrate how multi-layer designs enhance punching resistance without compromising initial stiffness, providing both higher ultimate capacity and improved damage tolerance compared to single-layer counterparts.

4.4.2 *Effect of Steel Contribution Ratio*

The steel contribution ratio ρ, defined as the thickness proportion of steel plates relative to the total panel thickness, significantly influences the load-displacement behavior of double-layer panels (see Fig. 4.17b). Specimens D140-4, D140-6, and D140-8 exhibit nearly identical first peak resistances (P_1) as this stage is governed by concrete and connector properties. However, their post-peak responses diverge markedly: higher ρ values correlate with greater secondary stiffness during membrane action development and enhanced ultimate capacities (P_2), where both top and middle plates contribute to shear resistance. The final resistance phase (P_3), controlled solely by the middle plate and connectors after top plate failure, demonstrates a direct relationship with ρ (D140-8 > D140-6 > D140-4), quantitatively confirming the steel proportion's critical role in determining both peak resistances and post-failure performance.

4.4.3 *Effect of Shear Connector Type*

Experimental results in Fig. 4.17c reveal that double-layer SCS panels with different shear connectors (J-hooks, headed studs, and hybrid) exhibit nearly identical load-displacement behavior. The specimens D140-6(J), D140-6(H), and D140-6 show comparable stiffness and load capacities at all stages (P_1-P_3), demonstrating that connector type has negligible effect on structural performance when minimum shear connection requirements are met. This suggests the load transfer mechanism depends more on overall composite action than specific connector configurations.

4.4.4 *Effect of Shear Connector Spacing*

Figure 4.17d compares the performance of double-layer panels with varying shear connector spacings (100 mm, 150 mm, and 200 mm). The specimen with closest spacing (D140-6) demonstrates full composite action, achieving 25% higher P_1 and

18% greater P_2 resistances than the partial composite panels (D140-6(S150)) and D140-6(S200)), due to enhanced shear transfer across concrete cracks. Reduced connector density decreases initial stiffness by approximately 30%, but increases ductility through controlled bond-slip, which redistributes stresses and improves energy absorption capacity by 15–20%. While both partial composite specimens exhibit similar load-displacement patterns, their performance differences highlight an optimal balance between structural efficiency (full composite) and construction economy (partial composite) in sandwich panel design.

4.4.5 Effect of Loading Area

The influence of loading area size on structural performance is evident in Fig. 4.17e, comparing specimens D140-6 (100 mm × 100 mm) and D140-6(A150) (150 mm × 150 mm). While both panels exhibit similar load-displacement trends indicating identical failure mechanisms, the larger loading area significantly enhances resistance capacities—increasing P_1 by 22%, P_2 by 18%, and P_3 by 15% due to expanded punching cone development. This performance improvement stems from greater material engagement, where the enlarged failure surface mobilizes additional concrete volume, shear connectors, and steel plate area to resist punching forces. The consistent behavioral patterns confirm that loading area dimensions primarily affect resistance magnitude rather than altering fundamental load transfer mechanisms in double-layer sandwich panels.

4.5 Analytical Model of Punching Shear

4.5.1 Analysis on Stiffness and Resistance

The energy absorption capacity represents a key performance metric for sandwich structures, closely tied to structural stiffness and deformation characteristics. Given the substantial computational demands associated with numerical modeling of double-layer SCS panels, this study develops an efficient analytical approach to predict both elastic and plastic stiffness behavior. The proposed methodology establishes an idealized load-displacement relationship that captures the essential mechanical response while significantly reducing analysis time and resource requirements compared to detailed FE simulations.

4.5.1.1 Stiffness at the Elastic Stage

During the elastic phase, the double-layer SCS panel demonstrates primarily global flexural deformation behavior, with the structural stiffness representing the panel's

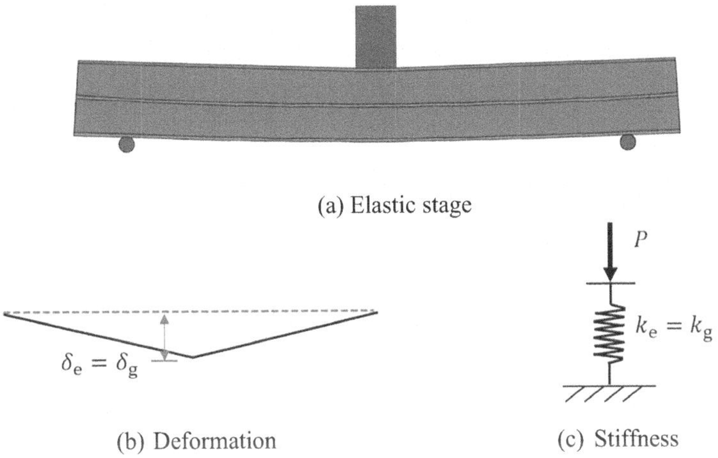

Fig. 4.18 Deformation and stiffness of SCS panel at elastic stage. (**a**) Elastic stage. (**b**) Deformation. (**c**) Stiffness. (Reprinted from Zhang et al. [27], copyright 2025, with permission from Elsevier)

overall performance (see Fig. 4.18). According to plate theory [16], the deformation response of simply supported rectangular panels under central concentrated loading can be theoretically predicted. For a panel with dimensions L_1 (length) × L_2 (width), the global deformation can be calculated using the following theoretical formulation:

$$\delta_g = \alpha \frac{PL_1^2}{D} = \frac{P}{k_g} \tag{4.3}$$

The bending stiffness D characterizes the plate's resistance to deformation while the numerical coefficient α varies with the aspect ratio L_2/L_1. For square plates ($L_1 = L_2$), α takes a value of 0.0116 as established in reference [16]. By substituting the span length L ($L = 2L_s$) between support rollers for L_1, we derive the following expression for the global deformation stiffness k_g in the current experimental configuration:

$$k_g = \frac{D}{0.0116L^2} \tag{4.4}$$

The composite bending stiffness D of the double-layer sandwich panel is determined by summing the contributions from both concrete (D_c) and steel plate (D_p) components. For this symmetric configuration, featuring equal-height upper/lower concrete cores and identical-thickness top/middle/bottom steel plates, the stiffness calculation simplifies significantly as all elements are evaluated about the composite section's neutral axis. This symmetrical design allows for simplified expressions of the total bending stiffness D and its constituent components D_c and D_p, which can be expressed as:

$$D = D_c + D_p \tag{4.5}$$

$$D_c = \frac{E_c}{1 - v_c^2} \cdot 2\left[\frac{h_c^3}{12} + h_c\left(\frac{h_c + t_p}{2}\right)^2\right] \tag{4.6}$$

$$D_p = \frac{E_p}{1 - v_p^2} \cdot \left\{\frac{t_p^3}{12} + 2\left[\frac{t_p^3}{12} + t_p(h_c + t_p)^2\right]\right\} \tag{4.7}$$

The bending stiffness calculation incorporates key material properties: concrete properties (elastic modulus E_c, Poisson's ratio v_c, core height h_c) and steel plate characteristics (elastic modulus E_p, Poisson's ratio v_p, thickness t_p). These parameters determine the individual stiffness contributions of each component in the composite section.

The conventional bending stiffness calculation assumes perfect bond between concrete and steel plates, leading to overestimated results. To account for interfacial slip effects, two FE models were developed for each design in Table 4.5—one with "Tie" constraints and another with "surface contact" interactions. Elastic material simulations revealed that the "surface contact" model's stiffness averaged only 20% of the "Tie" model's (ranging 11–29%), as shown in Table 4.6. This significant reduction prompted the introduction of a composite action parameter η, calculated as the ratio between connector shear capacity and steel plate tensile strength within triangular shear spans, as shown in Fig. 4.19. For simplification, the analysis uses half-span length ($L/2$) to determine the average cross-section when evaluating the steel plate's tensile capacity, thereby quantifying the degree of composite interaction more realistically. The composite action parameter η is expressed as:

$$\eta = \frac{n_s V}{f_{yp} t_p L/2} \leq 1.0 \tag{4.8}$$

The shear span region contains n_s connectors (one-quarter of the total), with individual capacities V_H for studs and V_J for J-hooks calculated per established equations in [17, 18], as given below:

$$V_H = 0.29 \alpha d_H^2 \sqrt{f_{ck} E_c} \leq 0.8 f_{u_H} A_H \tag{4.9}$$

$$V_J = 0.855 f_{ck}^{0.265} E_c^{0.469} A_J (h_J / d_J)^{0.154} \leq 0.8 f_{u_J} A_J \tag{4.10}$$

Table 4.5 FE models in the parametric study

FE models	L_s (mm)	H_c (mm)	t_p (mm)	a (mm)	S (mm)	ξ	⍰	f_{ck} (MPa)	f_{yp} (MPa)
SP-reference	500	140	6	100	100	0.16	1.00	45	275
SP-L425	425	140	6	100	100	0.19	1.00	45	275
SP-L575	575	140	6	100	100	0.14	1.00	45	275
SP-L650	650	140	6	100	100	0.12	1.00	45	275
SP-L725	725	140	6	100	100	0.11	1.00	45	275
SP-H100	500	100	6	100	100	0.12	1.00	45	275
SP-H120	500	120	6	100	100	0.14	1.00	45	275
SP-H160	500	160	6	100	100	0.18	1.00	45	275
SP-H180	500	180	6	100	100	0.20	1.00	45	275
SP-t4	500	140	4	100	100	0.15	1.00	45	275
SP-t8	500	140	8	100	100	0.16	1.00	45	275
SP-t10	500	140	10	100	100	0.17	1.00	45	275
SP-a50	500	140	6	50	100	0.16	1.00	45	275
SP-a150	500	140	6	150	100	0.16	1.00	45	275
SP-S125	500	140	6	100	125	0.16	0.83	45	275
SP-S150	500	140	6	100	150	0.16	0.65	45	275
SP-S175	500	140	6	100	175	0.16	0.51	45	275
SP-S200	500	140	6	100	200	0.16	0.37	45	275
SP-fc35	500	140	6	100	100	0.16	1.00	35	275
SP-fc55	500	140	6	100	100	0.16	1.00	55	275
SP-fc65	500	140	6	100	100	0.16	1.00	65	275
SP-fc75	500	140	6	100	100	0.16	1.00	75	275
SP-fp235	500	140	6	100	100	0.16	1.00	45	235
SP-fp355	500	140	6	100	100	0.16	1.00	45	355
SP-fp420	500	140	6	100	100	0.16	1.00	45	420
SP-fp460	500	140	6	100	100	0.16	1.00	45	460

Notes: "SP-Reference" indicates the reference model; "SP-L425" indicates model with $L_s = 425$ mm; "SP-H100" indicates model with $H_c = 100$ mm; "SP-t4" indicates model with $t_p = 4$ mm; "SP-a50" indicates model with $a = 50$ mm; "SP-S125" indicates model with $S = 125$ mm; "SP-fc35" indicates model with $f_{ck} = 35$ MPa; "SP-fp235" indicates model with $f_{yp} = 235$ MPa; $\xi = (H_c + 3t_p)/(2L_s)$ is the thickness/side length ratio; ⍰ is the degree of composite action

The coefficient $\alpha = 0.2(h_H/d_H + 1)$ for aspect ratios 3–4, and 1.0 for ratios >4. Connector capacities depend on their dimensions (height h_J, diameter d_J, area A_J) and ultimate strength f_{u_J}, with subscripts H and J denoting headed studs and J-hooks, respectively.

To account for partial composite action effects on stiffness k_g, the bending stiffness is scaled by a reduction factor of 0.2η. This modification yields values closely matching the $k_{g\_contact}/k_{g\_tie}$ ratios observed in Table 4.6. Consequently, the

Table 4.6 Comparison of k_g between the model using "Tie" constraint and "Surface contact"

Specimen	k_{g_tie} (kN/mm)	$k_{g\_contact}$ (kN/mm)	$k_{g\_contact}/k_{g\_tie}$	0.2[illegible]
SP-reference	851.4	163.3	0.18	0.20
SP-L425	876.0	185.5	0.21	0.20
SP-L575	596.8	133.2	0.22	0.20
SP-L650	508.7	110.1	0.22	0.20
SP-L725	416.5	96.6	0.23	0.20
SP-H100	476.1	87.6	0.18	0.20
SP-H120	614.5	124.5	0.20	0.20
SP-H160	837.6	203.3	0.24	0.20
SP-H180	959.7	278.1	0.29	0.20
SP-t4	672.2	120.7	0.19	0.20
SP-t8	988.8	190.6	0.19	0.20
SP-t10	983.0	222.7	0.23	0.20
SP-a50	540.2	145.3	0.27	0.20
SP-a150	899.0	186.9	0.21	0.20
SP-S125	742.9	110.4	0.15	0.17
SP-S150	747.8	105.4	0.14	0.13
SP-S175	737.6	81.6	0.11	0.10
SP-S200	752.4	75.9	0.10	0.07
SP-fc35	851.4	160.9	0.19	0.20
SP-fc55	851.4	164.8	0.19	0.20
SP-fc65	851.4	166.1	0.20	0.20
SP-fc75	851.4	168.3	0.20	0.20
SP-fp235	851.4	162.9	0.19	0.20
SP-fp355	851.4	165.0	0.19	0.20
SP-fp420	851.4	168.6	0.20	0.20
SP-fp460	851.4	169.0	0.20	0.20

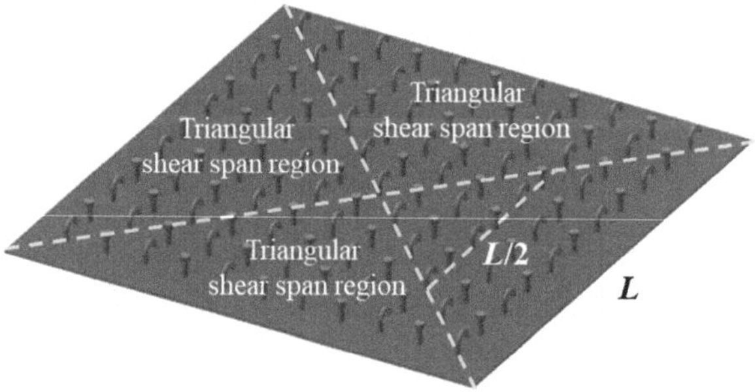

Fig. 4.19 Illustration of triangular shear span. (Reprinted from Zhang et al. [27], copyright 2025, with permission from Elsevier)

original stiffness in Eq. (4.4) is revised to incorporate this composite behavior adjustment, resulting in the following modified formulation:

$$k_g = \frac{0.2\eta D}{0.0116L^2} \tag{4.11}$$

The derived stiffness equation assumes thin plate behavior (thickness/side length ratio $\xi \leq 0.1$), making it unsuitable for direct application to thick double-layer SCS panels [16]. To address this limitation, a thickness correction factor λ is introduced, representing the ratio between Eq. (4.13) predictions and FE-derived elastic stiffness values. Analysis of cases in Table 4.5 (with ξ ranging 0.11–0.20) yields correction factors from 0.11 to 0.20 (see Table 4.7), where identical ξ values share averaged λ factors. Figure 4.20 presents the empirical relationship between ξ and λ through a fitted curve, expressed as:

$$\lambda = -1.65\xi + 0.75 \tag{4.12}$$

The elastic stiffness of double-layer SCS panels is ultimately expressed as:

$$k_{\mathrm{e}} = \lambda k_{\mathrm{g}} = \frac{0.2\eta\lambda D}{0.0116L^2} = \frac{\eta\lambda D}{0.058L^2} \tag{4.13}$$

Table 4.7 Correction to thickness/side length ratio for FE models

ξ	λ
0.11	0.60
0.12	0.53
0.14	0.52
0.15	0.52
0.16	0.49
0.17	0.39
0.18	0.46
0.19	0.40
0.20	0.49

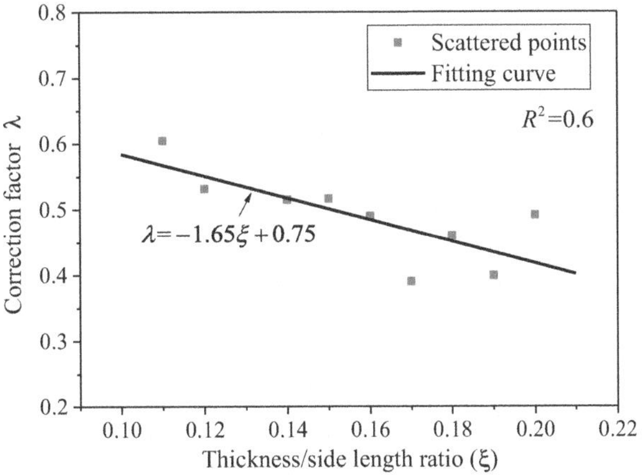

Fig. 4.20 Correction factor in relation to the thickness/side length ratio. (Reprinted from Zhang et al. [27], copyright 2025, with permission from Elsevier)

4.5.1.2 Stiffness at the Plastic Stage

During the plastic hardening stage, the double-layer SCS panel's load-bearing behavior follows a tandem spring model (see Fig. 4.21), where the total deformation δ comprises two components: global yielding (δ_g) and local indentation (δ_l). This decomposition ($\delta = \delta_g + \delta_l$) enables separate analysis of structural and localized responses, with the external load P exhibiting distinct relationships with each deformation mode as described by the following formulation:

$$P = k_g \delta_g = k_l \delta_l \tag{4.14}$$

The model combines global stiffness k_g (from Eq. (4.11)) with local indentation stiffness k_l to characterize the panel's plastic hardening behavior. The external load P can be equivalently expressed as a function of the total deformation δ through the following relationship:

$$P = k_p \delta \tag{4.15}$$

The plastic stiffness k_p combines global and local deformation effects during hardening. By combining Eqs. (4.14) and (4.15) with the deformation relationship between δ, δ_g, and δ_l, we derive the following unified expression for the panel's plastic behavior:

$$k_p = \frac{k_g k_l}{k_g + k_l} \tag{4.16}$$

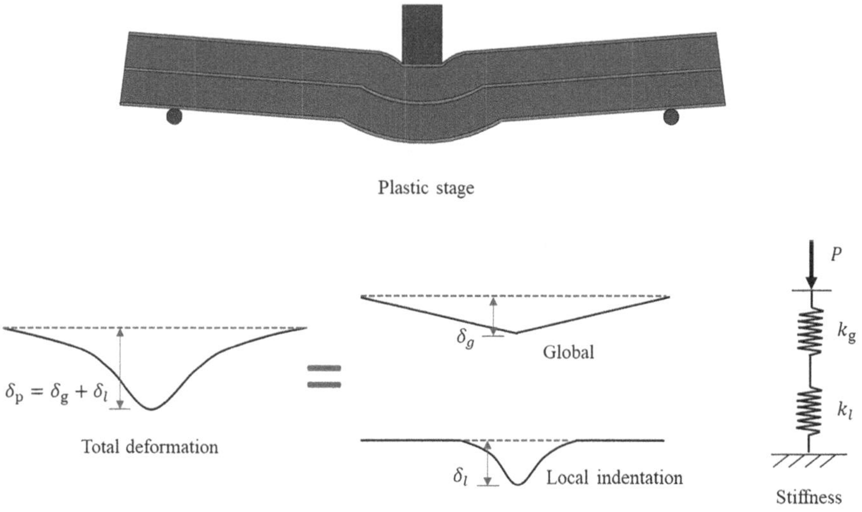

Fig. 4.21 Deformation and stiffness of SCS panel at plastic stage. (Reprinted from Zhang et al. [27], copyright 2025, with permission from Elsevier)

The local indentation response follows established solutions [17, 19] for steel plates under hemispherical loading:

$$P_s = 4\pi f_{yp} t_p \delta_l \tag{4.17}$$

where P_s is the steel plate load, f_{yp} the yield strength.

For cubic loading heads, the load-indentation relationship can be derived by modifying the hemispherical solution through geometric transformation. Considering the area ratio ($\pi/4$) between circular (diameter d) and square (side length d) contact surfaces, the governing equation adapts to:

$$P_s = 16 f_{yp} t_p \delta_l \tag{4.18}$$

The local stiffness of a simply supported steel plate under indentation loading is governed by the expression:

$$k_l = 16 f_{yp} t_p \tag{4.19}$$

The local stiffness k_l is adjusted for thickness effects using correction factor λ (Eq. (4.12)), giving $k_l = \lambda \times 16 f_{yp} t_p = 16\lambda f_{yp} t_p$ (Eq. (4.19)). Combining this with global stiffness k_g (Eq. (4.11)) and composite parameter η (Eq. (4.8)) yields the plastic stiffness for SCS panels:

$$k_p = \frac{k_g k_l}{k_g + k_l} = \frac{16\lambda\eta f_{yp} t_p D}{0.928 f_{yp} t_p L^2 + \eta D} \tag{4.20}$$

where D, η, and λ as given in Eqs. (4.5), (4.8), and (4.12) represent bending stiffness, composite action, and thickness correction, respectively.

4.5.1.3 Validation of Stiffness Models

The stiffness model's validity was assessed using eight additional double-layer SCS panel tests and 26 FE models (Tables 4.1 and 4.5). As shown in Tables 4.8 and 4.9, the proposed model demonstrates excellent agreement with experimental and numerical results for both elastic and plastic stages. The elastic stage predictions show a mean ratio of 1.03 (standard deviation 0.12) between calculated and measured values, whereas plastic stage predictions yield a mean ratio of 0.95 (standard deviation 0.13). These results confirm the model's accuracy across different deformation regimes, with close correspondence between theoretical predictions and empirical observations (Fig. 4.22).

Table 4.8 Comparison of stiffnesses at the elastic stage and plastic hardening stage between Test (or FE) and prediction model

Test models	k_e			k_p		
	Test (kN/mm)	Pred (kN/mm)	Pred/ Test	Test (kN/mm)	Pred (kN/mm)	Pred/ Test
D140-4	101.0	117.4	1.16	13.6	8.2	0.60
D140-6	155.2	163.3	1.05	15.1	11.9	0.79
D140-8	155.7	217.9	1.40	22.2	15.8	0.71
D140-6(J)	138.8	163.3	1.18	14.2	11.9	0.84
D140-6(H)	158.4	163.3	1.03	13.8	11.9	0.86
D140-6(S150)	86.2	105.5	1.22	10.7	11.5	1.07
D140-6(S200)	70.9	59.8	0.84	11.6	10.6	0.91
D140-6(A150)	154.4	163.3	1.06	16.4	11.9	0.73

Table 4.9 FE models in the parametric study

FE models	FE (kN/mm)	Pred (kN/mm)	Pred/ FE	FE (kN/mm)	Pred (kN/mm)	Pred/ FE
SP-Reference	163.3	163.3	1.00	12.2	11.9	0.97
SP-L425	185.5	203.0	1.09	13.2	10.9	0.83
SP-L575	133.2	131.8	0.99	12.9	12.5	0.97
SP-L650	110.1	109.7	1.00	11.3	12.9	1.14
SP-L725	96.6	90.8	0.94	11.1	12.8	1.16
SP-H100	87.6	94.9	1.08	12.5	12.6	1.01
SP-H120	124.5	127.8	1.03	12.3	12.5	1.01
SP-H160	203.3	200.4	0.99	12.0	11.3	0.94
SP-H180	278.1	238.0	0.86	11.8	10.6	0.90
SP-t4	120.7	117.4	0.97	7.2	8.2	1.14
SP-t8	190.6	217.9	1.14	15.3	15.8	1.04
SP-t10	222.7	268.2	1.20	19.4	19.2	0.99
SP-a50	145.3	163.3	1.12	10.1	11.9	1.17
SP-a150	186.9	163.3	0.87	14.6	11.9	0.81
SP-S125	110.4	134.3	1.22	12.8	11.7	0.91
SP-S150	105.4	105.5	1.00	12.3	11.5	0.93
SP-S175	81.6	82.6	1.01	12.7	11.1	0.87
SP-S200	75.9	59.8	0.79	12.1	10.6	0.88
SP-fc35	160.9	163.3	1.01	12.1	11.9	0.98
SP-fc55	164.8	163.3	0.99	12.2	11.9	0.97
SP-fc65	166.1	163.3	0.98	12.2	11.9	0.97
SP-fc75	168.3	163.3	0.97	12.3	11.9	0.96
SP-fp235	162.9	163.3	1.00	9.6	10.3	1.08
SP-fp355	165.0	163.3	0.99	15.5	15.1	0.97
SP-fp420	168.6	163.3	0.97	18.0	18.5	1.03
SP-fp460	169.0	163.3	0.97	19.5	20.1	1.03
Mean			1.03			0.95
Std. Dev.			0.12			0.13

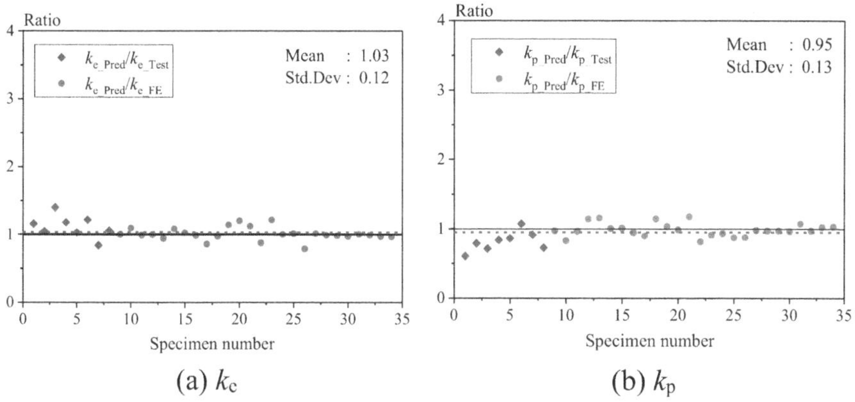

Fig. 4.22 Prediction to Test (FE) ratio of k_e and k_p. (**a**) k_e. (**b**) k_p. (Reprinted from Zhang et al. [27], copyright 2025, with permission from Elsevier)

4.5.1.4 Punching Shear Resistance

Huang et al. [1] proposed a punching cone model to predict the three characteristic peak resistances in double-layer sandwich panels under concentrated punching loads. These resistance peaks correspond to distinct failure mechanisms: concrete punching shear failure, top steel plate fracture, and middle steel plate fracture. The model accounts for potential bond-slip effects through composite action considerations. For practical design purposes, the first two peak resistances (concrete failure and top plate fracture) establish the critical lower and upper bounds of load-bearing capacity, making them particularly significant for structural evaluation. The occurrence of interfacial bond-slip depends on the degree of composite action between materials, which the model incorporates in its failure sequence predictions.

The first peak resistance P_1 in the punching cone model combines contributions from multiple components: the concrete layers (V_{c1} and V_{c2}), shear connectors (V_{s1} and V_{s2}), and the top steel plate (V_{pt}). This composite resistance is expressed as:

$$P_1 = V_c + V_s + V_{pt} = V_{c1} + V_{c2} + V_{s1} + V_{s2} + 0.33\sqrt{f_{ck}}\frac{E_p}{E_c}S_t t_p \tag{4.21}$$

where S_t is the perimeter of the loading patch.

The second peak resistance P_2 occurs after concrete punching cone formation, when diagonal cracks render the concrete ineffective. This resistance combines contributions from upper and lower shear connectors (V_{s1}, V_{s2}), the top steel plate (V_{pt}), and middle steel plate (V_{pm}), with composite action effects incorporated through parameter η. The resistance is expressed as:

$$P_2 = V_s + V_{pt} + V_{pm} = V_{s1} + V_{s2} + S_t t_p \frac{f_{up}}{\sqrt{3}} + \eta S_m t_p \frac{f_{yp}}{\sqrt{3}} \tag{4.22}$$

where S_t and S_m represent perimeters of the loading patch and cone-plate intersection respectively. The calculation methods for V_{c1}, V_{c2}, V_{s1} and V_{s2} follow Huang et al. [1]'s established procedures.

4.5.1.5 Idealized Load-Displacement Model

The load-displacement response of double-layer sandwich panels under punching loads can be idealized as a four-stage polyline model (see Fig. 4.23), incorporating elastic stiffness, plastic hardening, and two peak resistances (P_1, P_2). The post-P_1 plateau (δ_1') accounts for concrete punching failure and bond-slip effects, with its duration scaled by the composite action factor η'($\delta_1' = \delta_1/\sqrt{\eta'}$), following JEAG 4618-2005 [20], where $\sqrt{\eta'} = \sqrt{n_p/n_f}$ for connector arrangements. After P_2 (top plate fracture), the curve drops sharply, conservatively omitting the middle plate's membrane action observed in tests. Figure 4.24 demonstrates strong agreement between predicted, experimental, and FE results, validating the model's accuracy in capturing both elastic and plastic behaviors. This efficient four-stage approximation enables reliable assessment of energy absorption capacity for future dynamic analyses of SCS panels.

4.5.1.6 Single-Layer Sandwich Panel

Experimental results demonstrate that shear span-to-depth ratio (λ) governs the initial failure mode in single-layer panels, with high-λ specimens failing through flexural yielding and low-λ panels through concrete punching shear. However, both configurations ultimately reach capacity through identical top plate punching shear fracture at the second peak, which determines their ultimate resistance. This

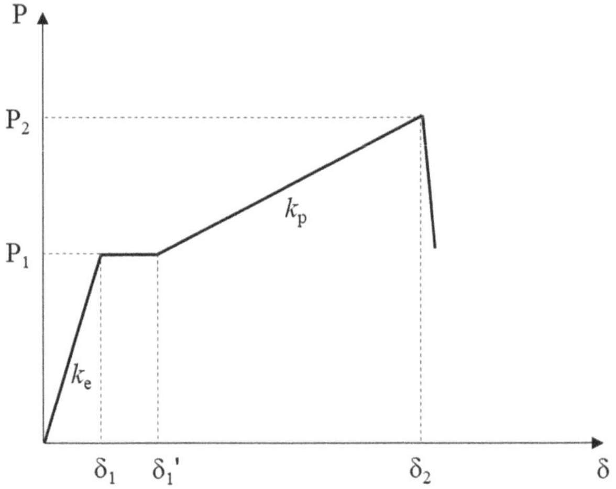

Fig. 4.23 Idealized load-displacement curve. (Reprinted from Zhang et al. [27], copyright 2025, with permission from Elsevier)

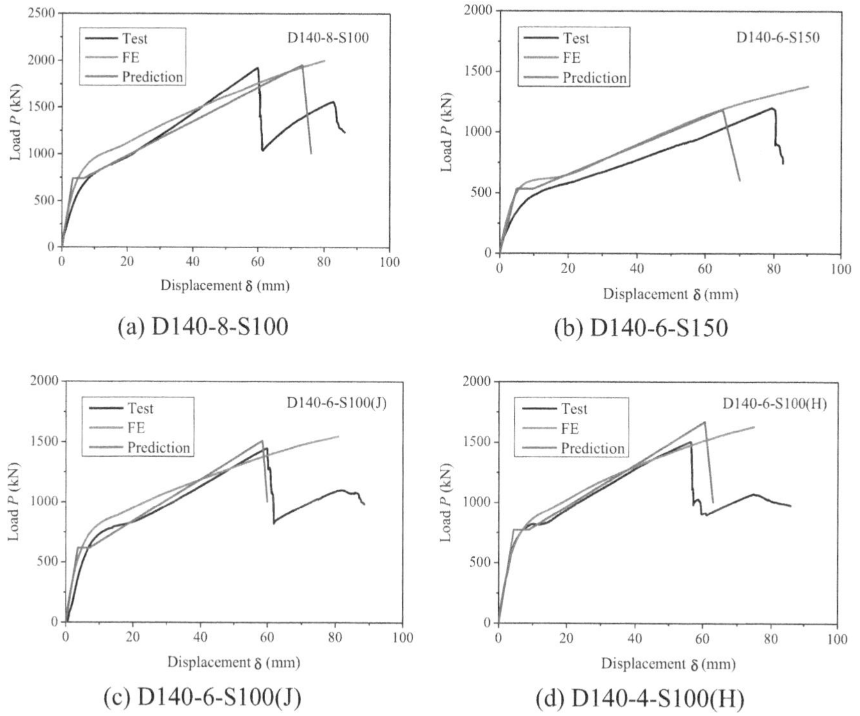

Fig. 4.24 Comparison of load-displacement curves among test, FE, and prediction. (**a**) D140-8-S100. (**b**) D140-6-S150. (**c**) D140-6-S100(J). (**d**) D140-4-S100(H). (Reprinted from Zhang et al. [27], copyright 2025, with permission from Elsevier)

behavioral transition reveals that while λ influences early-stage failure mechanisms, the final collapse mode remains consistently controlled by steel plate shear capacity regardless of panel geometry.

4.5.1.7 Flexural Yielding

The flexural capacity at first peak is determined using yield line theory, a well-established method for analyzing reinforced concrete and composite slabs [6, 21]. This approach applies the virtual work principle, equating external work from applied loads to internal energy dissipation through plastic hinge formation. As illustrated in Fig. 4.25a, the yield line pattern for sandwich panels consists of four diagonal lines radiating from the loading patch corners to the panel edges, plus four lines along the patch perimeter. The external work calculation considers $W_e = P_{1f} \times 1$.

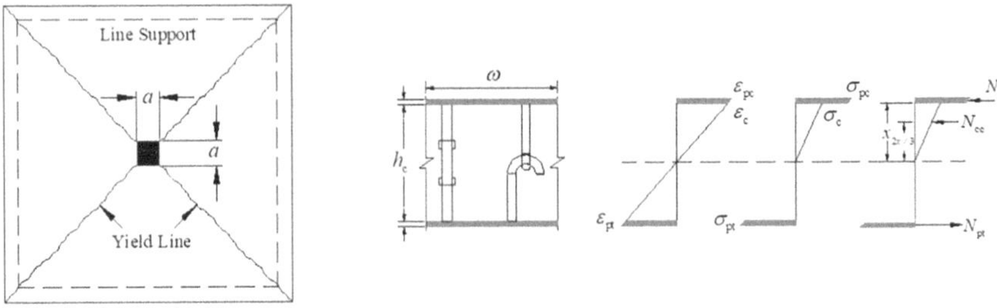

(a) Yield line pattern

(b) Elastic strain, stress, and internal force on the cross section

Fig. 4.25 Yield line model. (**a**) Yield line pattern. (**b**) Elastic strain, stress, and internal force on the cross section. (Reprinted from Huang et al. [1], copyright 2025, with permission from Elsevier)

The internal work done by the resisting moments equals:

$$W_{\mathrm{i}} = 8\left(m_{\mathrm{f}} \cdot \frac{a}{2} \cdot \frac{2}{L_{\mathrm{s}} - \mathrm{a}} + \left(\frac{L}{2} - \frac{a}{2}\right) \cdot m_{\mathrm{f}} \cdot \frac{2}{L_{\mathrm{s}} - \mathrm{a}}\right) = \frac{8 m_{\mathrm{f}} L}{L_{\mathrm{s}} - \mathrm{a}} \tag{4.23}$$

The flexural resistance at the first peak is obtained by equating the external work (We) with the internal work (W_i), expressed as:

$$P_{1\mathrm{f}} = \frac{8 m_{\mathrm{f}} L}{L_{\mathrm{s}} - a} \tag{4.24}$$

The flexural resistance per unit length (m_{f}) is determined through two established approaches: Yan and Liew [21] employ cross-sectional force equilibrium analysis considering elastic strain and stress distributions (see Fig. 4.25b), while Leng and Song [6] propose the simplified relation $m_{\mathrm{f}} \approx f_{\mathrm{py}} t_{\mathrm{b}} \cdot 0.9\ h_{\mathrm{c}}$. Both methods yield comparable m_{f} values, demonstrating consistent theoretical predictions despite differing analytical frameworks.

4.5.1.8 Punching Shear Failure of Concrete

The experimental results demonstrate the formation of a punching cone during loading, with the concrete's punching shear resistance at first peak load predicted using a 60° conical failure surface model, as shown in Fig. 4.26a, b. This angle assumption is justified by three key observations: (1) measured cone angles in tested panels ranged 45°–70°, with 60° representing a practical average; (2) prior studies on SCS beams [3, 22, 23] consistently show steeper crack angles than RC members,

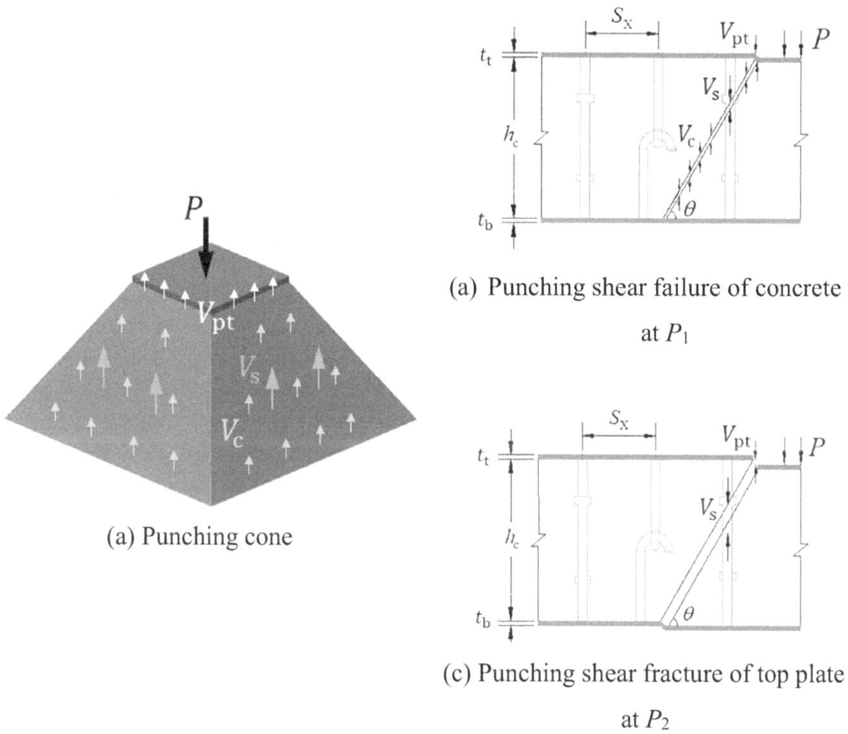

Fig. 4.26 Punching cone model for single-layer sandwich panel. (**a**) Punching cone. (**b**) Punching shear failure of concrete at P_1. (**c**) Punching shear fracture of top plate at P_2. (Reprinted from Huang et al. [1], copyright 2025, with permission from Elsevier)

with Zhang et al. [3] specifically validating 60° for SCS beam predictions; and (3) analogous load transfer mechanisms between SCS beams and panels support using the same angle. While this simplification provides satisfactory predictive accuracy, future work should establish more precise angle determinations through additional experimental and analytical investigations.

The punching shear resistance of single-layer panels comprises concrete (V_c), shear connector (V_s), and top steel plate (V_{pt}) contributions, reflecting the composite material action.

$$P_{1c} = V_c + V_s + V_{pt} \tag{4.25}$$

The study employs LHDCC with 0.7% PE fiber content, where the concrete contribution (V_c) is calculated following Eurocode 2 [24] provisions that account for both lightweight concrete properties and fiber reinforcement effects.

$$V_c = \left[C_c k_c \eta_1 (100\rho_1 f_{ck})^{\frac{1}{3}} + k_f \tau_{f,FRC}\right] S_c h_c \tag{4.26}$$

The concrete punching shear resistance calculation follows modified Eurocode 2 provisions incorporating material-specific coefficients: C_c takes values of $0.18/\gamma_c$ for normal-weight concrete and $0.15/\gamma_c$ for lightweight concrete, with the partial safety factor γ_c set as 1.0. The formulation includes the size effect factor $k_c = 1 + (200/h_c)^{0.5} \leq 2.0$ (h_c in mm), density correction $\eta_1 = 0.4 + 0.6\ u/2200 \leq 1.0$ (ρ in kg/m^3), and reinforcement ratio $\rho_1 = t_b/h_c \leq 0.02$. The model accounts for fiber reinforcement through the product of material coefficient k_f (0.216 for steel/0.290 for synthetic fibers) and fiber contribution $\tau_{f,FRC} = 4.23V_f$ (V_f = fiber volume percentage) as established by Mirsayah [25]. The critical perimeter S_c is evaluated at $h_c/2$ from the loading area, completing the comprehensive set of parameters governing the concrete's shear contribution in the composite system.

The contribution by the shear connectors V_s can be predicted as:

$$V_s = \sum_{i=1}^{n_c} T_i \sin \alpha_i \quad \text{and} \quad T_i = \min \begin{cases} T_{cb} = 0.33\sqrt{f_{ck}} A_N \\ T_{pl} = 0.9\varphi f_{ck} A_e + N_{jh} \\ T_u = \phi f_{su} A_e \\ T_{ps} = \dfrac{\pi d_s t_p f_{pu}}{\sqrt{3}} \end{cases} \tag{4.27}$$

The shear connector contribution (V_s) calculation involves several parameters: the number of connectors (n_c) within the punching cone surface, their inclination angles (α_i), and tensile capacities (T_i) taken as the minimum of concrete breakout (T_{cb}), pullout (T_{pl}), tensile fracture (T_u), and punching shear (T_{ps}) resistances [21]. The concrete breakout strength considers geometric parameters $A_N = \pi h^2_s(1 + d_s/h_s)$, where d_s and h_s represent connector diameter and height, along with the effective anchoring area A_e. The calculation includes hook straightening resistance (N_{jh}), a crack influence factor ($\varphi = 1.4$ for uncracked or 1.0 for cracked regions), strength reduction factor (ϕ), and ultimate strengths of both connector (f_{su}) and steel plate (f_{pu}).

The punching shear resistance contribution of the top steel plate (V_{pt}) at first peak load is determined through membrane action analysis using an equivalent section approach.

$$V_{\mathrm{pt}} = 0.33\sqrt{f_{\mathrm{ck}}}\frac{E_p}{E_\mathrm{c}}S_\mathrm{t}t_\mathrm{t} \tag{4.28}$$

where S_t represents the loading patch perimeter and $E_\mathrm{p}/E_\mathrm{c}$ denote the elastic moduli of steel plate and concrete, respectively.

4.5.1.9 Punching Shear Fracture of Top Steel Plate

At the second peak load (P_2), both single-layer sandwich panels experience punching shear fracture of the top plate, coinciding with full development of the concrete punching cone (see Fig. 4.26c). With the critical diagonal crack fully widened, the concrete no longer contributes to punching resistance, transferring the load entirely to the top plate (using its ultimate strength f_{pu}) and shear connectors. This mechanism determines the panel's ultimate resistance, calculated as:

$$P_2 = V_s + V_{\mathrm{pt}} = V_s + S_t t_t \frac{f_{\mathrm{pu}}}{\sqrt{3}} \tag{4.29}$$

where the calculation of V_s follows Eq. (4.27).

4.5.2 Double-Layer Sandwich Panel

Double-layer sandwich panels demonstrate two distinct failure sequences: full composite panels fail through concrete punching shear followed by steel plate fracture, while partial composite panels exhibit similar failure progression accompanied by interfacial bond-slip. Both mechanisms operate through the 60° punching cone model (see Fig. 4.27), with identical cone angles assumed for upper and lower layers based on the experimental justification presented previously. Figure 4.27b–d sequentially illustrates the punching cone development and corresponding load transfer mechanisms during progressive failure.

4.5.2.1 Punching Shear Failure of Concrete

The first peak load in double-layer sandwich panels results from concrete punching shear failure, where the middle steel plate's negligible stress contribution during elastic deformation justifies its exclusion from calculations. The total punching shear resistance combines three components (see Fig. 4.27b): concrete resistance ($V_\mathrm{c} = V_{\mathrm{c}1} + V_{\mathrm{c}2}$, calculated via Eq. (4.26) for each layer), shear connector capacity

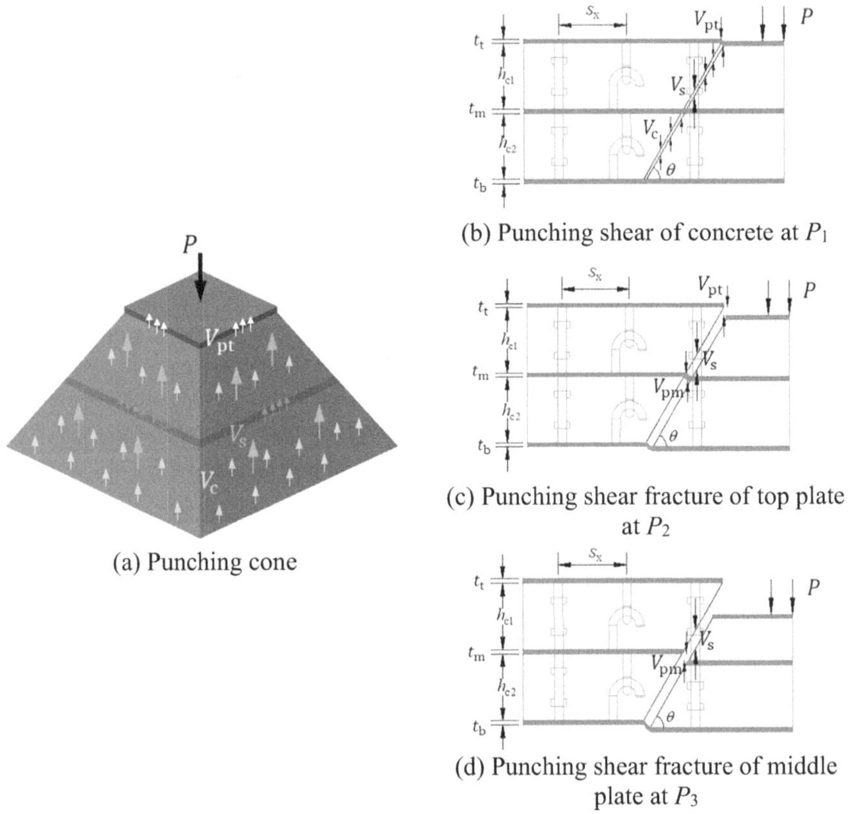

Fig. 4.27 Punching cone model for double-layer sandwich panel. (**a**) Punching cone. (**b**) Punching shear of concrete at P_1. (**c**) Punching shear fracture of top plate at P_2. (**d**) Punching shear fracture of middle plate at P_3. (Reprinted from Huang et al. [1], copyright 2025, with permission from Elsevier)

($V_s = V_{s1} + V_{s2}$, determined by Eq. (4.27)), and top plate resistance (V_{pt} from Eq. (4.28)), expressed as:

$$P_1 = V_c + V_s + V_{pt} = V_{c1} + V_{c2} + V_{s1} + V_{s2} + 0.33\sqrt{f_{ck}}\frac{E_p}{E_c}S_t t_t \quad (4.30)$$

4.5.2.2 Punching Shear Fracture of Top Steel Plate

The ultimate resistance (P_2) of double-layer sandwich panels occurs when the top plate undergoes punching shear fracture, as shown in Fig. 4.27c. At this stage, both concrete layers develop widened critical diagonal cracks, rendering them ineffective for load transfer, while shear connectors maintain functionality across these cracks. The top plate experiences rapid shear failure around loading perimeters, simultaneously with yielding of the middle plate at cone surface intersections—both exhibiting punching shear-dominated behavior.

The ultimate resistance of double-layer panels comprises three components: shear connector capacity (V_s,), top plate resistance (V_{pt}), and middle plate contribution (V_{pm}). The middle plate's shear resistance follows the same formulation as Eq. (27), but substitutes the plate's yield strength (f_{py}) for the ultimate strength (f_{pu}) used in top plate calculations.

$$V_{pm} = S_m t_m \frac{f_{py}}{\sqrt{3}} \tag{4.31}$$

The middle plate resistance (V_{pm}) uses the intersection perimeter (S_m) between the punching cone and plate, calculated similarly to V_{pt} but with yield strength (f_{py}).

For partial composite panels, bond-slip effects are accounted for by applying a reduction factor η to the middle plate's contribution in the ultimate resistance calculation as given below:

$$P_2 = V_s + V_{pt} + \eta V_{pm} = V_{s1} + V_{s2} + S_t t_t \frac{f_{pu}}{\sqrt{3}} + \eta S_m t_m \frac{f_{py}}{\sqrt{3}} \tag{4.32}$$

where V_{s1} and V_{s2} are calculated according to Eq. (4.27).

Figure 4.28 presents the triangular shear span concept and revised η definition for SCS panels, where the centrally-loaded, simply-supported configuration creates four triangular shear regions. The composite factor η represents the ratio between total connector shear capacity and tensile strength of the tapered plate section, using averaged cross-sectional properties to account for variable geometry. This formulation underpins the modified composite action equation as given in the equation below:

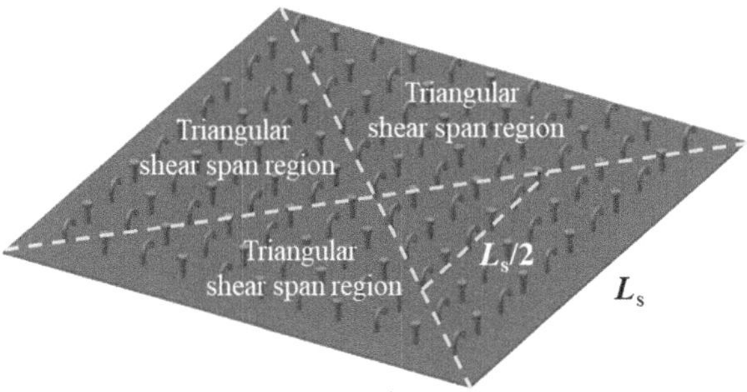

Fig. 4.28 Definition of degree of composite action. (Reprinted from Huang et al. [1], copyright 2025, with permission from Elsevier)

$$\eta = \frac{n_s P}{\frac{f_{py} t_p L_s}{2}} \leq 1.0 \tag{4.33}$$

The calculation incorporates three key parameters: the number of shear connectors (n_s) within each triangular shear span region, the plate thickness (t_p), and the shear capacity of individual connectors—either headed studs (P_H) determined by Eq. (4.34) [26] or J-hook connectors (P_J) calculated via Eq. (4.35) [19].

$$P_H = \frac{0.29\alpha d_H^2 \sqrt{f_{ck} E_c}}{\gamma_v} \leq \frac{0.8 f_{u\_H} A_H}{\gamma_v} \tag{4.34}$$

$$P_J = \frac{0.855 f_{ck}^{0.265} E_c^{0.469} A_J \left(\frac{h_J}{d_J}\right)^{0.154}}{\gamma_v} \leq \frac{0.8 f_{u\_J} A_J}{\gamma_v} \tag{4.35}$$

The calculation incorporates geometric and material parameters for both connector types: headed studs (with height h_H, diameter d_H, cross-section A_H, and ultimate strength f_{u_H}) and J-hooks (with corresponding dimensions h_J, d_J, A, and f_{u_J}). The reduction factor α equals $0.2(h_H/d_H + 1)$ for aspect ratios $3 \leq h_H/d_H \leq 4$ or $\alpha = 1.0$ for $h_H/d_H > 4$, while the partial safety factor γ_v is taken as 1.0.

4.5.2.3 Punching Shear Fracture of Top Steel Plate

Following the second peak load, the failed top plate transfers the concentrated load directly to the lower layer through compacted concrete. The third peak resistance develops from the combined action of lower-layer shear connectors (V_{s2}) and middle plate capacity (V_{pm}), assuming the original punching cone geometry remains unchanged in the lower layer. Thus,

$$P_3 = V_{s2} + V_{pm} = V_{s2} + S_m t_m \frac{f_{pu}}{\sqrt{3}} \tag{4.36}$$

The third peak resistance calculation uses the lower-layer shear connector capacity (V_{s2}) from Eq. (4.27), and the middle plate's effective perimeter (S_m) at the cone intersection.

4.5.3 Validations and Design Flow Chart

The analytical models demonstrate strong agreement with experimental results, as shown in Table 4.10 and Fig. 4.29, with mean test-to-prediction ratios of 1.12 ± 0.08

Table 4.10 Validation of the proposed model

Specimen	P_1				P_2				P_3			
	Failure mode	P_{1_test} (kN)	P_{1_prop} (kN)	$\frac{P_{1\_test}}{P_{1\_prop}}$	Failure mode	P_{2_test} (kN)	P_{2_prop} (kN)	$\frac{P_{2\_test}}{P_{2\_prop}}$	Failure mode	P_{3_test} (kN)	P_{3_prop} (kN)	$\frac{P_{3\_test}}{P_{3\_prop}}$
S90-6	FY	354	345	1.03	PST	927	880	1.05	/	/	/	/
S140-6	PSC	713	675	1.06	PST	956	904	1.06	/	/	/	/
D140-4	PSC	742	615	1.21	PST	1305	1138	1.15	PSM	868	678	1.28
D140-6	PSC	800	697	1.15	PST	1350	1590	0.85	PSM	1068	906	1.18
D140-8	PSC	806	736	1.09	PST	1923	1954	0.98	PSM	1552	1056	1.47
D140-6(J)	PSC	773	616	1.25	PST	1450	1510	0.96	PSM	1100	826	1.33
D140-6(H)	PSC	816	771	1.06	PST	1502	1669	0.90	PSM	1068	985	1.08
D140-6(S150)	PSC	573	534	1.07	PST + BS	1205	1185	1.02	/	/	/	/
D140-6(S200)	PSC	546	537	1.02	PST + BS	1235	994	1.24	/	/	/	/
D140-6(A150)	PSC	1194	984	1.21	PST	1827	2230	0.82	PSM	1332	1360	0.98
Mean				1.12				1.00				1.22
Std. Dev				0.08				0.13				0.18

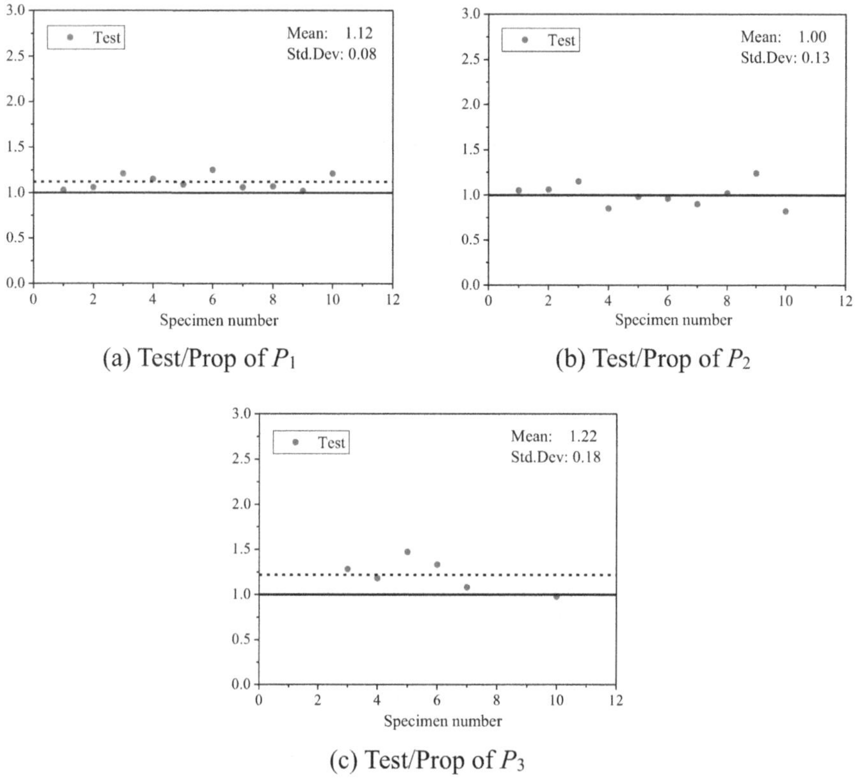

Fig. 4.29 Ratio of test to prediction results. (**a**) Test/Prop of P_1. (**b**) Test/Prop of P_2. (**c**) Test/Prop of P_3. (Reprinted from Huang et al. [1], copyright 2025, with permission from Elsevier)

for P_1, 1.00 ± 0.13 for P_2, and 1.22 ± 0.18 for P_3. Conservative P_1 predictions for double-layer panels result from neglecting middle plate contributions, while P_3 conservatism stems from assuming single-layer behavior after top plate failure. The models accurately capture bond-slip effects (evident in P_2 comparisons) and provide clear calculation procedures (see Fig. 4.30), with P_2 representing design-critical ultimate resistance and P_3 indicating post-failure robustness for double-layer configurations.

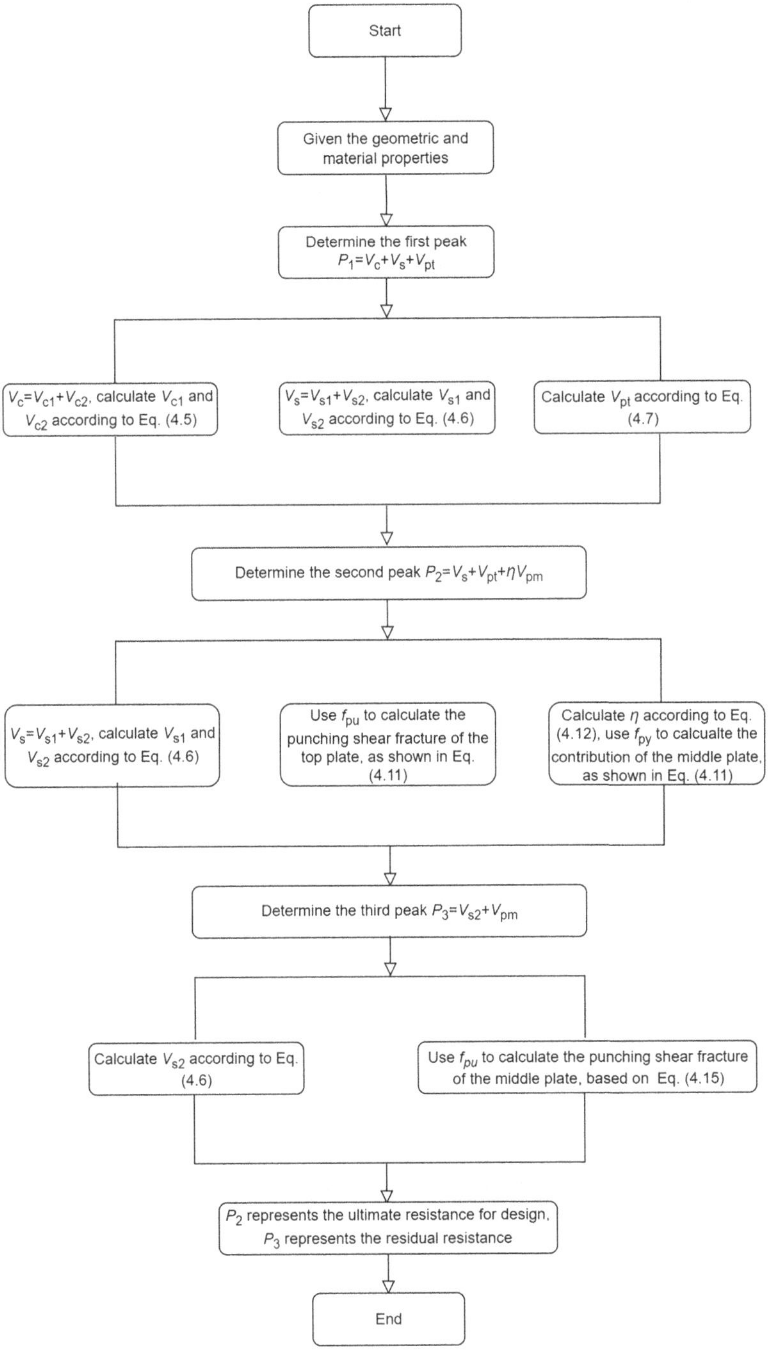

Fig. 4.30 Flow chart to determine load resistance of double-layer sandwich panels. (Reprinted from Huang et al. [1], copyright 2025, with permission from Elsevier)

References

1. Z. Huang, X. Zhao, W. Zhang, Z. Fu, Y. Zhou, L. Sui, Load transfer mechanism of novel double-layer steel–LHDCC–steel sandwich panels under punching loads, Eng. Struct. 226 (2021) 111427.
2. Y.-B. Leng, X.-B. Song, H. Wang, Failure mechanism and shear strength of steel–concrete–steel sandwich deep beams, J. Constr. Steel Res. 106 (2015) 89–98.
3. W. Zhang, Z. Huang, Z. Fu, X. Qian, Y. Zhou, L. Sui, Shear resistance behavior of partially composite steel–concrete–steel sandwich beams considering bond-slip effect, Eng. Struct. 210 (2020) 110394.
4. K.M.A. Sohel, J.Y.R. Liew, Steel–concrete–steel sandwich slabs with lightweight core—static performance, Eng. Struct. 33(3) (2011) 981–992.
5. J.-B. Yan, J.-Y. Wang, J.Y.R. Liew, X. Qian, Z.-X. Liu, Punching shear behavior of steel–concrete–steel sandwich composite plate under patch loads, J. Constr. Steel Res. 121 (2016) 50–64.
6. Y.-B. Leng, X.-B. Song, Flexural and shear performance of steel–concrete–steel sandwich slabs under concentrate loads, J. Constr. Steel Res. 134 (2017) 38–52.
7. Z. Huang, W. Du, S. Du, Z. Gu, W. Long, J. Ye, Effect of rubber particles and fibers on the dynamic compressive behavior of novel ultra-lightweight cement composites: Numerical simulations and metamodeling, Compos. Struct. 258 (2021) 113210.
8. X. Liu, T. Wu, Y. Liu, Stress-strain relationship for plain and fibre-reinforced lightweight aggregate concrete, Constr. Build. Mater. 225 (2019) 256–272.
9. W. Zhang, Y.S. Choo, L. Feng, Z. Huang, W. Shen, Representation of nonlinear behavior of fully grouted K joints in pushover analysis, J. Constr. Steel Res. 169 (2020) 106024.
10. Z. Huang, J.Y.R. Liew, Nonlinear finite element modelling and parametric study of curved steel–concrete–steel double skin composite panels infilled with ultra-lightweight cement composite, Constr. Build. Mater. 95 (2015) 922–938.
11. W. Zhang, Y.S. Choo, P. Yang, W. Shen, Z. Huang, Nonlinear behaviour of fully grouted CHS X joints and associated representation for overall frame analysis, Thin-Walled Struct. 152 (2020) 106761.
12. B. G. Rabbat, H. G. Russell, Friction coefficient of steel on concrete or grout, J. Struct. Eng. 111 (3) (1985) 505–515.
13. F.-X. Ding, T. Zhang, X.-M. Liu, Z.-H. Lu, Q. Guo, G.-S. Jiang, Behavior of steel-reinforced concrete-filled square steel tubular stub columns under axial loading, Thin-Walled Struct. 119 (2017) 737–748.
14. Z. Dai, S.D. Pang, J.Y.R. Liew, Axial load resistance of grouted sleeve connection for modular construction, Thin-Walled Struct. 154 (2020) 106883.
15. T. Chen, C. Chen, C. Zhang, X. Wang, K. Chen, G. Yuan, Numerical modeling and parametric analysis of grouted connections under axial loading, Thin-Walled Struct. 154 (2020) 106880.
16. S.P. Timoshenko, S. Woinowsky-Krieger, Theory of plates and shells, McGraw-Hill College, 1959.
17. K.M.A. Sohel, J.Y.R. Liew, Behavior of steel–concrete–steel sandwich slabs subject to impact load, J. Constr. Steel Res. 100(12) (2014) 163–175.
18. Q. Guo, W. Zhao, Displacement response analysis of steel–concrete composite panels subjected to impact loadings, Int. J. Impact Eng. 131 (2019) 272–281.
19. J.B. Yan, J.Y.R. Liew, K.M.A. Sohel, M.H. Zhang, Push-out tests on J-hook connectors in steel–concrete–steel sandwich structure, Mater. Struct. 47(10) (2014) 1693–1714.
20. JEAG 4618-2005, Technical guidelines for aseismic design of steel plate reinforced concrete structures—buildings and structures, Japan Electric Association Nuclear Standards Committee, Tokyo, 2005.
21. J.-B. Yan, J.Y.R. Liew, Design and behavior of steel–concrete–steel sandwich plates subject to concentrated loads, Compos. Struct. 150 (2016) 139–152.

22. F. Qin, Q. Kong, M. Li, Y.L. Mo, G. Song, F. Fan, Bond slip detection of steel plate and concrete beams using smart aggregates, Smart Mater. Struct. 24 (2015) 115039.
23. Y. Li, J. Yan, Z. Wang, F. Fan, Y. Yang, Z. Yu, Failure mechanism and failure patterns of SCS composite beams with steel-fiber-reinforced UHPC, Eng. Struct. 211 (2020) 110471.
24. Eurocode 2, Design of concrete structures: Part 1-1, general rules and rules for buildings, British Standards Institution, 2005.
25. A.A. Mirsayah, N. Banthia, Shear strength of steel fiber-reinforced concrete, Mater. J. 99 (2002) 473–479.
26. Eurocode 4, Design of composite steel and concrete structures part 1-1: General rules and rules for buildings, Thomas Telford, London, 2004.
27. W. Zhang, Z. Huang, R. Li, X. Zhao, J. Ye, Impact response of double-layer steel–RULCC–steel sandwich panels: Experimental, numerical and analytical approaches, J. Struct. Eng. 148 (10) (2022) 04022165.

Chapter 5
Impact Response and Post-impact Residual Flexural Resistance of Double-Layer Steel-RULCC-Steel Sandwich Panels: Experiment, Numerical Study and Analytical Prediction

5.1 Introduction

Compared to traditional reinforced concrete (RC) structures, which are prone to cracking and crushing under impact, SCS sandwich panels effectively utilize the tensile strength of steel and the compressive strength of concrete, preventing penetration and dissipating energy through the concrete core [1–3]. The composite action between steel plates and concrete is typically ensured by shear connectors, such as J-hooks, overlapped headed studs, or bolts, which enhance structural integrity and load transfer efficiency [4–6]. Recent innovations include the use of ultra-lightweight high-ductility cement composites (ULHDCC) and rubberized concrete (RULHDCC) to further improve impact resistance and energy absorption while addressing environmental concerns by incorporating recycled materials such as rubber powder [7–9].

The dynamic response of SCS sandwich structures under impact loading has been extensively studied. For instance, Liew et al. [10] and Remennikov et al. [3] investigated the behavior of SCS beams and panels under low-velocity impact, highlighting the role of tensile membrane action in steel plates and the energy dissipation capacity of lightweight concrete cores. Jung et al. [11] explored high-velocity impact resistance, demonstrating that SCS walls could mitigate projectile penetration and reduce debris. To optimize performance, Huang et al. [12] and Zhang et al. [13, 14] proposed double-layer SCS panels with hybrid shear connectors and RULHDCC cores, which exhibited enhanced ductility, load-bearing capacity, and impact resistance compared to single-layer designs. Numerical simulations using LS-DYNA and analytical models, such as single-degree-of-freedom (SDOF) and two-degree-of-freedom (TDOF) systems, have been developed to predict impact responses, including peak forces, deformations, and failure modes [15–17].

Despite these advancements, the residual performance of SCS structures after impact remains underexplored. While studies on RC components, such as beams and

Z. Huang, Y. Zhou, *Innovative Lightweight High-performance Steel-concrete Sandwich Composite for Marine Structures*,
https://doi.org/10.1007/978-981-95-4681-7_5

columns, have shown that residual strength and stiffness are influenced by impact energy and reinforcement ratios [18–20], similar investigations for SCS panels are limited. Recent work by Huang et al. [21] examined the residual flexural performance of double-layer Steel-RULHDCC panels post-impact, revealing that parameters such as connector spacing, concrete layer count, and rubber content significantly affect residual resistance. Finite element (FE) simulations validated these findings, emphasizing the importance of steel plate thickness and strength in maintaining post-impact integrity [22]. Additionally, repeated impact tests demonstrated that double-layer configurations and hybrid connectors reduce damage accumulation, while increased rubber content enhances energy dissipation [23–25].

Overall, SCS sandwich structures offer exceptional mechanical and impact-resistant properties. This chapter integrates experimental, numerical, and analytical approaches to comprehensively evaluate the impact, repeated impact and post impact static performance of double-layer Steel-RULHDCC panels, providing insights for their application in resilient and sustainable infrastructure.

5.2 Impact Test Program of Double-Layer Sandwich Panels

5.2.1 Specimen Design

For single and repeated impact test as well as residual flexural performance test, this study conducts a full-scale experimental program to examine the dynamic behavior of double-layer SCS panels under impact loading. Figure 5.1a, b show the configuration of single−/double-layer SCS panels, respectively. A total of 9 full-scale SCS sandwich panels were designed and fabricated, with key experimental variables included connector type (J-hook or hybrid), connector spacing (100 mm, 150 mm, or 200 mm), core layer configuration (single or double), rubber powder content (0%, 5%, or 10% by volume replacement), and impact loading conditions (1, 2, or 3 repeated impacts). Table 5.1 summarizes their geometric parameters, which were selected based on practical structural applications. The specimen identification system encoded these parameters systematically. For instance, SR10LHC(D)-150 (H)-6(2) denotes a double-layer panel with 10% rubber-modified lightweight high-ductility cement composite (LHDCC) core, hybrid connectors at 150 mm spacing, 6 mm steel plates, subjected to two impacts. Most specimens were tested under three impacts, with selected configurations subjected to fewer impacts to evaluate damage accumulation. Complete geometric specifications of impact test and post-impact static test specimens are provided in Tables 5.1 and 5.2. The shear connector spacing directly influenced the degree of composite action (η), calculated as the ratio of the total shear strength of connectors within a shear span to the steel plate's average tensile strength, as expressed in Eq. (5.1) below:

$$\eta = \frac{n_s V_s}{f_{yp} t_p L_s / 2} \tag{5.1}$$

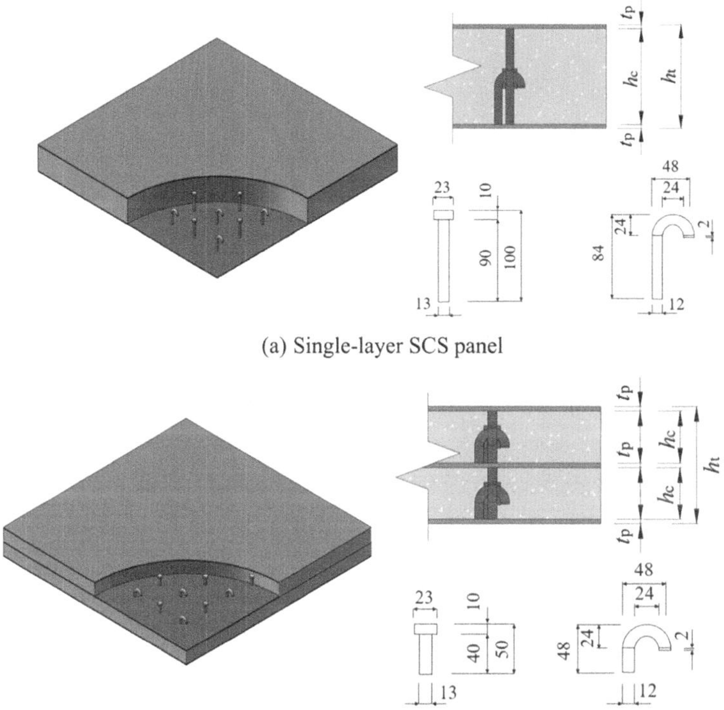

Fig. 5.1 Configuration of single-layer and double-layer SCS panels. (**a**) Single-layer SCS panel. (**b**) Double-layer SCS panel. (Reprinted from Zhang et al. (2022), copyright 2025, with permission from American Society of Civil Engineers)

where, η_s is the number of shear connectors within the shear span; V_s is the shear strength of a single shear connector; f_{yp} is the yield strength steel plates.

For impact-loading test, the study examines three ULCC mix designs: a baseline ULCC without rubber, and two modified versions incorporating 5% and 10% rubber powder by volume (designated R5ULCC and R10ULCC, respectively). The rubber powder replaces an equivalent volume of fine aggregates (fly ash cenospheres) in the mixture. Table 5.3 details the mix proportions for all three composites. To enhance ductility, each mix includes 5.8 kg/m^3 (0.6% by volume) of polyethylene (PE) fibers. A high-range water-reducing admixture was incorporated to maintain workability, achieving a flowability range of 220–230 mm as per BS EN 1015-3. Material properties of the ULCC, R5ULCC, and R10ULCC are summarized in Table 5.4. Compressive strength was evaluated using cube tests (GB/T 50081) and cylinder tests (ASTM C39/C39M-2021), while tensile strength followed JSCE (2008) standards. Results indicate that both compressive strength and elastic modulus decline as rubber content increases, confirming that rubber powder enhances flexibility at the

Table 5.1 Geometric details of impact test specimens. (Reprinted from Zhang et al. (2022), copyright 2025, with permission from American Society of Civil Engineers)

Specimen	Core material	Core layer	h_c (mm)	t_s (mm)	s (mm)	J-hook	Headed stud	ρ	η	Impact times
SULCS-150-9(SH)	LHDCC	1	141	9.2	150	ϕ12@300	ϕ13@300	11.5	0.43	3
SULCS-100-6(DH)	LHDCC	2	70*2	5.8	100	ϕ12@200	ϕ13@200	11.1	1.00	3
SULCS-150-6(DH)	LHDCC	2	70*2	5.8	150	ϕ12@300	ϕ13@300	11.1	0.64	3
SULCS-200-6(DH)	LHDCC	2	70*2	5.8	200	ϕ12@400	ϕ13@400	11.1	0.37	3
SULCS-150-6(DJ)	LHDCC	2	70*2	5.8	150	ϕ12@150	–	11.1	0.64	3
SR5ULCS-150-6(DH)	RLHDC(5%)	2	70*2	5.8	150	ϕ12@300	ϕ13@300	11.1	0.64	3
SR10ULCS-150-6(DH1)	RLHDC(10%)	2	70*2	5.8	150	ϕ12@300	ϕ13@300	11.1	0.64	1

Notes: h_c is the total thickness of the concrete core; t_s represents the thickness of the steel plate; s is the connector spacing; ρ indicates the steel ratio, $\rho = nt_s/(nt_s + h_c)$, n is the number of steel layers; η represents the composite degree of the composite panel [19] and is related to the number of connectors, $\eta = n_s P_s/(f_y t_s L/2)$, and n_s, P_s, L are the connector number in the shear span, shear resistance per connector and clear span of the panel

Table 5.2 Geometric details of post-impact static test specimens

Specimen	Core material	Core layer	h_c (mm)	t_s (mm)	S (mm)	J-hook	Headed stud	ρ	η	Impact times
SLHC(S)-150(H)-9	LHDCC	1	141	9.2	150	ϕ12@300	ϕ13@300	11.5	0.43	3
SLHC(D)-100(H)-6	LHDCC	2	70*2	5.8	100	ϕ12@200	ϕ13@200	11.1	1.00	3
SLHC(D)-150(H)-6	LHDCC	2	70*2	5.8	150	ϕ12@300	ϕ13@300	11.1	0.64	3
SLHC(D)-200(H)-6	LHDCC	2	70*2	5.8	200	ϕ12@400	ϕ13@400	11.1	0.37	3
SLHC(D)-150(J)-6	LHDCC	2	70*2	5.8	150	ϕ12@150	–	11.1	0.64	3
SR5LHC(D)-150(H)-6	RLHDC(5%)	2	70*2	5.8	150	ϕ12@300	ϕ13@300	11.1	0.64	3
SR10LHC(D)-150(H)-6(1)	RLHDC(10%)	2	70*2	5.8	150	ϕ12@300	ϕ13@300	11.1	0.64	1
SR10LHC(D)-150(H)-6(2)	RLHDC(10%)	2	70*2	5.8	150	ϕ12@300	ϕ13@300	11.1	0.64	2
SR10LHC(D)-150(H)-6(3)	RLHDC(10%)	2	70*2	5.8	150	ϕ12@300	ϕ13@300	11.1	0.64	3

Table 5.3 Mix proportion of ULCC and RULCC (kg/m^3)

Materials	W	OPC	SF	GGBFS	R	F	HWRA	SRA
ULCC	259.0	702.0	78.0	339.9	/	5.8	12.0	9.0
R5ULCC	259.0	702.0	78.0	322.9	18.8	5.8	12.4	9.0
R10ULCC	259.0	702.0	78.0	305.9	37.7	5.8	12.8	9.0

Notes: *W* water, *OPC* ordinary Portland cement, *SF* silica fume, *GGBFS* mineral powder, *R* rubber powder, *F* steel fiber, *HWRA* high Water reducing agent, *SRA* shrinkage reducing agent

Table 5.4 Material properties of ULCC and RULCC

Mix	Density (kg/m^3)	Cube compressive strength (MPa)	Cylinder compressive strength (MPa)	Elastic modulus (GPa)	Poisson's ratio	Tensile strength (MPa)	Maximum tensile strain (με)
ULCC	1388	49.2	35.6	11.8	0.28	2.9	0.0029
R5ULCC	1296	44.3	33.2	9.9	0.30	3.1	0.0031
R10ULCC	1203	40.2	31.0	8.5	0.33	3.2	0.0039

Table 5.5 Material properties of steel

Steel component	Yield strength (MPa)	Yield strain (με)	Ultimate strength (MPa)	Ultimate strain (με)	Elastic modulus (GPa)	Poisson's ratio
Steel plate ($t_s = 5.8$ mm)	288	2034	440	0.27	202	0.30
Steel plate ($t_s = 9.2$ mm)	292	1994	460	0.31	208	0.30
J-hook	463	2034	690	0.15	204	0.30
Headed stud	772	4972	916	0.11	205	0.30

expense of mechanical strength. Additionally, Table 5.5 presents the material properties of steel components, tested in accordance with ASTM E8/E8M.

5.2.2 Impact Test Setup and Instrumentation

The impact tests were performed using an STLH-50000 drop hammer impact test machine, which has a maximum counterweight capacity of 1012 kg. The experimental setup, illustrated in Fig. 5.2a, b, consists of a drop hammer released from a height of 5 m, delivering a maximum impact energy of 50,000 J. The hammer features a hemispherical impact head with a diameter of 100 mm to ensure consistent contact conditions. Figure 5.2c presents the clamping system, including the frame-rigid supporting floor and the specimen-support frame, where the SCS panel is simply supported on four circular steel bars welded to the foundation. All SCS panels were subjected to the same impact energy (50,000 J) to ensure comparative

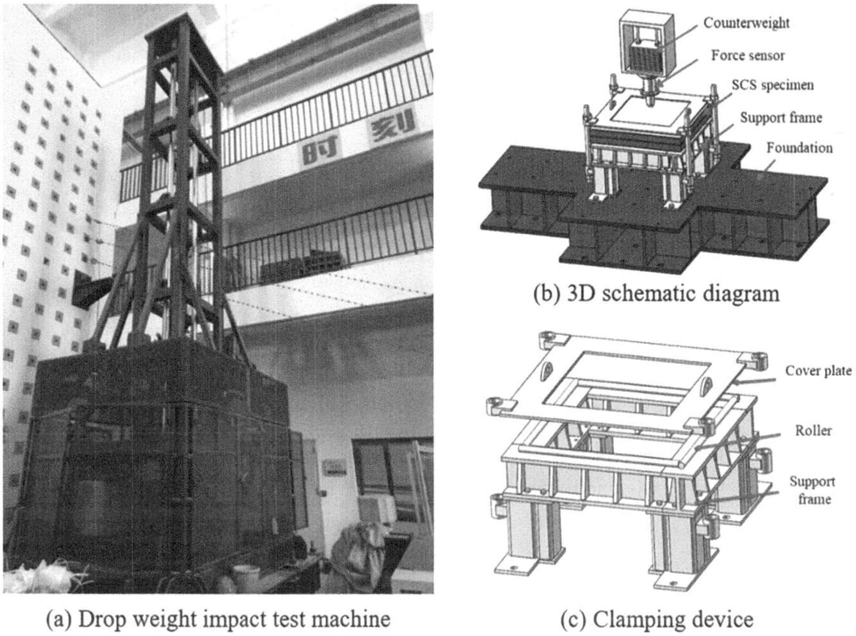

Fig. 5.2 Test set-up of impact test. (**a**) Drop weight impact test machine. (**b**) 3D schematic diagram. (**c**) Clamping device. (Reprinted from Zhang et al. (2022), copyright 2025, with permission from American Society of Civil Engineers)

analysis. During testing, key parameters such as impact force, panel deformation, steel plate strains, and high-speed impact progression were measured and recorded. The instrumentation and data acquisition setup are detailed in Fig. 5.3.

The repeated-impact test was conducted using the same drop-weight impact testing machine. Each specimen was subjected to three impact tests. The first two impacts were performed at a height of 5 m. After the second impact, the hammerhead had significantly penetrated the composite slab. To prevent damage to the force sensor above the hammerhead, the third impact height for the same specimen was reduced to 3 m, resulting in impact energies of 50 kJ, 50 kJ, and 30 kJ for the three impacts, respectively. The boundary conditions could be approximately considered as simply supported on all four edges.

The parameters measured during the test included impact force, displacement, impact velocity, steel plate strain, and high-speed video of the entire impact process. The impact force was recorded using a force sensor mounted on the hammerhead, with a range of 2000 kN. Displacement was measured using a draw-wire displacement transducer with a range of 150 mm. The impact velocity was captured by a built-in laser velocimeter in the testing machine. Strain was acquired via strain gauges. The impact process was recorded using a high-speed camera with an

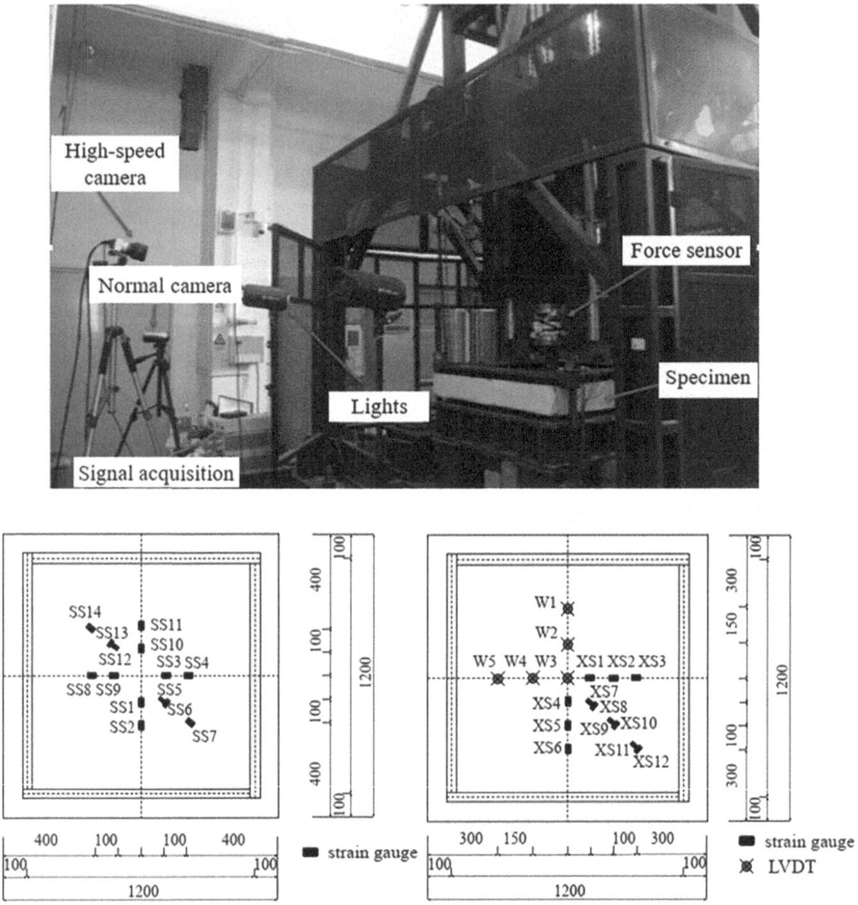

Fig. 5.3 Measurement scheme of impact test. (Reprinted from Zhang et al. (2022), copyright 2025, with permission from American Society of Civil Engineers)

exposure rate of 1000 frames per second. The impact force, displacement, and strain data were collected using a high-speed data acquisition system with a sampling frequency of 500 kHz.

5.2.3 Post-Impact Static Test Setup

This section aims to examine the residual static flexural performance of impact-damaged composite panels through systematic experimental investigation. To facilitate the bending tests, the original 1200 × 1200 mm impacted panels were precisely bisected using high-pressure water jet cutting technology, resulting in specimens with dimensions of 1200 × 600 mm. This approach was adopted for several critical reasons: the central impact zone exhibited significant permanent deformation from the hammer impact, making direct testing of intact panels impractical. The sectioned specimens provided clear visualization of internal damage patterns and failure mechanisms, which proved invaluable for subsequent FE model calibration. The water jet cutting method was specifically chosen to minimize additional specimen disturbance, with any cutting-induced damage being intentionally localized to the central region to preserve the integrity of the panel edges for reliable flexural testing. This methodology enabled comprehensive evaluation of post-impact structural performance while maintaining the representativeness of the damage conditions.

The flexural testing employed the Bangwei loading system (refer to Fig. 5.4) under displacement control at a constant rate of 0.5 mm/min, continuing until specimen failure or attainment of the system's maximum displacement capacity. Comprehensive instrumentation captured critical response parameters: load capacity was recorded via a force transducer, while mid-span deflection at the specimen's bottom surface was monitored using Linear Variable Displacement Transducers (LVDTs). Strain development in both steel plates and concrete was measured through strategically positioned strain gauges, with concrete gauges installed on both the section-cut profile (front view) and end surface (back view) as illustrated in Fig. 5.5. Interfacial slip behavior was quantified using a dedicated pair of LVDTs affixed separately to the steel plate and concrete at the specimen's end, where the relative displacement difference between these measurements provided direct quantification of slip magnitude between the composite constituents.

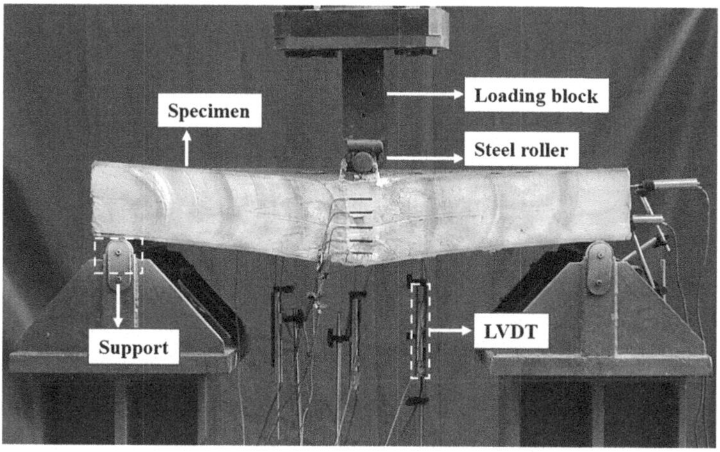

Fig. 5.4 Test setup for flexural test [21]

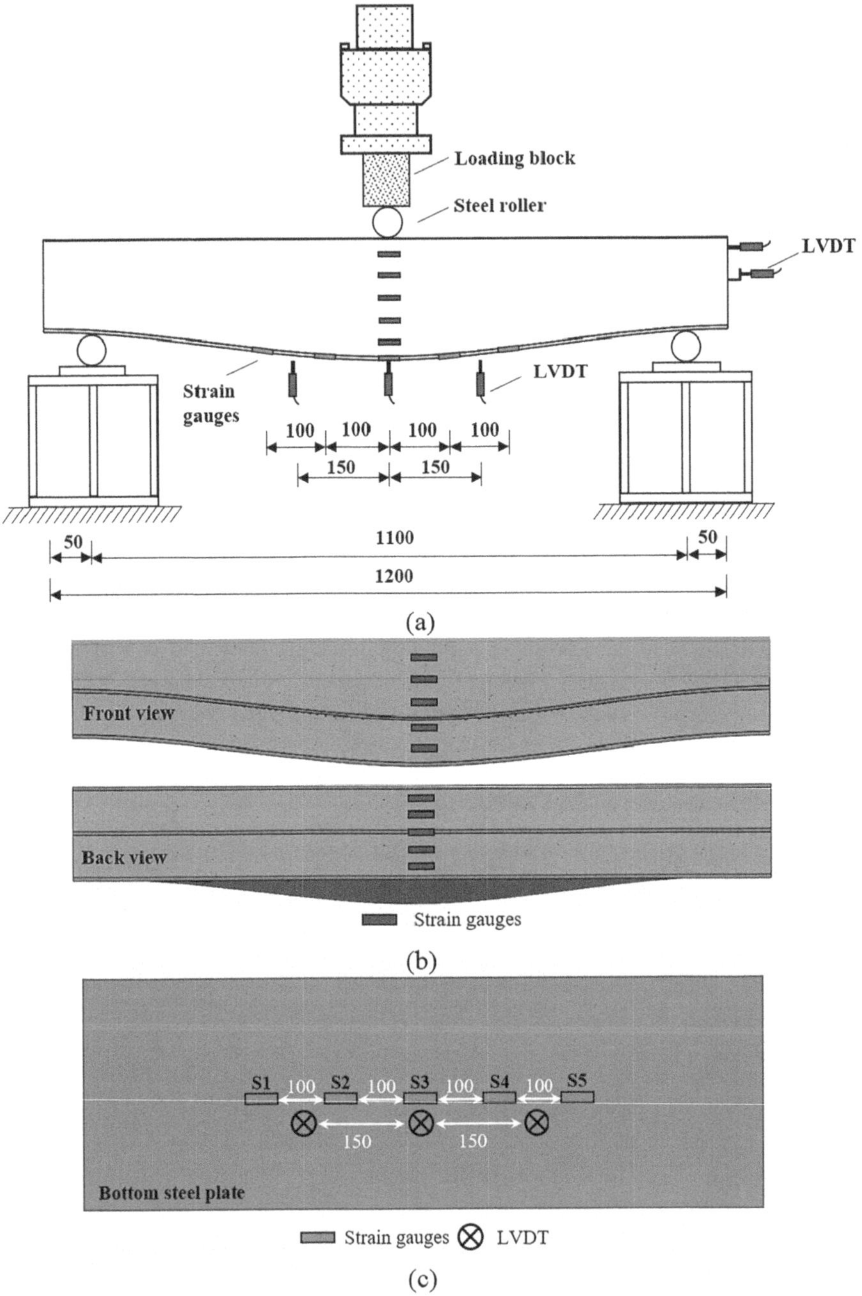

Fig. 5.5 Layout of measuring points (**a**) measurement points arrangement; (**b**) measurement points on concrete; (**c**) measurement points on steel [21]

5.3 Failure Mechanism and Dynamic Responses

5.3.1 Failure Modes

The impact process was systematically analyzed and divided into four distinct stages based on the hammer's dynamic interaction with the panel:

5.3.1.1 Stage I: Free Fall

Before contact, the hammer undergoes free-fall acceleration, reaching maximum velocity just prior to impact. This stage represents the initial kinetic energy input into the system.

5.3.1.2 Stage II: Contact and Energy Transfer

Upon striking the panel, the hammer transfers its momentum, inducing global bending and localized indentation. The panel's deformation increases until the hammer's velocity diminishes to zero, marking the point of maximum structural displacement. During this phase, impact energy is dissipated through elastic and inelastic deformation mechanisms.

5.3.1.3 Stage III: Hammer Rebound

Following peak deformation, the hammer remains in contact as the panel's elastic recovery initiates rebound. The hammer accelerates upward until separation occurs, with rebound velocity decaying to zero due to energy dissipation. This stage highlights the system's elastic restoration capacity.

5.3.1.4 Stage IV: Secondary Impact (Negligible Energy Contribution)

Subsequent hammer rebounds may lead to follow-on impacts, but their reduced drop height results in minimal energy transfer compared to the initial strike. Consequently, secondary impacts have negligible influence on the panel's overall response and are typically disregarded in analysis.

Experimental observations revealed distinct failure mechanisms between single-layer and double-layer SCS panels under impact loading (vid. Fig. 5.6). The single-layer configuration (SULCS-150-9(SH)) demonstrated significant vulnerability, with the impact zone showing severe indentation accompanied by steel plate fracture extending approximately 20% around the indentation perimeter. This failure mode indicates a concentration of stresses that exceeded the material's ductility limits in the single-layer arrangement.

In contrast, double-layer panels (exemplified by SULCS-150-6(DH)) exhibited markedly improved impact resistance, developing local indentations without any

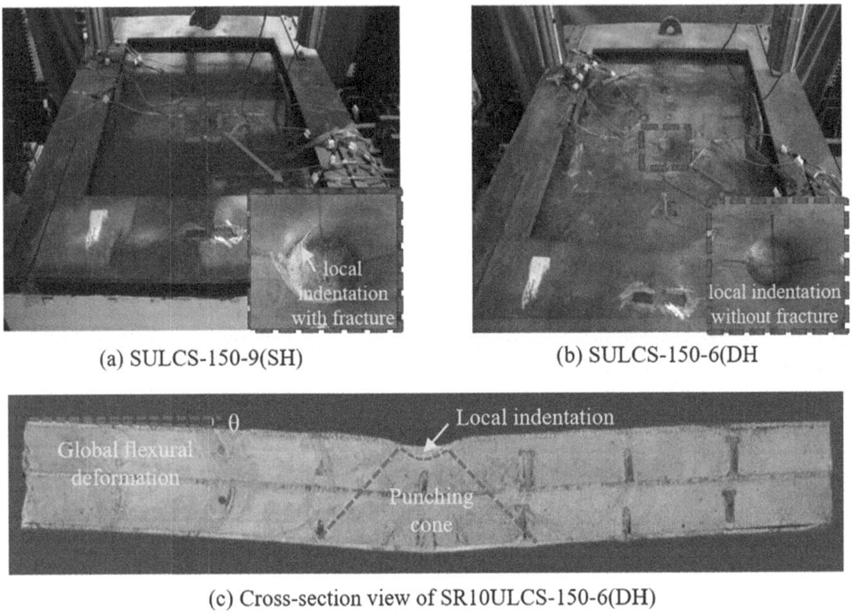

Fig. 5.6 Failure mode. (**a**) SULCS-150-9(SH). (**b**) SULCS-150-6(DH). (**c**) Cross-section view of SR10ULCS-150-6(DH). (Reprinted from Zhang et al. (2022), copyright 2025, with permission from American Society of Civil Engineers)

steel fracture. The enhanced performance stems from the composite action of the additional steel layer, which effectively redistributes impact forces across a broader panel area and engages more material in energy dissipation. This behavior was consistently observed across all tested double-layer specimens. Post-impact examination of a sectioned double-layer panel (SR10ULCS-150-6(DH)) through waterjet cutting provided further insights (vid. Fig. 5.6c). The cross-section revealed combined global flexural deformation and localized yielding at the impact zone, with the top steel plate undergoing substantial plastic deformation while maintaining structural continuity. The absence of perforation or cracking in the double-layer configuration confirms its superior damage tolerance compared to single-layer designs. These findings highlight the critical role of layered construction in impact-resistant design, where the additional steel layer in double-panel systems transforms the structural response from localized failure to distributed energy absorption. The results demonstrate that multi-layer SCS configurations offer significantly enhanced protective performance against impact loading scenarios.

5.3.2 *Effect of Number of Layers*

The impact performance of the specimens is summarized in Table 5.6, with Fig. 5.7 comparing the dynamic responses across four parameter variations: layer configuration, composite action, connector type, and rubber content. Notably, Fig. 5.7a

Table 5.6 Impact parameters and test results

Specimen	m_h (kg)	H (m)	v_{0h} (m/s)	P (kg·m/s)	E (J)	F_p (kN)	δ_p (mm)	δ_r (mm)	Failure mode
SULCS-150-9(SH)	1012	5	9.8	9948.0	48894.2	941.9	43.6	33.8	Fracture
SULCS-100-6(DH)	1012	5	9.8	9920.6	48626.0	1063.4	32.8	25.4	Indentation
SULCS-150-6(DH)	1012	5	9.9	10001.6	49422.9	960.2	43.3	32.4	Indentation
SULCS-200-6(DH)	1012	5	9.9	9998.6	49392.9	907.7	47.4	35.1	Indentation
SULCS-150-6(DJ)	1012	5	9.9	9978.3	49193.1	874.3	45.5	34.3	Indentation
SR5ULCS-150-6(DH)	1012	5	9.9	10008.7	49492.9	884.4	46.5	34.5	Indentation
SR10ULCS-150-6(DH)	1012	5	9.9	10008.7	4942.9	938.3	41.1	30.4	Indentation

Notes: m_h = hammer mass; H = drop height of the hammer; v_{0h}, P, E = impact velocity, impact momentum and impact energy of the hammer right before contacting the panel; F_p = peak load on the panel; δ_p = peak deformation at the bottom center of the panel; δ_r = residual deformation at the bottom center of the panel

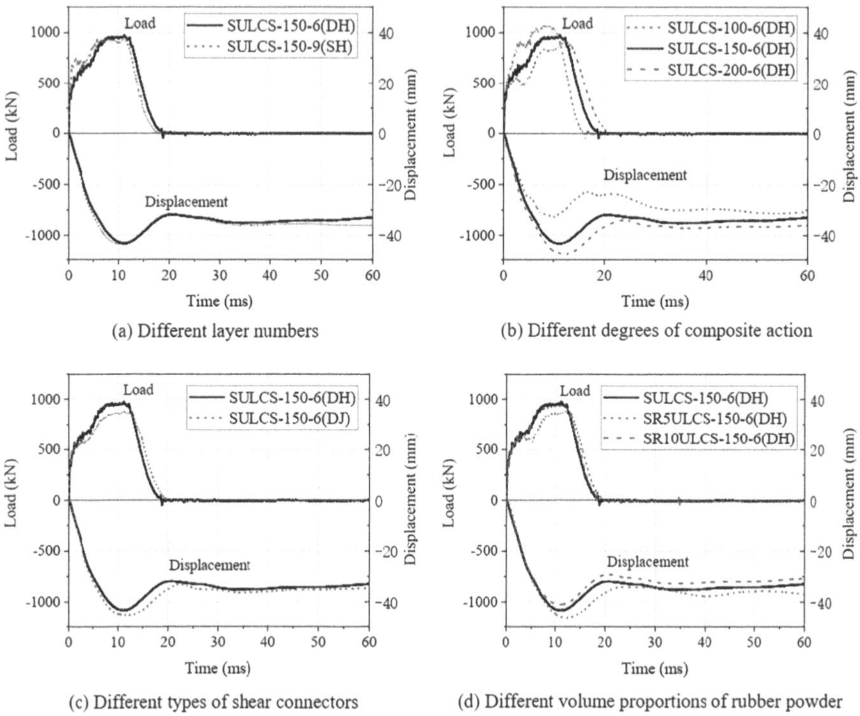

Fig. 5.7 Load-time and displacement-time responses. (**a**) Different layer numbers. (**b**) Different degrees of composite action. (**c**) Different types of shear connectors. (**d**) Different volume proportions of rubber powder. (Reprinted from Zhang et al. (2022), copyright 2025, with permission from American Society of Civil Engineers)

reveals nearly identical load-time and displacement-time curves for SULCS-150-6 (DH) (double-layer) and SULCS-150-9(SH) (single-layer), despite their differing damage patterns. The single-layer panel exhibited more severe damage, including top steel plate fracture with a crack extending 20% around the indentation perimeter—a failure mode posing significant safety and repair concerns. In contrast, the double-layer configuration demonstrated superior impact resistance through distributed damage absorption without fracture, proving both more structurally reliable and economically viable for protective applications. These findings clearly establish the performance advantages of double-layer SCS panels in impact scenarios, where their multi-layer design effectively mitigates localized failure while maintaining comparable initial resistance characteristics to single-layer systems.

5.3.3 Effect of Degree of Composite Action

The degree of composite action (η), governed by shear connector spacing, significantly influences the impact response of SCS panels. Figure 5.7b presents a comparative analysis of three double-layer specimens (SULCS-100-6(DH), SULCS-150-6(DH), and SULCS-200-6(DH)) with varying η values of 1.00, 0.64, and 0.37, respectively. The experimental results reveal an inverse relationship between composite action and structural flexibility: as η decreases from 1.00 to 0.37, the peak impact force reduces from 1063.4 kN to 907.7 kN, while the maximum displacement increases from 32.8 mm to 47.4 mm. This behavior stems from the bond-slip mechanism between steel and concrete components. Panels with lower composite action (wider connector spacing) exhibit enhanced flexibility and ductility, allowing greater energy absorption through deformation. Consequently, when subjected to identical impact energy, such panels demonstrate superior deformation capacity but sustain reduced impact forces compared to their strongly composite counterparts. The findings underscore the critical trade-off between impact force mitigation and displacement control in SCS panel design, highlighting how connector spacing can be optimized to achieve desired performance characteristics under dynamic loading conditions.

5.3.4 Effect of Shear Connectors

The impact performance of SCS panels demonstrates notable variations depending on shear connector type, as evidenced by comparative testing of specimens with different connection systems. The panel employing exclusively J-hook connectors (SULCS-150-6(DJ)) demonstrates effective impact resistance through a distinctive interlocking mechanism between J-hook pairs. When contrasted with its hybrid-connector counterpart (SULCS-150-6(DH)), the J-hook configuration exhibits a 9% reduction in peak impact force (874.3 kN versus 960.2 kN) coupled with a modest 5% increase in maximum displacement (45.5 mm versus 43.3 mm). These comparable energy absorption capacities between both connector types suggest similar overall performance under impact loading conditions. However, practical considerations favor the hybrid connector system for engineering applications. As established by Zhang et al. [5], the hybrid configuration offers superior transverse shear resistance compared to J-hooks alone. This enhanced shear capacity, combined with significantly simpler fabrication requirements, positions the hybrid connector system as an optimal solution for real-world implementations. The slightly higher impact forces observed in hybrid-connected panels are offset by their construction advantages and improved shear performance, making them particularly suitable for scenarios demanding both impact resistance and structural efficiency. These findings provide valuable guidance for engineers selecting connection systems in SCS panel design, where the balance between impact performance, shear capacity, and constructability must be carefully evaluated for specific application requirements.

5.4 Analytical Model

5.4.1 Resistances Under Quasi-Static Loading

The punching shear behavior of double-layer SCS panels under concentrated loading exhibits a well-defined failure progression, as illustrated in Fig. 5.8. Experimental observations reveal the formation of a characteristic 60° punching cone (refer to Fig. 5.8a), with the three-dimensional load transfer mechanism (refer to Fig. 5.8b) demonstrating how resistance is shared between steel plates, concrete core, and shear connectors. The panel's deformation response combines both global bending and local bulging components, as shown in Fig. 5.8c, which can be effectively modeled using a tandem spring approach accounting for global and local stiffness contributions. The load-displacement response follows a distinct three-phase failure sequence established in previous studies [12], as shown in Fig. 5.8d. The initial peak resistance (Point A) corresponds to concrete core punching failure, followed by sequential fracture of the top (Point C) and middle (Point E) steel plates. This failure progression highlights the composite action between materials and the importance of considering both global and local deformation mechanisms in design. The consistent 60° punching cone angle observed across tests provides a valuable geometric parameter for analytical modeling of such systems. These findings offer important

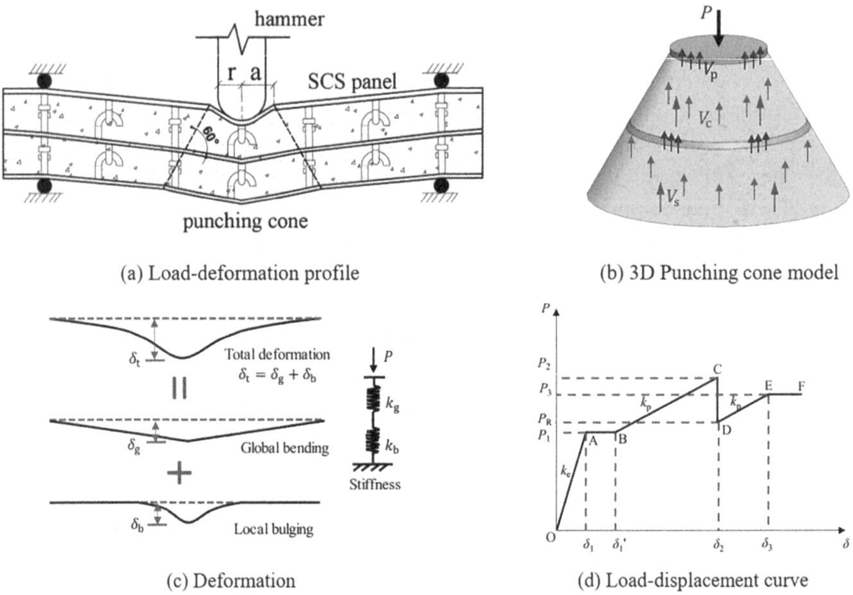

Fig. 5.8 Double-layer SCS panel under punching load. (**a**) Load-deformation profile. (**b**) 3D punching cone model. (**c**) Deformation. (**d**) Load-displacement curve. (Reprinted from Zhang et al. (2022), copyright 2025, with permission from American Society of Civil Engineers)

insights for optimizing the punching shear resistance of multi-layer SCS panels in practical applications.

At point A, the punching cone is initially formed and the first peak resistance P_1 consists of the contributions of the top steel plate V_{pt}, the upper layer concrete V_{c1}, the lower layer concrete V_{c2}, the upper layer shear connectors V_{s1}, and the lower layer shear connectors V_{s2} as given below:

$$P_1 = V_c + V_s + V_p = V_{c1} + V_{c2} + V_{s1} + V_{s2} + V_{pt} = V_{c1} + V_{c2} + V_{s1} + V_{s2} + 0.33\sqrt{f_c}\frac{E_p}{E_c}S_t t_p \tag{5.2}$$

where f_c is the compressive strength of concrete; E_p and E_c are the elastic modulus of steel plate and concrete, respectively; S_t is the perimeter of loading hammer; t_p is the thickness of steel plate. The details of V_{c1}, V_{c2}, V_{s1}, and V_{s2} can be referred to [12].

At the onset of punching shear fracture in the top steel plate (Point C), the fully developed punching cone renders the concrete core ineffective due to critical diagonal cracking, eliminating its load transfer capacity. The secondary resistance peak (P_2) consequently comprises four key components: the resistance contributions of the top steel plate V_{pt}, the middle steel plate V_{pm}, the upper layer shear connectors V_{s1}, and the lower layer shear connectors V_{s2}, which is given below:

$$P_2 = V_s + V_p = V_{s1} + V_{s2} + V_{pt} + V_{pm} = V_{s1} + V_{s2} + S_t t_p \frac{f_{up}}{\sqrt{3}} + \eta S_m t_p \frac{f_{yp}}{\sqrt{3}} \tag{5.3}$$

where η is the degree of composite action; f_{yp} and f_{up} are the yield strength and ultimate strength of steel plate, respectively; S_m is the perimeter of the intersection between the punching cone and the middle steel plate.

At point D, the top steel plate is completely punched through and loses its load-carrying capacity. The load is then directly applied to the lower layer. Thus, the residual load resistance at point D, P_R, is calculated by subtracting V_{pt} and V_{s1} from P_2, as given below:

$$P_R = P_2 - V_{pt} - V_{s1} = V_{s2} + \eta S_m t_p \frac{f_{yp}}{\sqrt{3}} \tag{5.4}$$

At point E, the middle steel plate starts to fail by punching shear fracture. The third peak load resistance, P_3, consists of the contributions of the middle steel plate and the lower layer shear connectors, as given below:

$$P_3 = V_s + V_p = V_{s2} + V_{pm} = V_{s2} + S_m t_p \frac{f_{pu}}{\sqrt{3}} \tag{5.5}$$

5.4.2 Stiffness Under Quasi-Static Loading

During the elastic stage (O-A), the double-layer SCS panel primarily undergoes global flexural deformation. While classical thin plate theory [26] provides the basis for calculating global stiffness (k_g), two critical modifications are required for accurate thick plate application: (1) incorporation of a thickness-to-side length ratio correction factor, and (2) accounting for steel-concrete bond-slip effects. These considerations yield the following modified stiffness formulation for double-layer SCS panels:

$$k_e = k_g = 0.2\eta\lambda \cdot \frac{D}{0.0116L_s^2} = \frac{\eta\lambda D}{0.058L_s^2} \tag{5.6}$$

where, $\lambda = -1.65\xi + 0.75$ and ξ is thickness-to-side length ratio of the panel. D is bending stiffness of the panel and is calculated as the sum of the concrete part D_c and the steel part D_s, as given by Eqs. (5.7)–(5.9) below:

$$D = D_c + D_p \tag{5.7}$$

$$D_c = \frac{E_c}{1 - v_c^2} \cdot 2\left[\frac{h_c^3}{12} + h_c\left(\frac{h_c + t_p}{2}\right)^2\right] \tag{5.8}$$

$$D_p = \frac{E_p}{1 - v_p^2} \cdot \left[\frac{t_p^3}{12} + 2\left(\frac{t_p^3}{12} + t_p\left(h_c + t_p\right)^2\right)\right] \tag{5.9}$$

where h_c is height of a single concrete layer; v_c and v_p are Poisson's ratio of the concrete and steel plate, respectively.

At the first plastic stage B-C, the total deformation δ_t consists of the global bending deformation δ_g and the local bulging deformation δ_b. The plastic stiffness k_{p1} is calculated by using as a tandem spring model of global bending stiffness k_g and local bugling stiffness k_b, as given below:

$$k_{p1} = \frac{k_g k_b}{k_g + k_b} \tag{5.10}$$

where, k_g is calculated according to Eq. (5.6), while k_b is calculated by Eq. (5.11) below:

$$k_b = 4\pi\lambda f_{yp} t_p \tag{5.11}$$

According to experimental results reported in [12], the slopes observed during the first (B-C) and second (D-E) plastic stages are comparable. This similarity arises because both stages are dominated by the membrane behavior of the steel plates.

Therefore, within the analytical model, the stiffness during the second plastic phase may be reasonably approximated using the value established for the first plastic stage, as expressed below:

$$k_{p2} = k_{p1} \tag{5.12}$$

The load-displacement response illustrated in Fig. 5.8d can be derived from the relationships among peak resistances and stiffness values. In partially composite panels, the plateau region (A–B) is longer than that in fully composite configurations. To account for this behavior, the displacement parameter δ_1' is defined as $\delta_1' = \delta_1 / \sqrt{\eta'}$, where η' represents a stiffness reduction factor. This factor is computed as $\eta' = \sqrt{n_p / n_f}$, with n_p denoting the actual number of shear connectors in the panel and n_f the number required for full composite action [27].

5.4.3 SDOF Model

In accordance with the single degree of freedom (SDOF) model, the drop hammer and the SCS panel are assumed to remain bonded following impact, as illustrated in Fig. 5.9a. The initial velocity of the combined system is derived using the principle of conservation of momentum, given by Eq. (5.13):

$$v_0 = \frac{m_h}{m_e} v_{0h} \tag{5.13}$$

where v_{0h} is the impact velocity of the drop hammer; $m_e = m_s + m_h$, in which m_h is the physical mass of the hammer, and m_s is the effective mass of the SCS panel

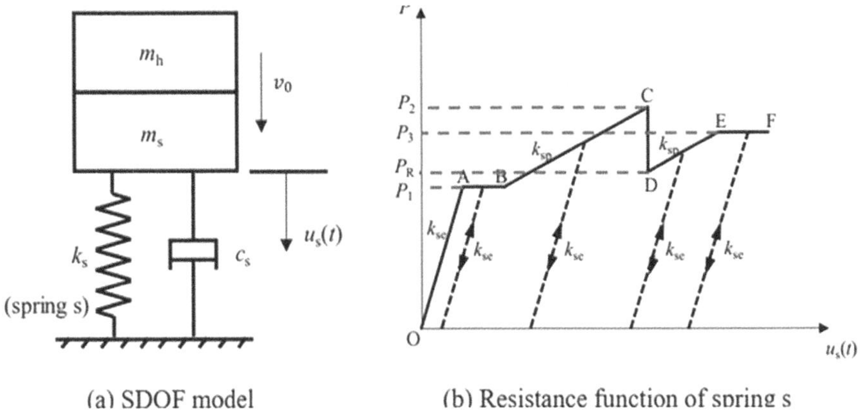

Fig. 5.9 Single Degree of Freedom (SDOF) model. (**a**) SDOF model. (**b**) Resistance function of spring s. (Reprinted from Zhang et al. (2022), copyright 2025, with permission from American Society of Civil Engineers)

calculated by multiplying the total mass of the panel with a transformation factor, which is dependent on the member geometry, support conditions, and expected response (elastic, elasto-plastic, or plastic). This factor is determined as 0.2 by integrating the square of shape function within the area of the panel, and the detailed procedure refers to the literature [16, 28].

The dynamic equation of motion of the SDOF model is then given as:

$$m_e \ddot{u}_s(t) + c_s \dot{u}_s(t) + k_s u_s(t) = 0 \tag{5.14}$$

The dynamic response of the two contacting objects is governed by their acceleration $\ddot{u}_s(t)$, velocity $\dot{u}_s(t)$, and displacement $u_s(t)$. The damping term c_s in this system can be omitted for modeling the initial loading, as it has a negligible effect on the maximum deformation during high-speed impact events [28]. In contrast, damping is a dominant factor in the unloading and free vibration phases, where it efficiently diminishes the amplitude of oscillation. Therefore, the coefficient is defined to be significant primarily for these phases.

$$c_s = \begin{cases} 0 & 0 \le t \le t(u_{s,\max}) \\ \sqrt{m_e k_{se}} & t > t(u_{s,\max}) \end{cases} \tag{5.15}$$

where $t(u_{s,max})$ is the time when the maximum deformation is reached. The term $k_s u_s(t)$ represents the load resistance function of the spring s, $P[u_s(t)]$, as shown in Fig. 5.9b. Thus, Eq. (5.14) can be modified as:

$$m_e \ddot{u}_s(t) + c_s \dot{u}_s(t) + P[u_s(t)] = 0 \tag{5.16}$$

The dynamic equation of motion represents an initial value problem, which can be numerically solved using the finite difference method. The displacement, velocity, and acceleration at each time step i are computed as follows:

$$u_s(t_i) = u_{s,i} \tag{5.17}$$

$$\dot{u}_s(t_i) = \dot{u}_{s,i} = \frac{1}{\Delta t}(u_{s,i} - u_{s,i\text{-}1}) \tag{5.18}$$

$$\ddot{u}_s(t_i) = \ddot{u}_{s,i} = \frac{1}{\Delta t^2}(u_{s,i+1} - 2u_{s,i} + u_{s,i\text{-}1}) \tag{5.19}$$

Substituting Eqs. (5.17)–(5.19) to Eq. (5.16) gives:

$$u_{s,i+1} = \left[2 - \frac{c_s \Delta t}{m_e}\right] u_{s,i} - \left[1 - \frac{c_s \Delta t}{m_e}\right] u_{s,i-1} - \frac{\Delta t^2}{m_e} P(u_{s,i}) \tag{5.20}$$

The initial conditions of the dynamic equation of motion are:

$$u_{s,0} = 0 \qquad \dot{u}_{s,0} = v_0 \tag{5.21}$$

The solution of $u_{s,1}$ needs $u_{s,0}$ and $u_{s,-1}$, which can be determined by considering $i = 0$ in Eq. (5.18) as given below:

$$u_{s,-1} = u_{s,0} - \Delta t\dot{u}_{s,0} = -\Delta t v_0 \tag{5.22}$$

5.4.4 TDOF Model

The TDOF analytical model represents the impact system through two discrete masses: one for the drop hammer and another for the SCS panel, interconnected through a coupled spring-damper element that simulates their contact interaction (refer to Fig. 5.10a). In this formulation, spring '*s*' characterizes the panel's structural resistance, maintaining an identical force-deformation relationship to the SDOF model (refer to Fig. 5.10b), while spring '*h*' models the contact stiffness between the impacting masses (refer to Fig. 5.10c). The deformation of spring '*h*' is defined by the relative displacement ($u_h - u_s$) between the hammer and panel. Following Wang et al. [16], the elastic stiffness coefficient k_{he} is derived as:

$$k_{he} = \frac{k_b k_i}{k_b - k_i} \tag{5.23}$$

where, k_b is the local bulging stiffness and is calculated according to Eq. (5.9). k_i is the local indentation stiffness, and is calculated by Eq. (5.24) given below:

$$k_i = 2.5\pi f_{yp} t_p + 2\pi r f_c \tag{5.24}$$

where, r is the radius of the hammer.

The plastic stiffness k_{hp} and unloading stiffness k_{hu} are modeled as the product of the elastic stiffness and two distinct coefficients, α and β, respectively. Based on [16], these coefficients are assigned values of 0.4 and 5.0. The significantly higher value of β accounts for the limited recovery of local indentation during the unloading phase.

The dynamic equation of motion of the TDOF model is then given as:

$$\begin{cases} m_s\ddot{u}_s(t) + c_s\dot{u}_s(t) + P_s[u_s(t)] = c_h[\dot{u}_h(t) - \dot{u}_s(t)] + P_h[u_h(t) - u_s(t)] \\ m_h\ddot{u}_h(t) + c_h[\dot{u}_h(t) - \dot{u}_s(t)] + P_h[u_h(t) - u_s(t)] = 0 \end{cases} \tag{5.25}$$

In this model, m_h denotes the physical mass of the hammer, while m_s represents the effective mass of the SCS panel, taken as 0.2 times its physical mass—consistent

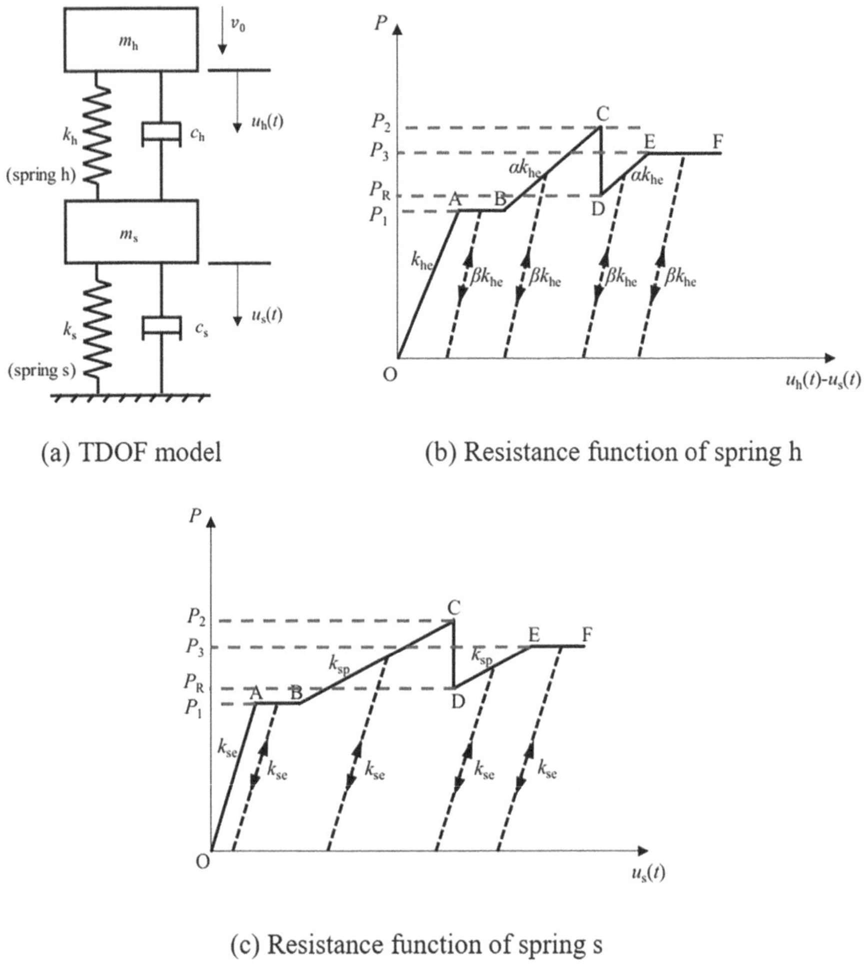

Fig. 5.10 Total Degree of Freedom (TDOF) model. (**a**) TDOF model. (**b**) Resistance function of spring h. (**c**) Resistance function of spring s. (Reprinted from Zhang et al. (2022), copyright 2025, with permission from American Society of Civil Engineers)

with the value adopted in the SDOF model. The TDOF model enables prediction of the contact force occurring between the drop hammer and the SCS panel, as expressed by the following equation:

$$F(t) = m_{\mathrm{h}}\ddot{u}_{\mathrm{h}}(t) = -c_{\mathrm{h}}[\dot{u}_{\mathrm{h}}(t) - \dot{u}_{\mathrm{s}}(t)] - P_{\mathrm{h}}[u_{\mathrm{h}}(t) - u_{\mathrm{s}}(t)] \tag{5.26}$$

The dynamic equation of motion is solved numerically via the finite difference method, yielding the following expressions:

$$\begin{cases} u_{s,i+1} = 2u_{s,i} - u_{s,i-1} + \dfrac{\Delta t}{m_s}\left[c_h\left(u_{h,i} - u_{h,i-1} - u_{s,i} + u_{s,i-1}\right) - c_s\left(u_{s,i} - u_{s,i-1}\right)\right] \\ \qquad + \dfrac{\Delta t^2}{m_s}\left[P_h\left(u_{h,i} - u_{s,i}\right) - P_s\left(u_{s,i}\right)\right] \\ u_{h,i+1} = 2u_{h,i} - u_{h,i-1} - \dfrac{\Delta t}{m_h} c_h\left(u_{h,i} - u_{h,i-1} - u_{s,i} + u_{s,i-1}\right) - \dfrac{\Delta t^2}{m_h} P_h\left(u_{h,i} - u_{s,i}\right) \\ F_{i+1} = -\dfrac{c_h}{\Delta t}\left(u_{h,i+1} - u_{h,i} - u_{s,i+1} + u_{s,i}\right) - P_h\left(u_{h,i+1} - u_{s,i+1}\right) \end{cases} \tag{5.27}$$

The resistance functions $P_h(u_{h,i-}u_{s,i})$ and $P_s(u_{s,i})$ are defined according to the relationships illustrated in Fig. 5.10b, c, respectively. Damping coefficients c_s and c_h are omitted during the initial loading phase and activated solely in subsequent unloading stages. Furthermore, c_h is disregarded once the hammer detaches from the top steel plate. The expressions for c_s and c_h are given as follows:

$$c_s = \begin{cases} 0 & 0 \le t \le t(u_{s,\max}) \\ \sqrt{m_s k_{se}} & t > t(u_{s,\max}) \end{cases} \tag{5.28}$$

$$c_h = \begin{cases} 0 & 0 \le t \le t(u_{h,\max}) \& P_h[u_h(t) - u_s(t)] > 0 \\ \sqrt{\dfrac{m_h m_s}{m_h + m_s}} k_{he} & t > t(u_{h,\max}) \& P_h[u_h(t) - u_s(t)] > 0 \\ 0 & t > t(u_{h,\max}) \& P_h[u_h(t) - u_s(t)] \le 0 \end{cases} \tag{5.29}$$

The TDOF model can be solved by imposing the following initial conditions.

$$\begin{cases} u_{s,0} = 0 & \dot{u}_{s,0} = 0 & u_{s,-1} = u_{s,0} - \Delta t \dot{u}_{s,0} = 0 \\ u_{h,0} = 0 & \dot{u}_{h,0} = v_{0h} & u_{s,-1} = u_{s,0} - \Delta t \dot{u}_{s,0} = -\Delta t v_{0h} \end{cases} \tag{5.30}$$

5.4.5 Verification

The comparative evaluation of SDOF and TDOF models for predicting the impact response of double-layer SCS panels reveals important insights into their respective capabilities and limitations. As summarized in.

Table 5.7 and illustrated in Fig. 5.11, both analytical approaches show reasonable agreement with experimental and FE results, though with distinct characteristics in their predictions. The displacement-time and load-time histories demonstrate that while the SDOF model provides adequate displacement estimates, it fundamentally cannot capture contact force variations due to its assumption of constant hammer-panel contact. This simplification leads to an overprediction of peak deformations

Table 5.7 Key parameters in the analytical models

Specimen	m_s	m_h	P_1	P_2	P_R	P_3	SDOF			TDOF		
							k_{se}	k_{sp}	c_s	k_{he}	c_s	c_h
SULCS-100-6(DH)	70.0	1012	648.9	1484.1	813.4	1060.4	158.0	12.5	413.5	51.4	105.2	58.0
SULCS-150-6(DH)	68.8	1012	620.2	1200.1	798.6	953.8	102.6	11.9	333.0	51.4	84.0	57.5
SULCS-200-6(DH)	68.0	1012	600.6	983.2	787.1	913.7	58.4	11.1	251.1	51.4	63.0	57.2
SULCS-150-6(DJ)	68.8	1012	617.6	1189.5	793.3	948.9	102.6	11.9	333.0	51.4	84.0	57.5
SR5ULCS-150-6(DH)	66.2	1012	611.4	1191.6	794.4	951.8	99.9	11.9	328.2	51.0	81.3	56.3
SR10ULCS-150-6(DH)	63.6	1012	605.5	1184.5	790.8	950.8	97.9	11.8	324.5	50.7	78.9	55.1

Notes: The unit of mass is kg, the unit of load is kN, the unit of stiffness is kN/mm, and the unit of damping is kN·s/m

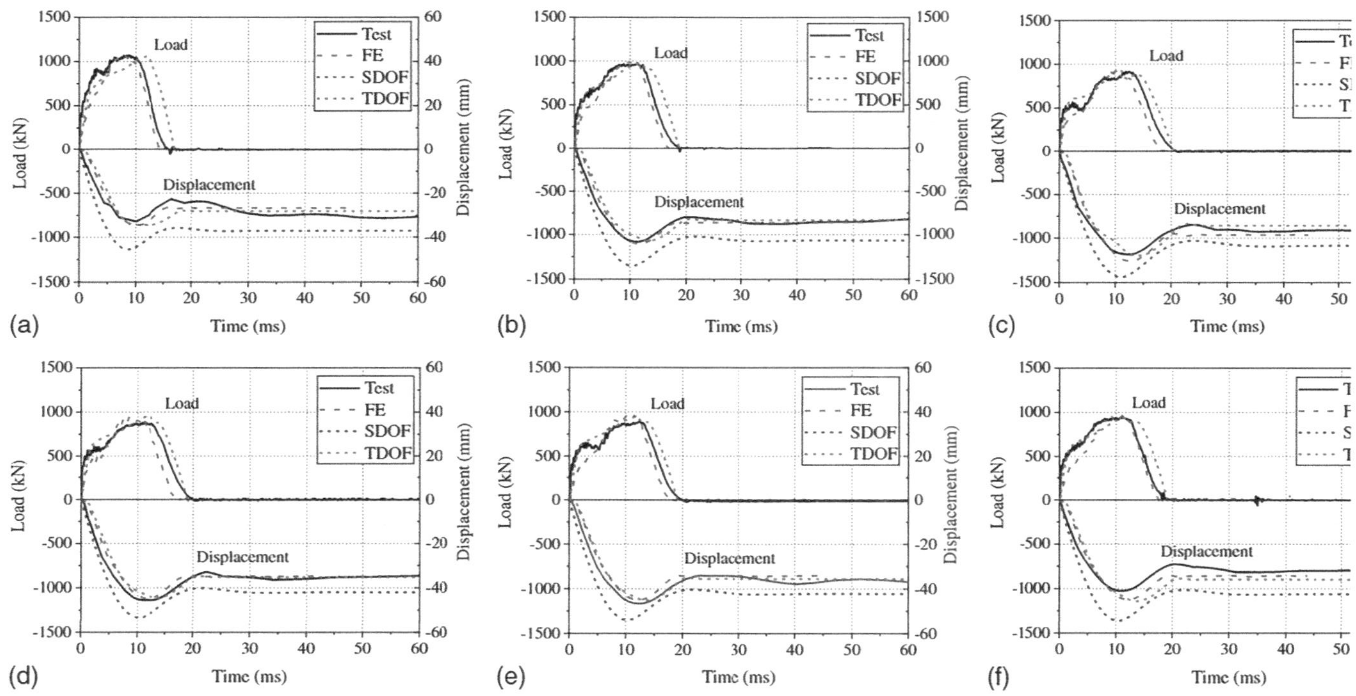

Fig. 5.11 Load-time and displacement-time responses: test versus FE versus prediction models: (**a**) SULCS-100-6(DH); (**b**) SULCS-150-6(DH); (**c**) SULCS-200-6(DH); (**d**) SULCS-150-6(DJ); (**e**) SR5ULCS-150-6(DH); (**f**) SR10ULCS-150-6(DH). (Reprinted from Zhang et al. (2022), copyright 2025, with permission from American Society of Civil Engineers)

compared to the more sophisticated TDOF approach. The observed differences between models stem primarily from their treatment of energy distribution in the system. With a substantial hammer-to-panel mass ratio of approximately 15, both models begin with equivalent initial kinetic energy due to negligible energy loss during impact. However, their divergent approaches to energy conversion yield different results: the SDOF model concentrates all energy in a single spring representing the panel, while the TDOF model properly partitions energy between separate springs representing both the hammer and panel components. This fundamental difference explains why the SDOF model's effective panel stiffness appears artificially increased, resulting in greater predicted deformations. The comparison of key response parameters in Table 5.8, including peak deformation, residual deformation, and peak impact force, further highlights the TDOF model's superior predictive capability, particularly for force-related quantities. These findings emphasize that while simplified SDOF approaches may suffice for preliminary displacement estimates, the TDOF framework provides more comprehensive and accurate predictions for systems involving significant mass disparities and complex energy transfer mechanisms. The results underscore the importance of selecting appropriate modeling techniques based on both the desired level of accuracy and the specific response parameters of interest in impact scenarios.

The numerical and analytical models demonstrate robust predictive capability for the impact behavior of double-layer SCS panels, with both FE and TDOF models showing excellent agreement with experimental results. As evidenced by the displacement-time and load-time curves, these computational approaches not only match the test data within acceptable statistical variations (as indicated by mean values and standard deviations of prediction-to-test ratios), but also provide enhanced clarity in identifying material behavior trends that were less distinct in physical testing. Particularly noteworthy is the consistent demonstration by both FE and TDOF models of how rubber powder modification influences panel performance—a relationship that physical tests suggested but could not conclusively establish. The computational results uniformly show that increasing the rubber content from 0% (SULCS-150-6(DH)) to 10% (SR10ULCS-150-6(DH)) produces a measurable increase in peak deformation alongside a corresponding reduction in peak impact force. This systematic variation, clearly quantified in Table 5.8, confirms that rubber powder incorporation enhances the panels' energy absorption capacity through greater deformation capability, while simultaneously reducing the maximum transmitted forces—a valuable combination for impact-resistant design. The strong correlation between numerical predictions and experimental measurements, coupled with the models' ability to elucidate subtle material effects, validates their utility for both performance evaluation and parametric studies of SCS panel systems.

Table 5.8 Verification of the FE model, SDOF model and TDOF model

Specimen	Test	FE	FE/ Test	SDOF	SDOF/ Test	TDOF	TDOF/ Test
	δ_p **(mm)**						
SULCS-100-6(DH)	32.8	34.7	1.06	45.4	1.38	34.8	1.06
SULCS-150-6(DH)	43.3	44.4	1.03	54.1	1.25	42.9	0.99
SULCS-200-6(DH)	47.4	50.4	1.06	57.7	1.22	48.0	1.01
SULCS-150-6(DJ)	45.5	44.9	0.99	53.3	1.17	44.6	0.98
SR5ULCS-150-6 (DH)	46.5	45.0	0.97	54.1	1.16	45.3	0.97
SR10ULCS-150-6 (DH)	41.1	45.3	1.10	54.7	1.33	45.7	1.11
Mean			1.03		1.25		1.02
Std. Dev.			0.05		0.09		0.05
	δ_r **(mm)**						
SULCS-100-6(DH)	25.4	26.7	1.05	36.9	1.45	28.3	1.11
SULCS-150-6(DH)	32.4	34.3	1.06	42.9	1.32	33.3	1.03
SULCS-200-6(DH)	35.1	38.6	1.10	43.5	1.24	34.2	0.97
SULCS-150-6(DJ)	34.3	34.9	1.02	41.7	1.22	35.2	1.03
SR5ULCS-150-6 (DH)	34.5	34.2	0.99	42.3	1.23	35.4	1.03
SR10ULCS-150-6 (DH)	30.4	34.5	1.13	42.5	1.40	36.1	1.19
Mean			1.06		1.31		1.06
Std. Dev.			0.05		0.10		0.08
	F_p **(kN)**						
SULCS-100-6(DH)	1063.4	1075.7	1.01	/	/	1065.6	1.00
SULCS-150-6(DH)	960.2	995.2	1.04	/	/	973.4	1.01
SULCS-200-6(DH)	907.7	911.1	1.00	/	/	932.5	1.03
SULCS-150-6(DJ)	874.3	950.8	1.09	/	/	957.8	1.10
SR5ULCS-150-6 (DH)	884.4	988.5	1.12	/	/	948.1	1.07
SR10ULCS-150-6 (DH)	938.3	984.6	1.05	/	/	943.0	1.01
Mean			1.05				1.04
Std. Dev.			0.04				0.04

5.5 Numerical Investigation

5.5.1 FE Modelling of Multi-Impacts

The restart method is used for numerical simulations of multiple impacts. This approach requires using the deformation and stress-strain fields from the previous impact as the initial conditions for the next impact, followed by resubmitting the K-file for calculation, making the process relatively cumbersome. To simplify the computational procedure, this study defines multiple drop hammers in a single K-file, allowing them to simultaneously impact the composite slab from different heights to simulate multiple impacts.

Preliminary drop-weight impact tests revealed that the duration of a single impact is approximately 20 ms. The impact velocity of the drop hammer upon contacting the steel plate can be theoretically calculated. In the numerical model, parameters such as the height, initial velocity, and gravitational acceleration of each drop hammer must be set to ensure that the impact velocity matches the calculated contact velocity. Additionally, the time interval between successive drop hammers must be sufficiently large so that the subsequent hammer strikes only after the previous one has completed its impact and rebounded from the specimen. Since no contact is defined between the drop hammers, they do not interfere with each other.

The 3D modeling of the specimen was performed using SOLIDWORKS, followed by mesh refinement in HYPERMESH. The final K-file was then imported into LS-DYNA for computation. As shown in Fig. 5.12, due to the symmetry of the simulated structure, a quarter-model was adopted to reduce computational effort, with symmetric boundary conditions applied. In the model, the drop hammer, concrete, and connectors were modeled using solid elements (SECTION_SOLID), with minimum mesh sizes of 25 mm × 25 mm × 25 mm, 10 mm × 10 mm × 10 mm, and 2 mm × 2 mm × 2 mm, respectively. The steel plate was modeled using thick shell elements (SECTION_TSHELL), with a minimum mesh size of 10 mm × 10 mm.

The J-hooks and studs were simplified as straight rods of equivalent diameter in the finite element model. The rods and steel plate were fixed using shared nodes, while the centers of the upper and lower rods were connected via spring elements. The spring elements employed Material S04 (SPRING-NONLINEAR-ELASTIC), requiring the keyword *DEFINE_CURVE to define a force-displacement curve [14]. This curve was obtained from pull-out tests conducted on sandwich specimens containing J-hooks and studs.

5.5.2 Calibration of Material Constitutive Model

LS-DYNA provides various constitutive models for concrete material characterization. For steel-concrete composite structures subjected to impact loading, the Concrete Damage Model (MAT_72R3) and Continuous Surface Cap Model (CSCM, MAT_159) [15, 16] are most frequently employed. Through comprehensive numerical trials and comparative analysis, the CSCM model (MAT_159) was ultimately selected to simulate the mechanical behavior of both ULHDCC and RULHDCC. This advanced model features a smooth transition surface connecting the shear yield surface and hardening cap, effectively incorporating material hardening, damage evolution, and strain-rate effects. Its demonstrated reliability in simulating low-velocity impact scenarios on reinforced concrete structures has led to widespread adoption in the field.

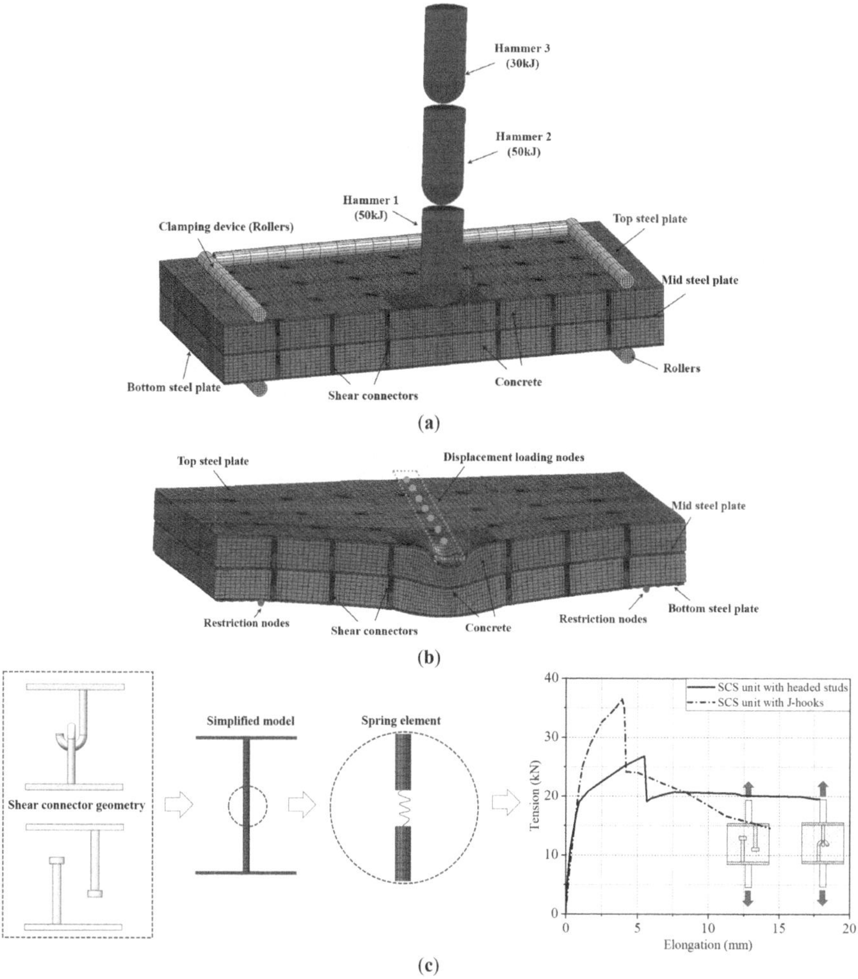

Fig. 5.12 FE model of SCS composite panel (**a**) impact stage; (**b**) post-impact static stage; (**c**) simplified model for shear connectors [21]

Fig. 5.13 Yield surface of CSCM model [30]. (Reprinted from Huang et al. (2023), copyright 2025, with permission from Journal of Building Structures (In Chinese))

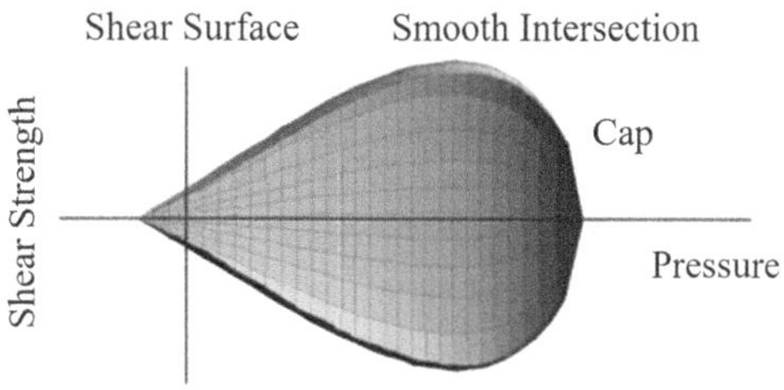

Figure 5.13 illustrates the yield surface morphology of the CSCM model, mathematically described by Eq. (5.31) below using first stress invariant (I_1), second/third deviatoric stress invariants (J_2, J_3), and cap hardening parameter (k):

$$f(J_1, J_2, J_3, k) = J_2 - R^2 F_f^2 F_c \quad (5.31)$$

where, R is Reduction factor, F_f is Shear failure surface function, F_c is Cap hardening function.

The shear failure surface is defined as:

$$F_f(J_1) = \alpha - \lambda \exp^{-\beta J_1} + \theta J_1 \quad (5.32)$$

Parameters α, β, λ, and θ are derived from cylindrical concrete triaxial compression tests. Under high hydrostatic pressure, material behavior is governed by the movable hardening cap surface, described by Eq. (5.33), which characterizes pressure-dependent strain hardening:

$$F_c(J_1, k) = 1 - \frac{[J_1 - L(k)][|J_1 - L(k)| + J_1 - L(k)]}{2[X(k) - L(k)]^2} \quad (5.33)$$

The cap evolution function $L(k)$ is given by:

$$L(k) = \begin{cases} k & (k > k_0) \\ k_0 & (k \leq k_0) \end{cases} \quad (5.34)$$

where: k is Current cap position, k_0 is Initial cap position

As an elastoplastic damage model, CSCM incorporates a damage algorithm to simulate material softening. The effective stress is computed via:

$$\sigma_{ij}^{d} = (1 - d)\sigma_{ij}^{vp} \quad (5.35)$$

where, d is Damage variable (0 = intact, 1 = fully damaged), σ_{ij}^{vp} is Effective undamaged strain tensor.

The CSCM model involves a total of 37 input parameters. Parameters associated with the yield and cap surfaces are automatically derived based on fundamental concrete material properties. These core properties consist of density, compressive strength, tensile strength, shear modulus, and bulk modulus—the latter two being obtained from Eqs. (5.36) and (5.37), respectively. The remaining parameters are adopted from reference [29] and further refined through iterative trial-and-error calibration. Table 5.9 presents the full set of CSCM parameters calibrated specifically for Ultra-High Performance Ductile Cementitious Composite.

$$G = E/(2 + 2\nu) \quad (5.36)$$

$$K = E(1 - 2\nu)/3 \quad (5.37)$$

Table 5.9 Input parameters for ULHDCC

Parameter	Value	Parameter	Value
Density/$t{\cdot}mm^{-3}$	1.388e-9	Maximum plastic strain, W	0.05
Shear modulus, G/MPa	4610	Linear hardening exponent, D_1/MPa	2.5e-4
Bulk modulus, K/MPa	1810	Nonlinear hardening exponent, D_2	3.492
Triaxial compression meridian constant, α/GPa	14.72	Compression softening parameter, B	100
Triaxial compression meridian linear term, θ/MPa^{-1}	0.302	Compression fracture energy, G_{fc}	7.107
Triaxial compression meridian nonlinear, λ/MPa	10.51	Tension/shear softening parameter, D	0.1
Triaxial compression meridian exponent, β/MPa^{-1}	1.929e-2	Tension fracture energy, G_{ft}	7.107
Torsion meridian constant, α_1	0.7474	Pure shear fracture energy, G_{fs}/MPa·mm	7.107e-2
Torsion meridian linear term, θ_1/MPa^{-1}	1.106e-3	PWRC	5
Torsion meridian nonlinear term, λ_1	0.17	PWRT	1
Torsion meridian exponent, β_1/MPa^{-1}	6.886e-2	PMOD	-3
Triaxial tension meridian constant, α_2	0.66	Compressive strain rate effect parameter, η_{oc}	1.024e-04
Triaxial tension meridian linear term, θ_2/MPa^{-1}	1.333e-3	Compressive strain rate exponent, NC	0.78
Triaxial tension meridian nonlinear term, λ_2	0.16	Tensile strain rate effect parameter, η_{ot}	6.33e-5
Triaxial tension meridian exponent, β_2/MPa^{-1}	6.886e-2	Tensile strain rate exponent, NT	0.48
Initial hardening parameter, N_h	1	OVERC/MPa	49.2
Hardening rate, C_h	0	OVERT/MPa	2.9
Cap surface aspect ratio, R	5	SRATE	1
Initial cap position, X_0/MPa	91.34	REPOW	1

The steel components, including face plates, stud connectors, and J-hooks (reinforcement bars), were modeled using the bilinear kinematic hardening material model *MAT_PLASTIC_KINEMATIC. Based on experimental observations of failure modes, the erosion strains were specified as 0.1 for stud connectors and 0.3 for J-hooks. For the steel face plates, parametric studies suggested an optimal erosion strain of 0.45, which is slightly higher than the 0.3 fracture strain obtained from material tests to account for the constraining effects in the composite system. The drop hammer, manufactured from high-strength cast iron, was modelled as a purely elastic material (*MAT_ELASTIC) considering its demonstrated negligible deformation during multiple impact tests. This simplification, which includes only the basic parameters of density, elastic modulus, and Poisson's ratio, effectively captures the hammer's rigid-body dynamics while maintaining computational efficiency. The elastic assumption is justified by the hammer's exceptionally high strength and hardness, as evidenced by its minimal deformation in repeated experimental impacts.

5.5.3 Contact Definition and Boundary Conditions

LS-DYNA offers exceptional performance in modeling nonlinear contact interactions, surpassing other nonlinear finite element analysis software. The program's contact algorithms can be classified into four primary categories: one-way contact, two-way contact, single-surface contact, and tied contact. In this study's drop-weight impact analysis of steel-concrete composite panels, the *CONTACT_AUTOMATIC_SURFACE_TO_SURFACE algorithm with penalty-based enforcement was utilized. This advanced contact formulation simplifies the modeling process by only requiring identification of potential interacting component pairs, while automatically determining contact locations through computational detection. The contact mechanics are principally controlled by several critical parameters: static friction coefficient (FS), kinetic friction coefficient (FD), along with penalty scaling factors for both primary (SFS) and secondary (SFM) surfaces. Table 5.10 details the specific parameter values implemented in the current impact simulation. Importantly, supplementary contact definitions were established between the impactor and concrete elements to accurately represent the steel plate perforation behavior witnessed in experimental impact testing. These additional contact parameters maintain simulation precision when the upper steel plate experiences failure, ensuring reliable results under severe deformation scenarios.

5.5.4 FE Results and Discussions

The numerical simulation results obtained from LS-DYNA show remarkable consistency with experimental observations in terms of failure modes and dynamic responses. Both approaches successfully captured the characteristic damage patterns, including upper steel plate tearing, interfacial slip between steel and concrete

Table 5.10 Contact definition

Contact	FS	FD	SFS	SFM
Drop hammer-steel plate	0	0	1	1
Drop hammer-concrete	0.35	0.3	0.15	0.15
steel plate-concrete	0.7	0.5	0.15	0.15
Stud-concrete	0.7	0.5	0.15	0.15
J-hook-concrete	0.7	0.5	0.15	0.15
Steel plate-support	0	0	1	1

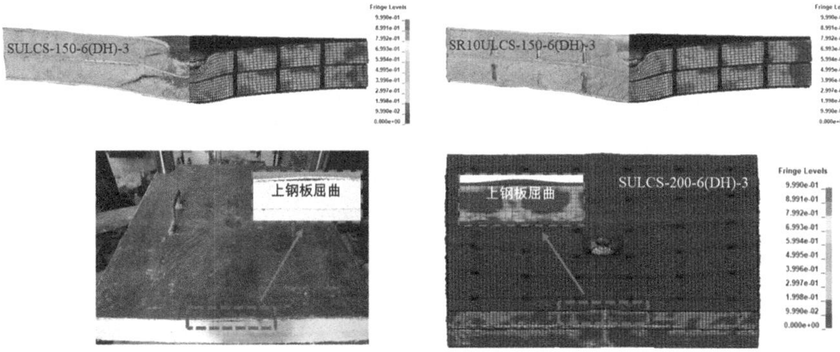

Fig. 5.14 FE vs. Test: failure mode. (Reprinted from Huang et al. (2023), copyright 2025, with permission from Journal of Building Structures (In Chinese))

components, and localized buckling of the top plate in specimen SULCS-200-6 (DH) following three successive impacts, as illustrated in Fig. 5.14. For specimen SULCS-150-6(DH), the numerical predictions of impact force-time histories and midspan displacement-time curves closely match the experimental measurements across all three impact events as shown in the Fig. 5.15 Quantitative analysis presented in Table 5.11 indicates that the maximum differences between simulated and experimental results for peak impact forces and maximum displacements remain within an acceptable 15% margin for all tested specimens.

The minor discrepancies observed can be primarily attributed to inherent limitations in simulating progressive damage accumulation and geometric idealizations in the numerical model. While the physical specimens exhibited localized, asymmetric damage patterns (such as partial circumferential tearing of the upper steel plate), the quarter-symmetry assumption in the simulation necessarily produces perfectly symmetric damage propagation. This fundamental difference becomes increasingly significant with successive impacts, as the physical specimens may experience eccentric loading due to prior asymmetric damage, whereas the numerical model maintains strict symmetry. Additionally, the challenge of precisely replicating the evolving material state after each impact, particularly the cumulative effects of concrete cracking and steel plate tearing, contributes to the observed variations between numerical and experimental results.

The overall strong correlation between simulation and experiment demonstrates the model's effectiveness in capturing the complex behavior of steel-concrete composite panels under multiple impacts. The validation confirms the reliability of the numerical approach for engineering applications while identifying specific areas where further refinement could enhance predictive accuracy, particularly in addressing asymmetric damage progression and improving cumulative damage modeling techniques.

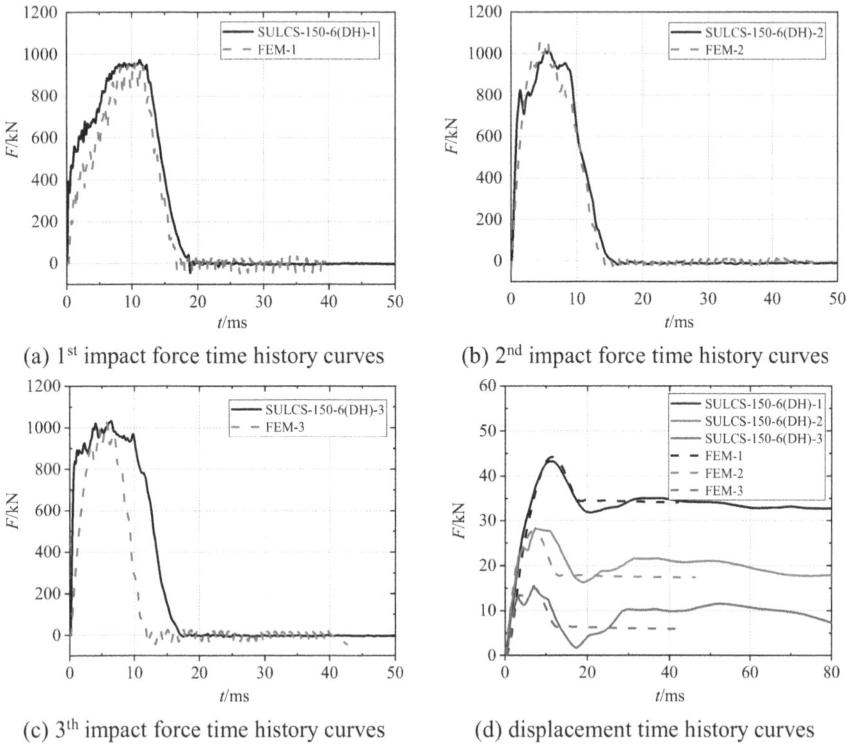

Fig. 5.15 FE vs. Test: impact force and displacement history curves. (**a**) First impact force time history curves. (**b**) Second impact force time history curves. (**c**) 3th impact force time history curves. (**d**) Displacement time history curves. (Reprinted from Huang et al. (2023), copyright 2025, with permission from Journal of Building Structures (In Chinese))

5.6 Post-Impact Residual Flexural Performance

5.6.1 Failure Modes

The post-impact flexural tests revealed that all damaged specimens maintained significant residual ductility, achieving midspan displacements between 30–60 mm before failure. As shown in Fig. 5.16, the panels ultimately failed through a combined flexural and bond-slip mechanism, demonstrating consistent energy dissipation capacity despite prior impact damage. This failure mode highlights the composite system's ability to sustain substantial deformation while maintaining structural integrity after impact events.

The post-impact flexural behavior of specimens SLHC(S)-150(H)-9, SLHC(D)-100(H)-6, SLHC(D)-150(H)-6, and SLHC(D)-150(J)-6 exhibited consistent damage progression patterns. As illustrated in Fig. 5.16a, these panels developed prominent main cracks during initial impact testing [23], which subsequently widened during

Table 5.11 Comparison between test results and numerical results

Specimen	Impact times	$F_{t,max}$/ kN	$F_{n,max}$/ kN	η_F/%	$\Delta_{t,max}$/ kN	$\Delta_{n,max}$/ kN	η_Δ/%
SULCS-150-9(SH)	1	941.9	964.6	+2.4	43.6	45.3	+4.7
	2	887.5	957.2	+7.9	26.1	22.8	−12.6
	3	645.6	637.2	−1.3	14.5	14.7	+1.4
SULCS-100-6(DH)	1	1063.4	1092.7	+2.8	32.8	34.9	+6.5
	2	1147.2	1177.3	+2.6	25.6	22.2	−13.3
	3	910.7	1038.8	+14.1	16.7	15.7	−6.0
SULCS-150-6(DH)	1	960.2	951.5	−0.9	43.3	44.2	+2.1
	2	1022.0	1062.2	+3.9	28.3	27.6	−2.5
	3	1013.0	1021.9	+0.9	15.5	15.3	−1.3
SULCS-200-6(DH)	1	907.7	936.2	+3.1	47.4	50.0	+5.5
	2	1070.1	1193.1	+11.5	31.4	30.2	−3.8
	3	961.1	1024.9	+6.6	18.5	16.9	−8.6
SULCS-150-6(DJ)	1	884.4	975.4	+10.3	46.5	45.2	−2.8
	2	1016.6	1133.2	+11.5	23.5	25.0	+6.4
	3	928.7	1052.8	+13.4	18.2	16.8	−7.7
SR5ULCS-150-6 (DH)	1	938.3	986.2	+5.1	41.1	44.1	+7.3
	2	1045.6	1141.1	+9.1	26.3	25.8	−1.9
	3	919.1	1031.2	+12.2	17.7	16.2	−8.6
SR10ULCS-150-6 (DH)	1	874.3	898.7	+2.8	45.5	48.0	+5.5
	2	893.0	1079.1	+20.8	26.7	27.2	+1.9
	3	795.2	936.3	+17.7	21.4	21.7	+1.4

Note: $F_{t,max}$ experimental peak impact force, $F_{n,max}$ numerical peak impact force, $\Delta_{t,max}$ experimental maximum midspan displacement, $\Delta_{n,max}$ numerical maximum midspan displacement, $\eta_F = (F_{n,max} - F_{t,max})/F_{t,max}$, $\eta_\Delta = (\Delta_{n,max} - \Delta_{t,max})/\Delta_{t,max}$

flexural loading. This crack propagation occurred concurrently with increasing interfacial slip at the specimen ends, reflecting progressive degradation of composite action. The failure mechanism evolved through three distinct phases: (1) stable crack widening accompanied by measurable end slip; (2) sequential fracture of shear connectors; and (3) complete interfacial separation with significant slippage at ultimate failure. This failure sequence demonstrates how impact-induced damage precipitates bond deterioration, ultimately leading to composite action breakdown and capacity reduction in the flexural loading stage.

The flexural behavior of specimens SLHC(D)-200(H)-6, SR5LHC(D)-150(H)-6, and SR10LHC(D)-150(H)-6(1–3) differed from other panels. These specimens developed only fine cracks during impact testing [23], with no dominant main cracks before flexural loading (vid. Fig. 5.16b). Under flexural loading, the microcracks gradually widened and coalesced into primary fractures. Concurrently, interfacial slip increased, followed by stud fracture and eventual steel-concrete separation, leading to complete loss of load capacity. The specimens with rubber additives or reduced composite action exhibited more distributed cracking patterns compared to conventional panels.

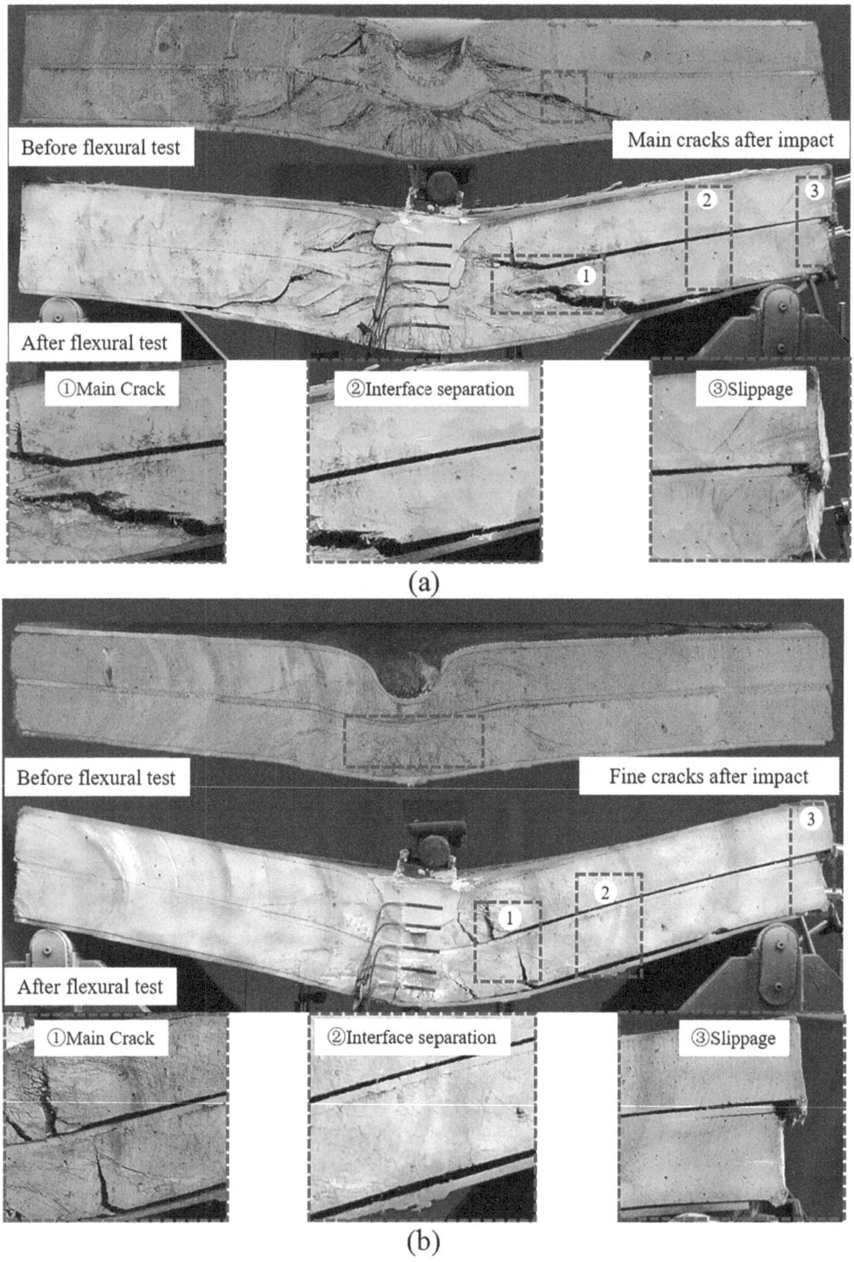

Fig. 5.16 Failure mode of specimens (**a**) SLHC(D)-100(H)-6; (**b**) SLHC(D)-200(H)-6 [21]

5.6.2 *Strain Development*

The strain development in the bottom steel plates, as shown in Fig. 5.17, reveals distinct behavioral patterns among the specimens. While most double-layer panels (SLHC(D)-100(H)-6, SLHC(D)-150(H)-6, SR5LHC(D)-150(H)-6, and SR10LHC (D)-150(H)-6 series) achieved yield strain levels, two specimens exhibited notably different responses. The single-layer SLHC(S)-150(H)-9 panel showed reduced strain capacity due to severe connector damage from impact loading, which compromised composite action and limited stress transfer to the steel plate during subsequent flexural loading. Similarly, the SLHC(D)-200(H)-6 specimen with wider connector spacing (200 mm) and consequently lower composite action demonstrated diminished strain development, reflecting less effective stress redistribution across the section.

The strain development in concrete was investigated through strategically positioned strain gauges on both the front (the profile exposed after the cut) and back (end surface) sides of impact-damaged specimens, as shown in Fig. 5.5b. Comparative analysis revealed markedly different strain distribution patterns between these surfaces, as exemplified by specimen SLHC(D)-100(H)-6 (see Fig. 5.18). The front surface exhibited minimal strain development, with only bottom measurement points registering significant values, indicating severely compromised load transfer capacity in the impact-affected zone. In contrast, the undisturbed end surface demonstrated classical linear strain distribution consistent with plane section theory. These

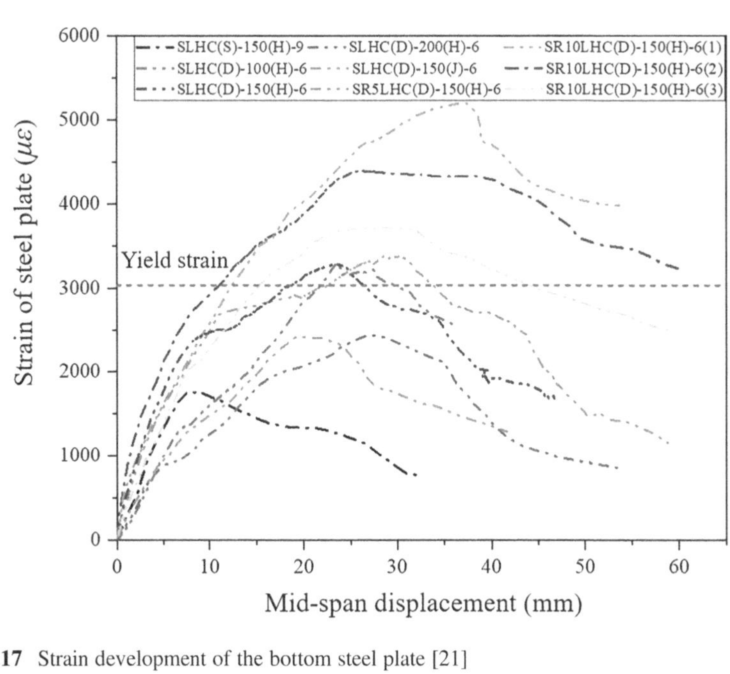

Fig. 5.17 Strain development of the bottom steel plate [21]

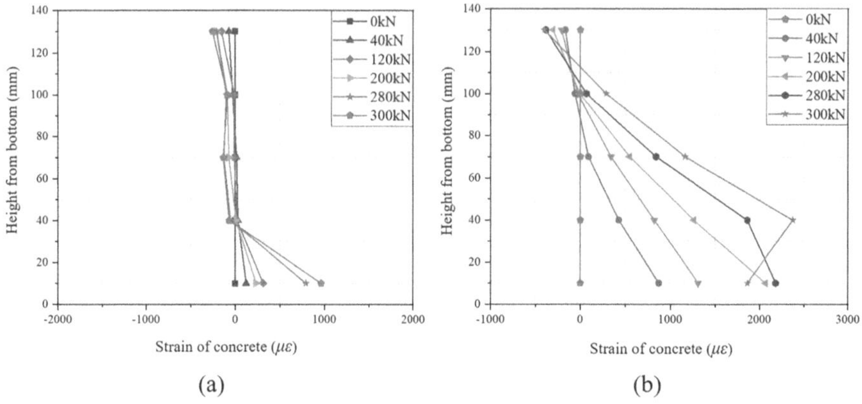

Fig. 5.18 Strain development of concrete (**a**) front surface; (**b**) end surface [21]

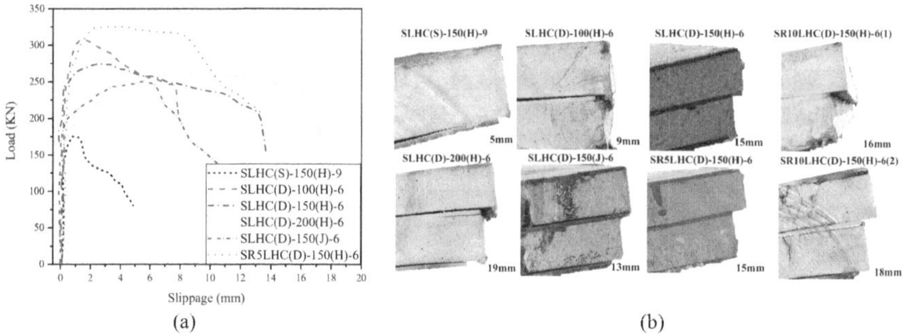

Fig. 5.19 Slip behaviour of the panel end (**a**) load-slip curves; (**b**) slippage magnitude [21]

findings establish a clear correlation between proximity to impact damage and deviation from ideal sectional behavior, demonstrating that mechanical assumptions must be adjusted based on damage localization in post-impact structural assessment.

The flexural tests revealed consistent failure modes characterized by flexural failure with interfacial bond-slip, evidenced by steel-concrete separation and end slippage. As shown in Fig. 5.19a, the load-slip curves demonstrate negligible slip during initial cracking (O-A) but significant development during the platform stage (A-B). Post-test measurements (see Fig. 5.19b) quantified terminal slippage, showing increased slip with wider stud spacing: 9 mm (100 mm spacing), 15 mm (150 mm), and 19 mm (200 mm). This correlation confirms that larger connector spacing reduces interface restraint capacity. Comparative analysis showed greater slip and displacement in double-layer SLHC(D)-150(H)-6 than single-layer SLHC (S)-150(H)-9, while connector type and impact repetition showed minimal influence on slip behavior.

5.6.3 Load-Deflection Curves

The residual load-displacement behavior of the impact-damaged specimens is characterized in Fig. 5.20, with Fig. 5.20a showing the overall response and Fig. 5.20b providing a representative curve. The post-impact behavior evolves through three distinct phases as shown below:

1. The crack development stage (O-A) is marked by internal damage within the composite panel caused by prior impact, resulting in a nonlinear rise in the load-displacement curve during the initial flexural loading phase. As this stage progresses, the panel's stiffness gradually diminishes due to the proliferation of fine cracks near the mid-span, along with the continued widening of pre-existing impact-induced primary cracks. The growing mid-span deflection eventually leads the panel to reach its peak residual resistance, P_y, signifying the culmination of this phase.
2. The slippage stage (A-B) initiates as deflection further increases, characterized by the load-displacement curve entering a plateau phase where the structural behavior is dominated by steel-concrete interface slip. Throughout this stage, progressive stud failure occurs, often accompanied by audible fracturing sounds, while the primary crack continues to expand. The applied load experiences a slight decrease, and displacement accelerates at a noticeably faster rate compared to the earlier crack development phase.
3. The unloading stage (B-C) begins when connector fractures and concrete cracking occur, causing a sharp decline in load-bearing capacity. Structural failure is officially recognized once the load diminishes to 85% of its maximum value (P_y), reaching the ultimate displacement δ_m. At this critical point, a distinct separation forms between the steel plate and concrete interface in the shear span region, resulting in total structural collapse and complete loss of load-bearing functionality in the composite panel.

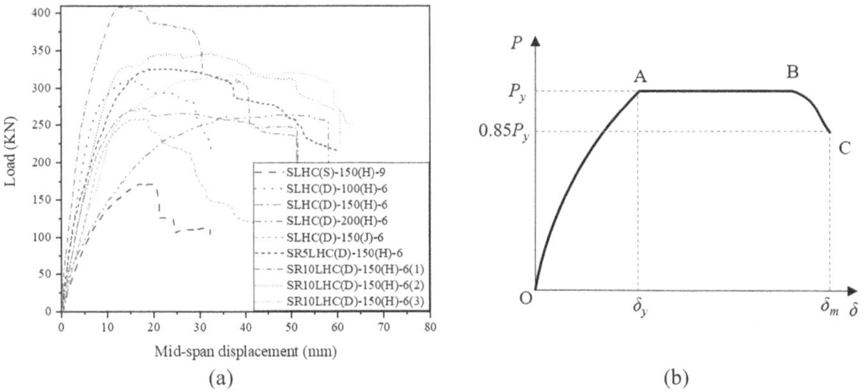

Fig. 5.20 Residual load-displacement curve (**a**) all specimens; (**b**) typical load-displacement curve [21]

5.6.4 Load-Slip Curves

The experimental findings revealed a uniform flexural failure pattern in all test specimens, characterized by pronounced bond-slip behavior manifested through visible steel-concrete interface separation and measurable end-panel displacement. The load-slip analysis, as shown in Fig. 5.21a delineated two primary response phases: an initial loading stage (O-A in Fig. 5.21b) with minimal slip during crack formation, followed by a distinct platform phase (A–B) featuring substantial interfacial slip until final loading conditions. Post-experiment measurements using precision vernier calipers, shown in Fig. 5.21b established a clear relationship between stud spacing and slip magnitude, with specimens SLHC(D)-100(H)-6, SLHC(D)-150(H)-6, and SLHC(D)-200(H)-6 exhibiting progressively larger terminal slips of 9 mm, 15 mm, and 19 mm corresponding to their respective 100 mm, 150 mm, and 200 mm stud spacings. This trend demonstrates the diminishing interfacial constraint effectiveness with increasing stud spacing. Comparative evaluation showed that double-layer core specimens (i.e., SLHC(D)-150(H)-6) developed approximately 25% greater slip and mid-span displacement than single-layer configurations (SLHC(S)-150(H)-9). Interestingly, variations in connector type and impact duration within the tested ranges exhibited negligible statistical influence on slip characteristics.

5.6.5 Post-Impact Static Simulation

The numerical investigation progresses from dynamic impact simulation to post-impact residual capacity analysis through LS-DYNA's advanced restart capabilities. The *STRESS_INITIALIZATION feature plays a pivotal role in this transition by precisely transferring the complete stress-strain history and damage state from the impact phase to the subsequent static analysis. The validation process employs a multi-faceted approach, combining non-destructive evaluation with destructive sectioning of test specimens to verify the model's ability to capture internal damage progression in the concrete core. The excellent correlation between numerical predictions and experimental measurements, establishes the model's credibility for subsequent residual capacity assessment. The following section details the specific

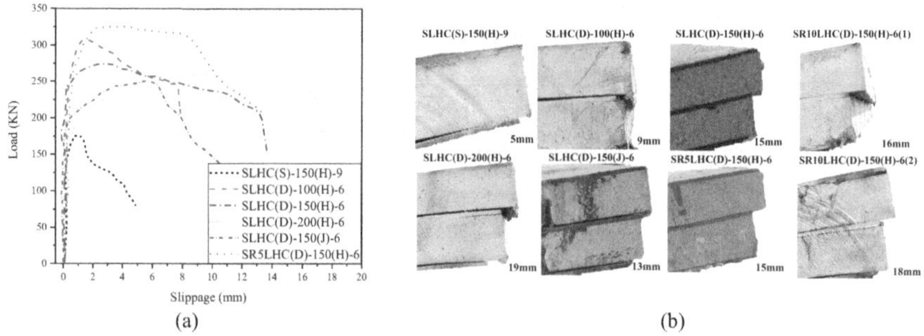

Fig. 5.21 Slip behaviour of the panel end (**a**) load-slip curves; (**b**) slippage magnitude [21]

implementation of these critical modifications, ensuring a seamless transition from impact simulation to residual capacity evaluation while maintaining the fidelity of the damage state accumulated during the impact phase. The transition from impact simulation to residual capacity analysis involves six critical modifications to the finite element model, as illustrated in Fig. 5.14b. First, all transient components including the drop hammer, support rollers, and their associated contact definitions and constraints are removed to streamline the model for static analysis. The loading regime is then converted to quasi-static conditions by implementing a controlled vertical displacement at the midspan's upper surface node through the *BOUNDARY_PRESCRIBED_MOTION_SET command. Revised boundary conditions are established by fully constraining nodal rows at both ends of the specimen using *BOUNDARY_SPC_SET to simulate simply supported conditions. For accurate load monitoring, nodal force output is activated via DATABASE_NODAL_FORCE_GROUP to capture the structural response during bending. The accumulated damage state from the impact phase is preserved through stress initialization (STRESS_INITIALIZATION), while strain rate effects are deactivated in material models (IRATE = 0 for concrete and SRC=CRP = 0 for steel) to reflect static loading conditions. These comprehensive modifications are implemented in a revised K-file, which along with the corresponding d3dump restart file, must be resubmitted to initiate the residual capacity computation.

The numerical simulations effectively characterize SCS panel damage through effective plastic strain contours, revealing critical strain localization zones. Figure 5.22 demonstrates strong correlation between simulated plastic damage patterns

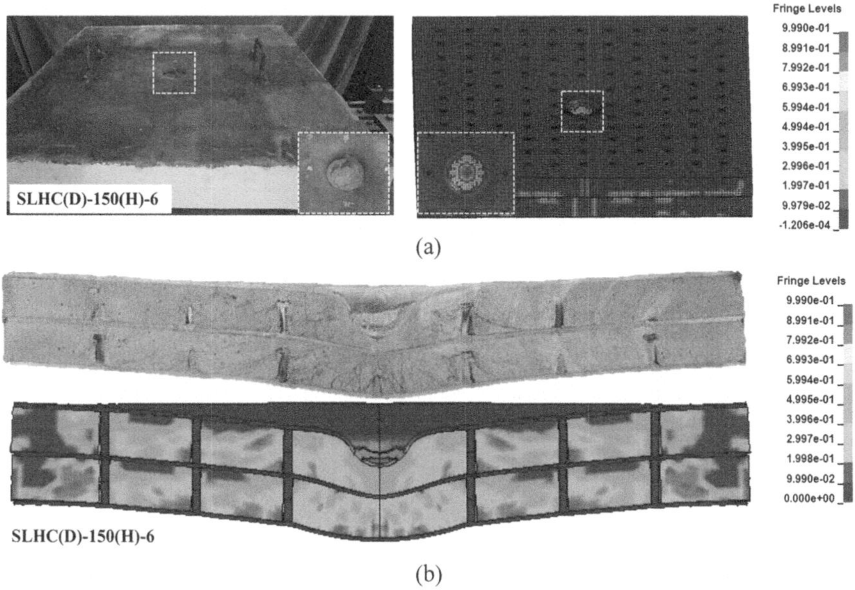

Fig. 5.22 Comparison of impact damage characteristics of composite panels (**a**) steel plate tearing; (**b**) cracks in concrete [21]

and experimental impact test results, accurately capturing both steel plate tearing and internal concrete cracking. Additional validation through comprehensive response comparisons (impact force, support reactions, midspan displacement) in prior research [23] confirms the explicit dynamic analysis' reliability for investigating SCS panel behaviour under repeated impacts. These results collectively verify the model's capability to simulate both damage progression and dynamic response.

The numerical simulation demonstrates good agreement with experimental observations for post-impact static behavior. Figure 5.23 shows the FE model accurately reproduces key failure characteristics, including end slip and midspan concrete damage patterns observed in tests. Load-displacement comparisons in Fig. 5.24 reveal slightly stiffer response in simulations (approximately 10% higher maximum resistance), though maintaining consistent overall trends with experimental data. Quantitative evaluation in Table 5.12 yields mean T/F ratios of 0.94 ± 0.07 for maximum resistance and 1.12 ± 0.14 for maximum displacement, indicating

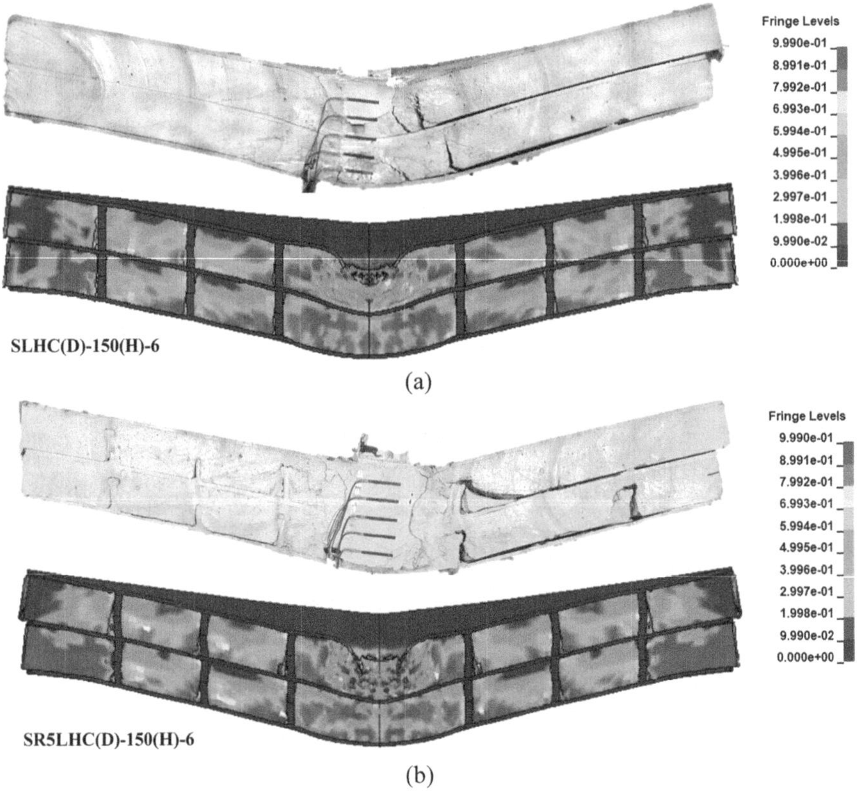

Fig. 5.23 Damage characteristics of composite panels at residual static loading stage (**a**) SLHC(D)-150(H)-6; (**b**) SR5LHC(D)-150(H)-6 [21]

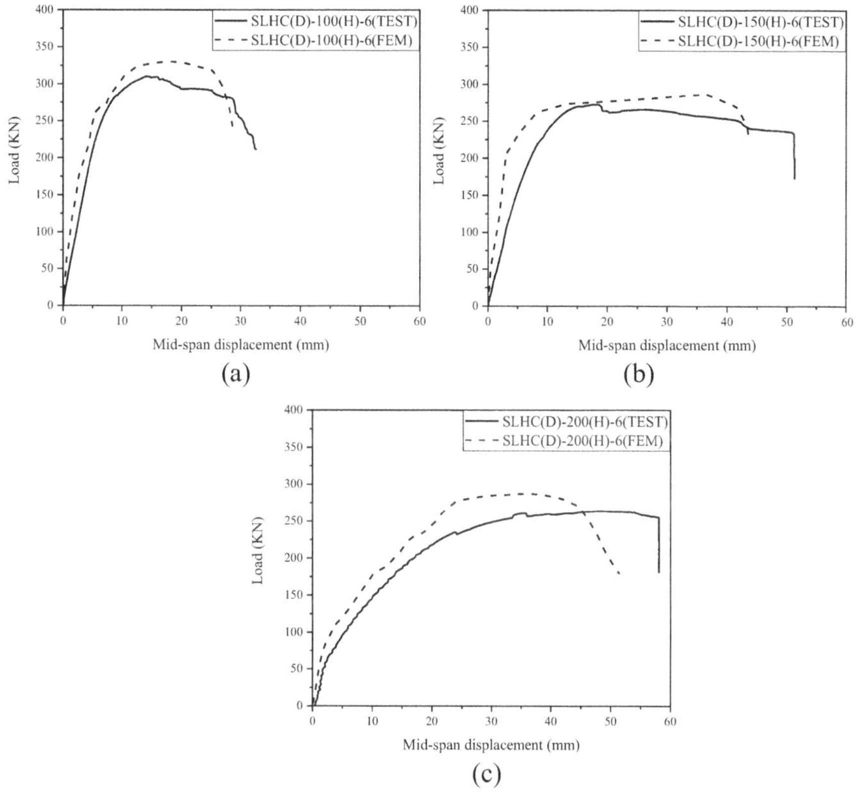

Fig. 5.24 Comparison of load-displacement curves: (**a**) SLHC(D)-100(H)-6; (**b**) SLHC(D)-150(H)-6; (**c**) SLHC(D)-200(H)-6 [21]

Table 5.12 Comparison of residual flexural tests and FE simulations

	Maximum residual resistance P_y (KN)			Ultimate displacement δ_m (mm)		
Specimens	Test	FE	Test/FE	Test	FE	Test/FE
SLHC(S)-150(H)-9	170	192	0.89	21	23	0.91
SLHC(D)-100(H)-6	308	330	0.93	28	26	1.08
SLHC(D)-150(H)-6	269	308	0.87	51	43	1.19
SLHC(D)-200(H)-6	259	285	0.91	58	51	1.14
SLHC(D)-150(J)-6	257	282	0.91	26	29	0.90
SR5LHC(D)-150(H)-6	324	310	1.05	46	41	1.12
SR10LHC(D)-150(H)-6(1)	407	381	1.07	50	45	1.11
SR10LHC(D)-150(H)-6(2)	344	366	0.94	61	48	1.27
SR10LHC(D)-150(H)-6(3)	317	354	0.90	58	43	1.35
Mean	–	–	0.94	–	–	1.12
Standard deviation	–	–	0.07	–	–	0.14

reasonable prediction accuracy. The modest overestimation of residual capacity likely stems from additional damage introduced during specimen preparation (pre-test cutting), which the restart analysis methodology successfully accounts for in simulating post-impact performance.

References

1. S.D. Adhikary, B. Li, K. Fujikake, Dynamic behavior of reinforced concrete beams under varying rates of concentrated loading, Int. J. Impact Eng. 47 (2012) 24–38.
2. X. Zhao, L. Han, Double skin composite construction, Prog. Struct. Eng. Mater. 8(3) (2006) 93–102.
3. A.M. Remennikov, S.Y. Kong, B. Uy, The response of axially restrained non-composite steel–concrete–steel sandwich panels due to large impact loading, Eng. Struct. 49 (2013) 806–818.
4. J.Y.R. Liew, K.M.A. Sohel, Lightweight steel–concrete–steel sandwich system with J-hook connectors, Eng. Struct. 31(5) (2009) 1166–1178.
5. W. Zhang, Z. Huang, Z. Fu, X. Qian, Y. Zhou, L. Sui, Shear resistance behavior of partially composite steel–concrete–steel sandwich beams considering bond-slip effect, Eng. Struct. 210 (2020) 110394.
6. C. Yan, Y. Wang, X. Zhai, Low velocity impact performance of curved steel–concrete–steel sandwich shells with bolt connectors, Thin-Walled Struct. 150 (2020) 106672.
7. Z. Huang, L. Sui, F. Wang, S. Du, Y. Zhou, J. Ye, Dynamic compressive behavior of a novel ultra-lightweight cement composite incorporated with rubber powder, Compos. Struct. 244 (2020) 112300.
8. M.A.E.M. Ali, A.M. Soliman, M.L. Nehdi, Hybrid-fiber reinforced engineered cementitious composite under tensile and impact loading, Mater. Des. 117 (2017) 139–149.
9. J. Xue, M. Shinozuka, Rubberized concrete: A green structural material with enhanced energy-dissipation capability, Constr. Build. Mater. 42 (2013) 196–204.
10. J.Y.R. Liew, K.M.A. Sohel, C.G. Koh, Impact tests on steel–concrete–steel sandwich beams with lightweight concrete core, Eng. Struct. 31(9) (2009) 2045–2059.
11. J.W. Jung, Y.C. Yoon, H.W. Jang, J.W. Hong, Investigation on the resistance of steel-plate concrete walls under high-velocity impact, J. Constr. Steel Res. 162 (2019) 105732.
12. Z. Huang, X. Zhao, W. Zhang, Z. Fu, Y. Zhou, L. Sui, Load transfer mechanism of novel double-layer steel–LHDCC–steel sandwich panels under punching loads, Eng. Struct. 226 (2021) 111427.
13. W. Zhang, Z. Huang, J. Ye, Youtam, Numerical-based analytical model of double-layer steel-LHDCC sandwich composites under punching loads, Compos. Struct. 286 (2022) 115271.
14. W. Zhang, Z. Huang, R. Li, X. Zhao, J. Ye, Impact response of double-layer steel–RULCC–steel sandwich panels: Experimental, numerical and analytical approaches, J. Struct. Eng. 148 (10) (2022) 04022165.
15. Q. Guo, W. Zhao, Displacement response analysis of steel–concrete composite panels subjected to impact loadings, Int. J. Impact Eng. 131 (2019) 272–281.
16. L. Wang, W. Zhao, G.-T. Yang, Q. Guo, TDOF model for evaluating the global and local impact response of steel-plate composite panels, Thin-Walled Struct. 164 (2021) 107879.
17. K. Fujikake, B. Li, S. Soeun, Impact response of reinforced concrete beam and its analytical evaluation, J. Struct. Eng. 135(8) (2009) 938–950.
18. S.D. Adhikary, B. Li, K. Fujikake, Residual resistance of impact-damaged reinforced concrete beams, Mag. Concr. Res. 67(7) (2015) 364–378.
19. W. Fan, B. Liu, G.R. Consolazio, Residual capacity of axially loaded circular RC columns after lateral low-velocity impact, J. Struct. Eng. 145(6) (2019) 04019039.

20. G. Dok, N. Caglar, A. Ilki, C.Yilmaz, Effect of impact loading on residual flexural capacity of high-strength reinforced concrete beams, Struct. 27 (2020) 2466–2480.
21. Z. Huang, X. Zhao, Y. Guo, X. Liu, Residual flexural performance of double-layer steel–RLHDC composite panels after impact, Build. 13(12) (2023) 80–92.
22. W. Zhao, Q. Guo, X. Dou, Y. Zhou, Y. Ye, Impact response of steel–concrete composite panels: Experiments and FE analyses, Steel Compos. Struct. 26(3) (2018) 255–263.
23. Z. Huang, W. Zhang, R. Li, Impact resistance of double-layer steel–RULHDCC sandwich panels subjected to repeated impact loads, J. Build. Eng. (2023) 80–92 (in Chinese).
24. F. Liu, G. Chen, L. Li, Y. Guo, Study of impact performance of rubber reinforced concrete, Constr. Build. Mater. 36 (2012) 604–616.
25. F. Wang, Dynamic performance of ultra-lightweight high ductility cement composite incorporated with rubber powder, Ph.D. thesis, Shenzhen University, 2020 (in Chinese).
26. S.P. Timoshenko, S. Woinowsky-Krieger, Theory of plates and shells, McGraw-Hill College, 1959.
27. JEAG 4618, Technical guidelines for aseismic design of steel plate reinforced concrete structures – buildings and structures, Japan Electric Association Nuclear Standards Committee, Tokyo, 2005.
28. J.C. Bruhl, A.H. Varma, J.M. Kim, Static resistance function for steel-plate composite (SC) walls subject to impactive loading, Nucl. Eng. Des. 295 (2015) 843–859.
29. GB50081-2002, Standard for test methods of mechanical properties of ordinary concrete, China Building Industry Press, Beijing, 2022 (in Chinese).
30. Y.D. Murray, A. Abu-Odeh, R. Bligh, Evaluation of LS-DYNA concrete material model 159 (Report No. FHWA-HRT-05-063), Federal Highway Administration, McLean, VA, 2007.

Chapter 6
Flexural and Repeated Impact Performance of Stainless Steel-Lightweight Concrete Sandwich Composites for Marine Structures

6.1 Introduction

Marine structures are frequently subjected to repeated low-velocity, high-mass impacts, such as collisions from ice floes, helicopter landings, and supply vessel docking. These scenarios pose significant challenges to structural integrity, necessitating materials and designs with superior impact resistance. Steel-concrete-steel (SCS) composite structures, consisting of external steel plates and an internal concrete core, have gained widespread use in offshore platforms, immersed tunnels, and protective structures due to their high strength, durability, and energy absorption capacity. However, conventional SCS composites face durability issues in harsh marine environments, where corrosion resistance is critical.

To address this, stainless steel has been proposed as an alternative to carbon steel due to its excellent corrosion resistance, mechanical properties, and fire resistance. Previous research on stainless steel-concrete composites has primarily focused on static behavior, with limited studies on their dynamic performance under impact loading—a crucial consideration for marine protective structures. Additionally, the choice of concrete core material significantly influences impact resistance. Advanced cementitious materials, such as engineered cementitious composites (ECC), ultra-high-performance concrete (UHPC), and lightweight concrete (LWC), offer tailored properties for different applications. Among these, lightweight high-ductility cement composites (LHDCC) and rubberized LHDCC (RLHDCC) have been developed to enhance energy absorption while reducing structural weight, making them ideal for marine applications.

Despite extensive research on SCS composites under single-impact loading, repeated impacts—which induce cumulative damage, residual stresses, and progressive failure—remain understudied. Existing studies have demonstrated that SCS structures with J-hook connectors, axial restraints, or optimized concrete cores exhibit improved impact resistance. However, the influence of repeated impacts on

Z. Huang, Y. Zhou, *Innovative Lightweight High-performance Steel-concrete Sandwich Composite for Marine Structures*,
https://doi.org/10.1007/978-981-95-4681-7_6

structural performance, particularly for stainless steel-LHDCC composites, requires further investigation. This chapter explores the flexural and repeated impact performance of a novel stainless steel-LHDCC composite beam designed for marine applications. The research integrates experimental and numerical approaches to:

1. Investigate the effects of key design parameters (connector spacing, type, concrete thickness, material type, and layering) on impact resistance.
2. Calibrate constitutive models for LHDCC using uniaxial, triaxial, and dynamic tests to ensure accurate finite element (FE) simulations.
3. Develop a validated FE model using LS-DYNA to simulate repeated impacts and assess cumulative damage.
4. Propose a modified single-degree-of-freedom (SDOF) model incorporating strain rate effects and composite action to predict dynamic responses.

By addressing these aspects, this work advances the understanding of stainless steel-LHDCC composites under repeated impacts, providing design insights for safer and more durable marine structures.

6.2 Flexural and Repeated Impact Test Program

The experimental investigation examined the flexural and impact performance of stainless steel-lightweight concrete composite beams through static and repeated impact tests. A total of eleven specimens were designed, including one for static loading and ten for impact testing. All specimens measured 1200 mm in length and 300 mm in width, with a clear span of 875 mm between supports. The specimens incorporated two types of shear connectors (overlapped headed studs and short headed studs) and two concrete layer configurations (single and double layers).

Material characterization included tensile testing of stainless steel plates and studs using a 3000 kN MTS machine following GB/T 228.1-2010, revealing yield strengths of 401.8 MPa and 489.7 MPa, respectively. Four concrete types were evaluated: lightweight high-ductility cement composite (LHDCC), rubberized LHDCC (RLHDC), ultra-high performance concrete (UHPC), and normal concrete (NC). LHDCC exhibited a compressive strength of approximately 50 MPa at half the density of NC, along with high tensile strength (4 MPa) and ductility (3.6% strain) with only 0.7% polyethylene (PE) fibers.

Static three-point bending tests were conducted using a 2000 kN machine under displacement control (0.5 mm/min), with instrumentation including load cells, LVDTs, and strain gauges. Impact tests employed a drop-hammer machine (maximum energy: 50,000 J) with a 512 kg hammer dropped from varied height. Each specimen was subjected to two consecutive impacts, with impact and reaction forces recorded by load cells, while LVDTs and strain gauges monitored displacement and strain evolution. High-speed cameras (1000 fps) captured crack propagation, and data were acquired at 500 kHz. The test setup ensured clamped supports to prevent rebound, enabling comprehensive evaluation of structural behavior under repeated impacts. The design details of the specimens are shown in Fig. 6.1 while Tables 6.1 and 6.2 lists the geometric parameters of each specimen.

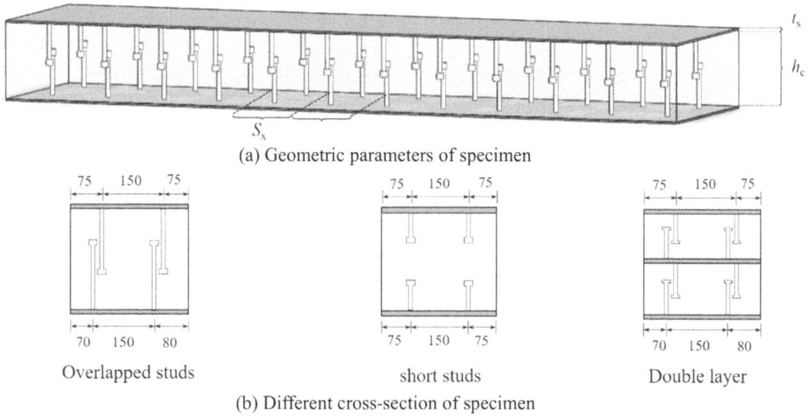

Fig. 6.1 Design details of the specimen. (**a**) Geometric parameters of specimen. (**b**) Different cross-section of specimen. (Reprinted from Zhao et al. [23], copyright 2025, with permission from Elsevier

Table 6.1 Parameters of the impact test specimens

Specimen	Loading type	h_c (mm)	S_x (mm)	t_s (mm)	Core material	Connector	λ	ρ (%)	η
SLHC-125-100(S)	Static	125	100	3	LHDCC	Overlapped	3.5	4.58	0.54
SLHC-125-100	Impact	125	100	3	LHDCC	Overlapped	3.5	4.58	0.54
SLHC-125-150	Impact	125	150	3	LHDCC	Overlapped	3.5	4.58	0.41
SLHC-125-200	Impact	125	200	3	LHDCC	Overlapped	3.5	4.58	0.34
SLHC-80-100	Impact	80	100	3	LHDCC	Overlapped	5.5	6.98	0.54
SLHC-160-100	Impact	160	100	3	LHDCC	Overlapped	2.7	3.75	0.54
SLHC-125-100(U)	Impact	125	100	3	LHDCC	Short	3.5	4.58	0.54
SRLHC-125-100	Impact	125	100	3	RLHDC	Overlapped	3.5	4.58	0.46
SUHC-125-100	Impact	125	100	3	UHPC	Overlapped	3.5	4.58	1.00
SNC-125-100	Impact	125	100	3	NC	Overlapped	3.5	4.58	0.91
SLHC-160-100(D)	Impact	80*2	100	2	LHDCC	Overlapped	2.7	3.75	0.54

h_c denotes the thickness of the concrete core, S_x represents the center-to-center spacing of shear connectors; and t_s refers to the thickness of the stainless steel plate. The parameter λ corresponds to the shear span-to-depth ratio, where the shear span is defined as half the clear span—i.e., the distance from the point of loading to the support—and the depth refers to the overall beam depth. The steel contribution ratio ρ is expressed as $\rho = at_s/(at_s + h_c)$, with a representing the number of steel layers. The composite degree η is defined as $\eta = nP_s/f_y t_s B$, where n is the number of studs within the shear span, P_s denotes the shear resistance per stud, and B is the width of the beam

Table 6.2 Parameters of the repeated impact test specimens

Specimen	h_c (mm)	S_x (mm)	t_s (mm)	Core material	Connector	λ	ρ (%)	η
L-H125-S100	125	100	3	LHDCC	Overlapped	3.5	4.58	0.54
L-H125-S150	125	150	3	LHDCC	Overlapped	3.5	4.58	0.41
L-H125-S200	125	200	3	LHDCC	Overlapped	3.5	4.58	0.34
L-H80-S100	80	100	3	LHDCC	Overlapped	5.5	6.98	0.54
L-H160-S100	160	100	3	LHDCC	Overlapped	2.7	3.75	0.54
L-H125-S100(U)	125	100	3	LHDCC	Short	3.5	4.58	0.54
R-H125-S100	125	100	3	RLHDC	Overlapped	3.5	4.58	0.46
U-H125-S100	125	100	3	UHPC	Overlapped	3.5	4.58	1.00
N-H125-S100	125	100	3	NC	Overlapped	3.5	4.58	0.91
L-H125-S100(D)	80*2	100	2	LHDCC	Overlapped	2.7	3.75	0.54

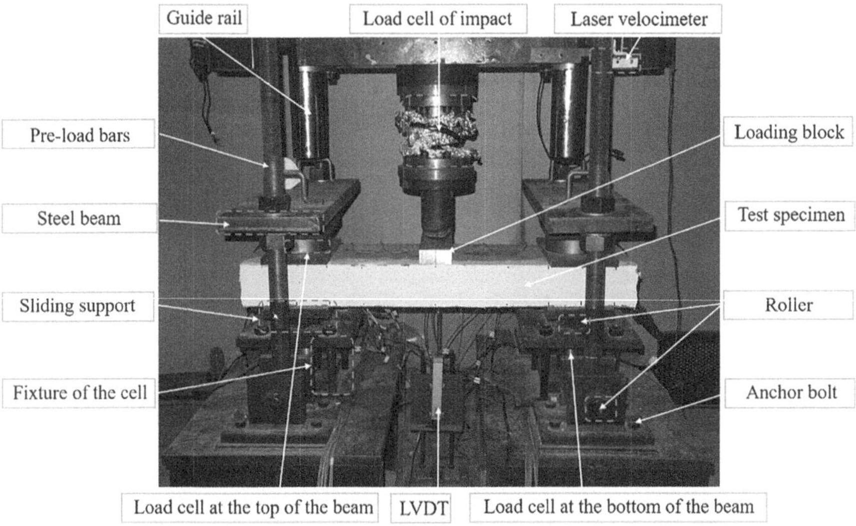

Fig. 6.2 Test set-up for multiple impact tests. (Reprinted from Zhao et al. [23], copyright 2025, with permission from Elsevier)

The test setup for the multiple impact tests is shown in Fig. 6.2. The impact forces are recorded by the load cell at the hammer head, while the reaction forces are recorded by the load cell at the bottom of the beam. A high-speed camera is used to record the whole process of impact. The LVDT and strain gauges measure the displacement of the beam and the strain of the steel plate.

6.3 Dynamic Responses of Stainless Steel-Lightweight Composites Under Impact

6.3.1 Failure Modes

The experimental investigation revealed three distinct failure modes in the stainless steel-concrete composite beams: (1) flexural failure with bond-slip, (2) pure flexural failure, and (3) shear failure. The failure mode was primarily governed by the concrete core material properties and shear connector configuration.

For specimens with LHDCC cores (L-H125-S100, L-H125-S150, L-H125-S200, L-H80-S100, L-H160-S100, L-H160-S100(D), L-H125-S100(U), R-H125-S100), flexural failure with bond-slip was observed, as shown in Fig. 6.3. After initial impact, flexural cracks concentrated in the mid-span region while interfacial slippage occurred at beam ends. Subsequent impacts widened existing cracks without forming new ones, with bond-slip magnitude increasing significantly from 5.3 mm to 22.3 mm as stud spacing increased from 100 mm to 200 mm. The double-layer configuration (L-H160-S100(D)) demonstrated superior performance, exhibiting only 2.8 mm and 5.7 mm slippage after first and second impacts respectively, due to enhanced composite action from the additional steel plate.

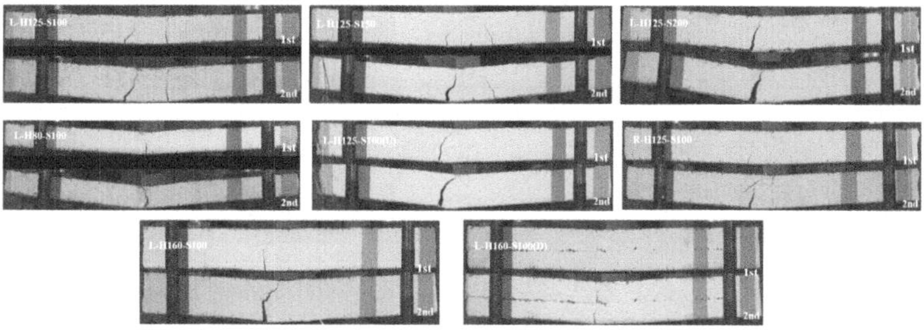

(a) Critical crack in specimen

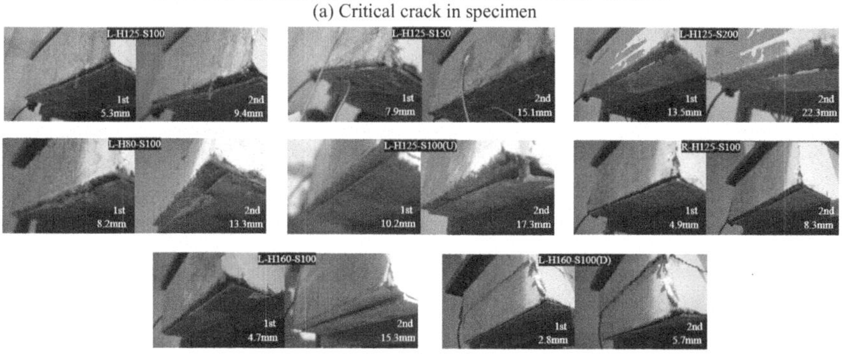

(b) Bond-slip mechanism in specimen

Fig. 6.3 Flexural failure with bond-slip. (**a**) Critical crack in specimen. (**b**) Bond-slip mechanism in specimen. (Reprinted from Zhao et al. [23], copyright 2025, with permission from Elsevier)

Shear failure dominated in normal concrete specimens (N-H125-S100, SNC-125-100), characterized by diagonal cracking propagating from impact zone to supports (see Fig. 6.4). The low shear strength of normal concrete caused complete section separation after second impact, accompanied by top steel plate buckling. Ultra-high performance concrete (UHPC) specimens (U-H125-S100, SUHC-125-100) exhibited pure flexural failure (see Fig. 6.5) with $\eta = 1.0$ composite action. A single critical flexural crack developed at mid-span without interfacial slippage, demonstrating UHPC's superior compressive strength and composite efficiency. The high shear resistance of UHPC prevented diagonal cracking, maintaining structural integrity through multiple impacts.

The failure mechanisms highlight critical material influences: LHDCC's fiber reinforcement (0.7% PE fibers) enhanced tensile strength and crack resistance but showed limited composite action ($\eta < 1.0$); normal concrete's low shear strength precipitated premature shear failure; while UHPC's balanced properties enabled full composite behavior. Connector configuration significantly affected performance, with overlapped studs outperformed short studs (10.2 mm vs 5.3 mm slippage at 100 mm spacing), and reduced stud spacing consistently improved interfacial bond performance.

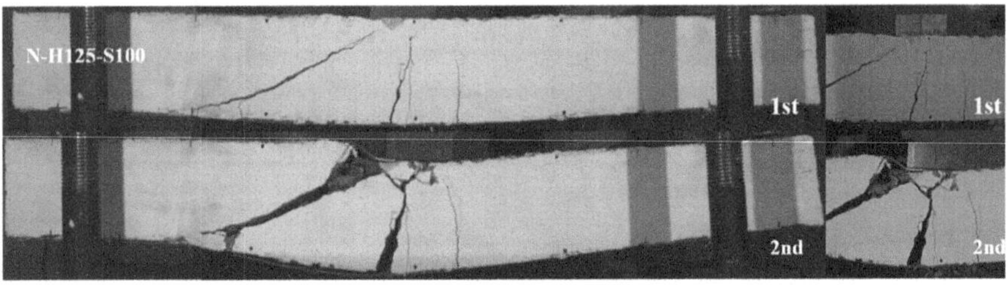

Fig. 6.4 Shear failure in normal concrete specimen. (Reprinted from Zhao et al. [23], copyright 2025, with permission from Elsevier)

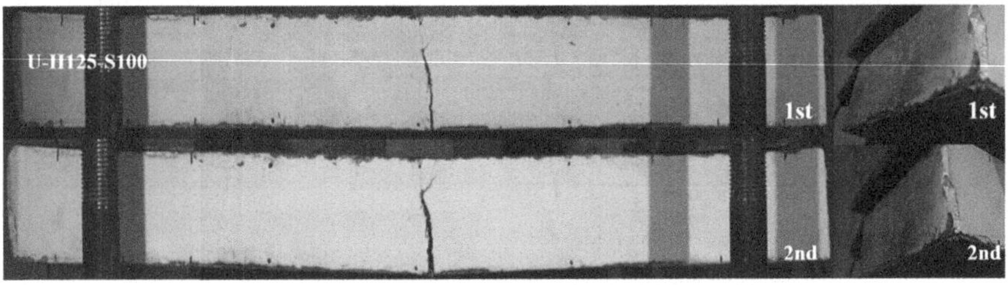

Fig. 6.5 Flexural failure in ultra-high performance concrete specimen. (Reprinted from Zhao et al. [23], copyright 2025, with permission from Elsevier)

6.3.2 Static Response

The static load-displacement response of the composite beam exhibits four distinct phases, as illustrated in Fig. 6.6. Initially, the curve demonstrates linear elastic behavior (oa) until flexural cracking initiates in the concrete core, causing stiffness reduction (ab). Subsequently, a prolonged plateau stage (bc) emerges due to progressive interfacial slip between the bottom steel plate and concrete. The final phase (cd) shows rapid load capacity deterioration resulting from critical crack formation in the concrete and shear failure of stud connectors. Strain measurements (see Fig. 6.7) reveal that the bottom steel plate reaches a maximum strain of 2380 με, significantly below the nominal yield strain of 4100 με, attributed to slip-induced strain limitation—a phenomenon consistent with previous findings [1]. The stud connectors demonstrate minimal initial strain until mid-span displacement exceeds 6 mm, after which rapid strain development occurs. This behavioral transition corresponds with bond-slip activation, where relative steel-concrete movement induces significant shear and bending deformation in the studs, particularly those connected to the bottom plate. The observed strain differential between top and bottom studs further confirms the predominant role of bottom plate slippage in governing the structural response.

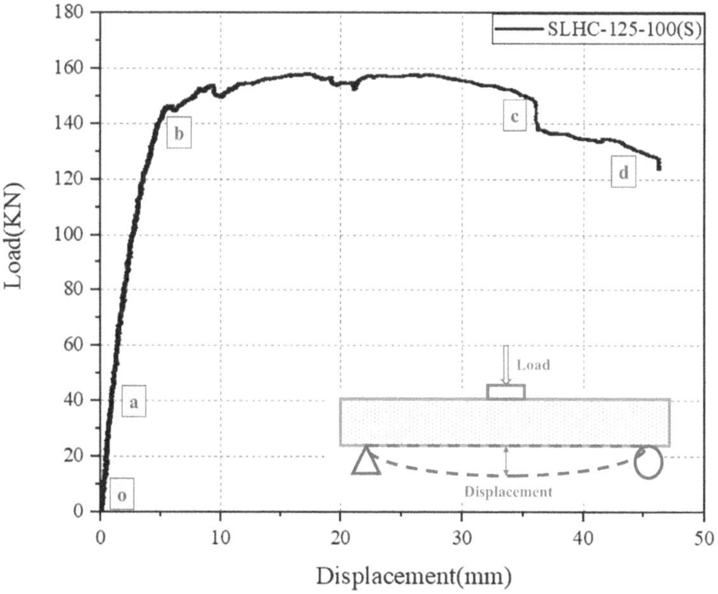

Fig. 6.6 Load-displacement curve of SLHC-125-100(S) under static test. (Reprinted from Huang et al. [15], copyright 2025, with permission from American Society of Civil Engineers)

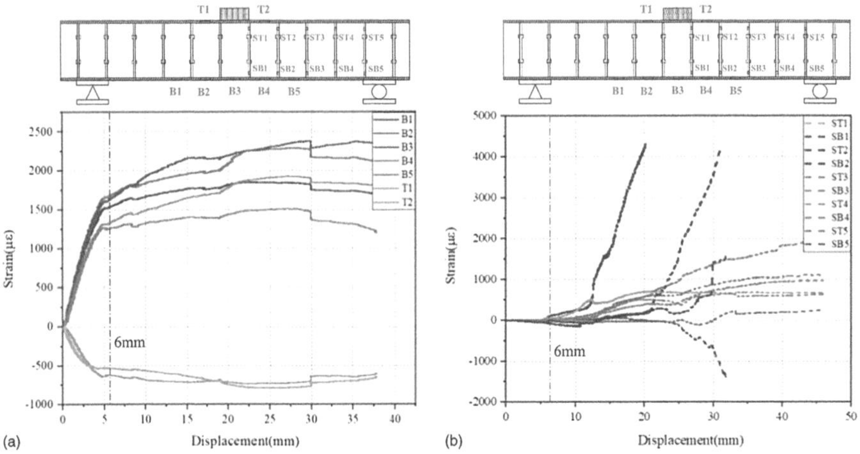

Fig. 6.7 Strain curves of SLHC-125-100(S) under static test: (**a**) steel plates; and (**b**) studs. (Reprinted from Huang et al. [15], copyright 2025, with permission from American Society of Civil Engineers)

6.3.3 Impact Force History

The impact response of the composite beams exhibits three distinct stages: loading, rebounding, and detaching, as illustrated in Figs. 6.8 and 6.9. During the initial loading stage (0–1 ms), the impact force rapidly peaks ($P_{\max}$) while being resisted primarily by inertial forces as given in Eq. (6.1), since the stress wave has not yet propagated to the supports. A temporal lag of 0.21–0.28 ms is observed between the impact and reaction forces, with the inertia force curve initially coinciding with the impact force curve. As the loading stage progresses, momentum transfer induces beam deformation, causing high-frequency oscillations in all force components (impact, reaction, and inertia) due to stress wave propagation. These vibrations gradually dampen, and the impact force stabilizes into a plateau, with the average force (P_{ave}) calculated via temporal integration, as given in Eq. (6.2). The mid-span displacement increases continuously, peaking at the end of this stage.

$$I(t) = -\int_0^l \overline{m}\ddot{u}(x,t)dx \approx \overline{m}X[a_1(t) + a_2(t) * 2] \tag{6.1}$$

$$P_{\text{ave}} = \frac{\int_{t_1}^{t_2} P(t)dt}{t_2 - t_1} \tag{6.2}$$

In the rebounding stage, the hammer begins to retract while maintaining contact, leading to a decline in impact and reaction forces. The inertia force diminishes to near-zero, and the reaction force closely mirrors the impact force. Concurrently, the beam undergoes elastic recovery, reducing mid-span displacement. Finally, during

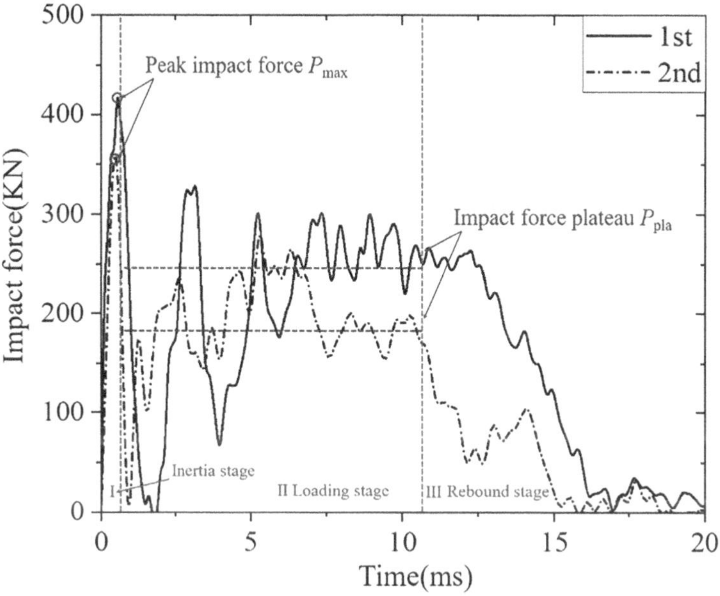

Fig. 6.8 Impact force history of L-H125-S100. (Reprinted from Zhao et al. [23], copyright 2025, with permission from Elsevier)

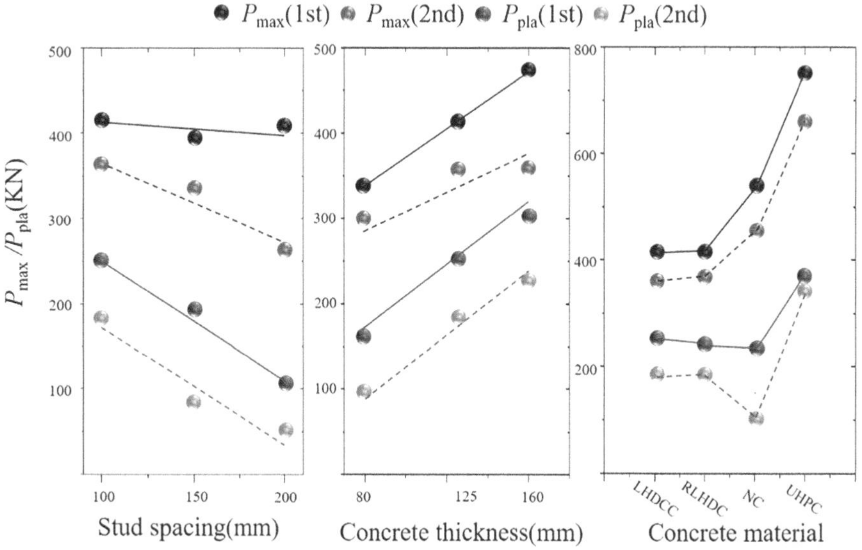

Fig. 6.9 Effect of different parameters on impact force,. Reprinted from Zhao et al. [23], copyright 2025, with permission from Elsevier)

the detaching stage, the hammer loses contact, causing both impact and reaction forces to drop to zero, with the remaining residual displacement reflecting permanent deformation.

6.3.4 Displacement and Velocity Response Analysis

The dynamic response characteristics of composite beams under multiple impacts are illustrated through mid-span displacement and velocity histories, as shown in Fig. 6.10. Upon initial contact, the impact point velocity surges instantaneously while displacement remains negligible. As energy transfers to the beam, displacement increases while velocity decays, exhibiting oscillatory behavior from hammer-beam interaction. Maximum displacement (D_{max}) coincides with zero velocity, after which negative velocity values indicate elastic recovery before stabilizing at residual displacement (D_{res}).

Figure 6.11 illustrates the parametric studies, which reveal significant displacement variations:

Stud spacing effects: Increasing spacing from 100 mm to 200 mm elevates D_{max} by 67% and D_{res} by 91% in first impacts, with greater amplification (102% D_{res} increase) in subsequent impacts due to cumulative damage.

Concrete thickness influence: Thicker sections (80 mm–160 mm) reduce D_{max} by 54% and D_{res} by 60% through enhanced stiffness.

Damage accumulation: Most specimens show increased D_{max} in second impacts (average $\Delta_{max} = 15\%$), except double-layer (L-H160-S100(D)) and UHPC-core beams (U-H125-S100) which maintain stable displacements through either intermediate steel plate reinforcement ($\eta = 0.92$) or full composite action ($\eta = 1.0$).

Material limitations: Normal concrete specimens (N-H125-S100) exhibit the highest Δ_{max} (38%) from brittle tensile failure and concrete spalling.

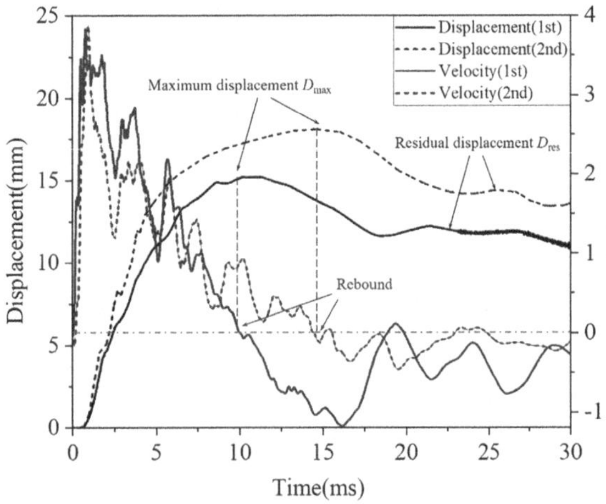

Fig. 6.10 Mid-span displacement and velocity–time history of L-H125-S100. (Reprinted from Zhao et al. [23], copyright 2025, with permission from Elsevier)

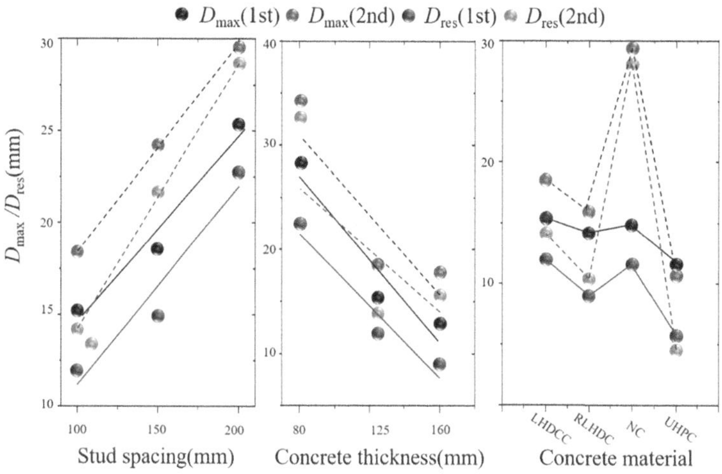

Fig. 6.11 Effect of main parameters on displacement. (Reprinted from Zhao et al. [23], copyright 2025, with permission from Elsevier)

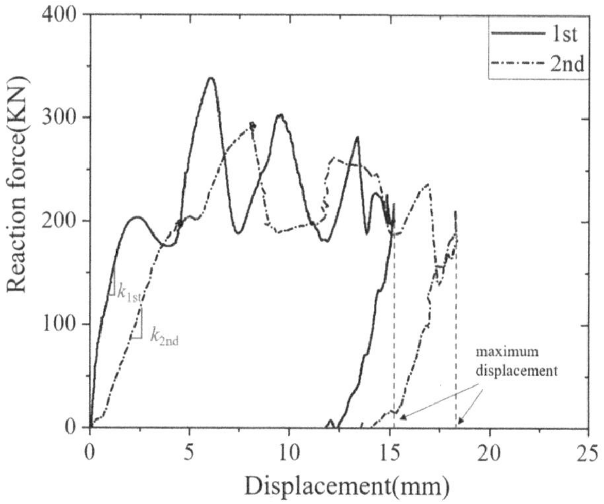

Fig. 6.12 Reaction force-displacement curves of L-H125-S100. (Reprinted from Zhao et al. [23], copyright 2025, with permission from Elsevier)

6.3.5 Energy Absorption Characteristics

Reaction force-displacement curves in Fig. 6.12 demonstrate three-phase behavior: (1) linear loading with dynamic stiffness k; (2) oscillatory plateau at maximum displacement; (3) elastic unloading with slope $\approx k$. The k, E_w, E_s of the composite beam under two impacts are shown in Table 6.2. With the increase of stud spacing and the decrease of concrete thickness, the dynamic global stiffness k of the beam under two impacts gradually decreases. k is more sensitive to concrete thickness. Energy analysis reveals: 48–94% of impact energy (E_k) converts to deformation energy (E_w), calculated via Eq. (6.3). Lower E_w/E_k ratios (48–60%) in L-H125-S200, L-H80-S100, and N-H125-S100 correlate with localized concrete crushing. Rubberized specimens (R-H125-S100) achieve maximum specific energy absorption ($E_s = 1.8$ J/kg) through rubber particle damping, while UHPC cores show minimal E_s (0.9 J/kg) due to higher mass despite superior stiffness.

$$E_{\mathrm{w}} = \int_0^{D_{\max}} R(\mathrm{w})d\mathrm{w} \tag{6.3}$$

6.3.6 Strain Evolution and Failure Mechanisms

Strain histories (see Fig. 6.13) differentiate failure modes. In flexural-slip specimens, maximum strains remain below the yield threshold of 4840 με due to partial composite action ($\eta < 1.0$), with second-impact strains reduced by 20–35% as a consequence from stud damage. Shear-critical beams, by contrast, exhibit limited steel strain development (<2200 με), governed by the low tensile strength of the material and the disruption of shear transfer caused by concrete cracking. UHPC specimens demonstrate a markedly different response, reaching yield strain during the first impacts and showing approximately 15% strain growth under second

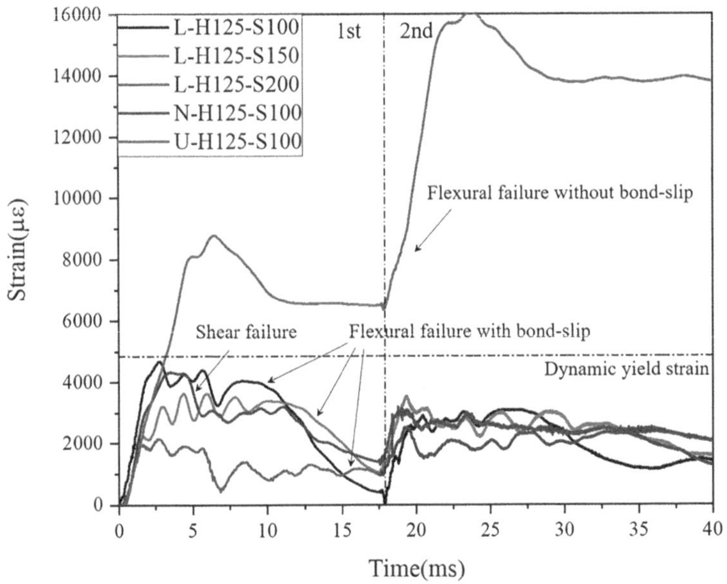

Fig. 6.13 Strain curves for different failure modes during impact. (Reprinted from Zhao et al. [23], copyright 2025, with permission from Elsevier)

impacts; owing to full composite action ($\eta = 1.0$), maintaining their structural integrity. These measurements quantify how connector design, section properties, and material selection collectively govern energy dissipation and damage progression under repeated impacts. The data provides critical benchmarks for optimizing impact-resistant marine structures.

6.3.7 Parametric Analysis of Impact Performance

6.3.7.1 Effect of Stud Spacing

The composite action of sandwich beams decreases with increasing stud spacing, leading to reduced structural stiffness. While specimens with varying stud spacings (100 mm, 150 mm, 200 mm) exhibit similar first peak impact forces (<5% difference), the post-peak response diverges significantly (see Fig. 6.14a). Beams with smaller spacings maintain higher impact forces during the loading stage (7–15 ms) due to enhanced composite action. The 200 mm-spaced specimen (SLHC-125-200) shows severe performance degradation, with stud fracture and steel-concrete interface separation causing 90% greater residual deformation than the 100 mm-spaced counterpart. This demonstrates the critical role of connector density in maintaining structural integrity under impact.

6.3.7.2 Effect of Stud Type

Comparative analysis of overlapped (SLHC-125-100) and short-headed studs (SLHC-125-100(U)) reveals distinct mechanical behaviors (see Fig. 6.14b). The overlapped configuration provides 11% higher peak impact force and 20% lower displacement by resisting both longitudinal shear (preventing bond-slip) and transverse shear (enhancing load resistance). Conversely, short studs promote flexural deformation, yielding greater energy absorption but reduced load capacity. This trade-off between strength and energy dissipation highlights the importance of connector design in impact-resistant applications.

6.3.7.3 Effect of Concrete Core Thickness

Increased core thickness (80–160 mm) improves structural performance through modified shear span-to-depth ratios (see Fig. 6.14c). Specimens with 125 mm and 160 mm cores exhibit 23–40% higher peak forces and 46–60% lower residual deformations compared to the 80 mm core. The thicker sections demonstrate greater stiffness, distributing impact energy more efficiently while limiting permanent deformation—a crucial consideration for structures requiring repeated impact resistance.

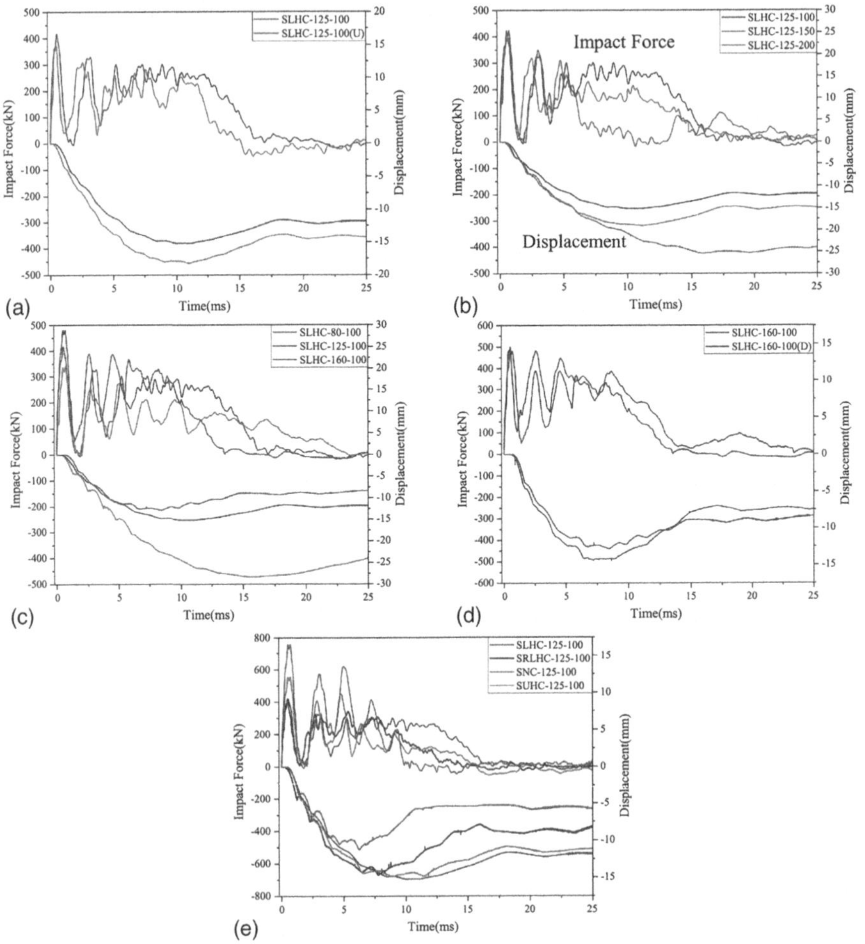

Fig. 6.14 Influences of design parameters: (**a**) Stud spacing, (**b**) Stud type, (**c**) Concrete thickness, (**d**) Layer number, (**e**) Type of concrete core. (Reprinted from Huang et al. [15], copyright 2025, with permission from American Society of Civil Engineers)

6.3.7.4 Effect of Layer Configuration

Despite similar peak responses (<4% difference), double-layer specimens (SLHC-160-100(D)) outperform single-layer designs in damage tolerance (see Fig. 6.14d). The intermediate steel plate in double-layer beams confines crack propagation, maintaining structural integrity where single-layer specimens develop full-depth flexural cracks. This illustrates how layered construction can enhance durability under repeated impacts without compromising initial stiffness.

6.3.7.5 Effect of Core Material

Material properties govern the stiffness hierarchy: UHPC > NC > LHDCC > RLHDC (see Fig. 6.14e). While UHPC specimens (SUHC-125-100) demonstrate superior strength and minimal displacement, rubberized LHDCC (SRLHC-125-100) shows unique advantages, as its 60% lower residual deformation than conventional LHDCC stems from rubber particles' energy absorption. Normal concrete (SNC-125-100) exhibits premature shear cracking due to lack of fiber reinforcement, underscoring the importance of material selection in balancing strength and deformation capacity.

These parametric studies collectively demonstrate that impact performance can be optimized through strategic combinations of connector design, section geometry, and advanced materials—with particular emphasis on maintaining composite action and controlling damage progression.

6.3.8 Repeated Impact Performance

Table 6.3 presents the dynamic stiffness (k), energy absorption ratio (E_w/E_k), and specific energy absorption (E_s) values for the composite beams subjected to two successive impacts. The results demonstrate an inverse relationship between stud spacing and dynamic stiffness, with k values decreasing as both stud spacing increases and concrete thickness reduces. Notably, the stiffness parameter exhibits greater sensitivity to variations in concrete thickness compared to stud spacing. Energy absorption analysis reveals that the beams dissipated between 48% and 94% of the input kinetic energy through structural deformation. Three specimens (L-H125-S200, L-H80-S100, and N-H125-S100) showed reduced energy absorption efficiency, attributable to localized failure modes where significant energy dissipation occurred through concrete crushing. The rubber-modified specimen (R-H125-S100) achieved the highest specific energy absorption, with the 10% rubber powder addition significantly improving energy dissipation capacity. Conversely, while specimen U-H125-S100 demonstrated optimal impact resistance (manifested through minimal maximum deflection and peak stiffness), its specific energy absorption was compromised by the substantial self-weight of the configuration.

6.4 Numerical Investigation

This study introduces an innovative multi-drop hammer modeling technique as an alternative to conventional restart methods for simulating repeated impacts. Unlike traditional approaches that require iterative modifications of input files between impacts [2], the proposed method simultaneously defines two independent hammers

Table 6.3 Repeated impact test results

Specimen	Impact Parameters i	M	v	E_k	P_{max}	P_{ave}	Δ_{pla} (%)	D_{max}	D_{res}	Δ_{max} (%)	k	Δ_k	E_w	$\frac{E_w}{E_k}$ (%)	E_s	Failure mode
L-H125-S100	1	512	3.85	3795	415.1	236.2	22	15.2	11.9	21	85.5	18	3497	92	44.2	I
	2	512	3.84	3775	356.4	184.6		18.4	14.2		70.3		3137	83	39.7	I
L-H125-S150	1	512	3.86	3814	393.4	212.3	23	18.7	14.9	29	80.1	12	3253	85	41.1	I
	2	512	3.86	3814	343.5	163.1		24.2	21.7		70.4		3084	81	39.0	I
L-H125-S200	1	512	3.86	3814	407.8	115.8	29	25.4	22.7	17	69.9	28	2567	67	32.5	I
	2	512	3.84	3775	265.2	82.3		29.6	28.7		50.1		2250	60	28.4	I + IV
L-H80-S100	1	512	3.83	3755	337.8	147.7	33	28.3	22.4	21	40.7	49	2358	63	41.7	I
	2	512	3.84	3775	298.5	98.3		34.2	33.0		20.7		2074	55	36.6	I + IV
L-H160-S100	1	512	3.83	3755	474.5	282.6	18	12.9	8.9	38	111.05	19	3340	89	34.5	I
	2	512	3.81	3716	359.6	230.5		17.8	15.7		90.4		2935	79	30.3	I
L-H125-S100(U)	1	512	3.82	3736	370.8	231.6	28	18.2	14.4	15	78.5	31	2786	75	35.2	I
	2	512	3.82	3736	304.5	167.8		20.9	18.1		54.3		2714	73	34.3	I + IV
R-H125-S100	1	512	3.84	3775	416.4	255.7	26	14.2	9.0	12	80.4	22	3515	93	49.6	I
	2	512	3.82	3736	368.2	189.8		15.9	10.4		62.6		3144	84	44.4	I
U-H125-S100	1	512	3.81	3716	751.6	369.6	8	11.2	5.4	−4	161.6	3	3483	94	24.7	II
	2	512	3.81	3716	658.3	341.3		10.8	5.0		156.3		3427	92	24.3	II
N-H125-S100	1	512	3.83	3755	540.2	141.8	19	14.8	11.6	86	76.8	39	2310	62	18.3	III
	2	512	3.84	3775	456.8	115.2		27.6	29.5		46.7		1795	48	14.2	III + V
L-H160-S100(D)	1	512	3.86	3814	495.1	308.6	10	14.2	7.6	−2	110.5	3	3575	94	36.9	I
	2	512	3.86	3814	414.2	277.3		13.9	7.1		106.7		3548	93	36.7	I

i impact number, M impact mass (kg), v impact velocity (m/s), Impact energy $E_k = 0.5MV^2$ (J), P_{max} peak impact force (kN), P_{ave} impact force plateau (kN), $\Delta_{pla} = (P_{ave,first} - P_{ave,1st})/P_{ave,1st} \times 100\%$, D_{max} maximum mid-span displacement (mm), D_{res} residual mid-span displacement (mm), $\Delta_{max} = (D_{res,second} - D_{res,1st})/D_{res,1st} \times 100\%$, k dynamic global stiffness (kN/mm), $\Delta_k = (k_{1st} - k_{2nd})/k_{1st} \times 100\%$, E_w energy absorbed by deformation (J), Specific energy absorption $E_s = E_w/M_b$ (J/kg), M_b mass of beam, *I* flexural failure with bond-slip, *II* flexural failure, *III* shear failure, *IV* shear stud rupture (attached to bottom plate), *V* top steel plate buckling

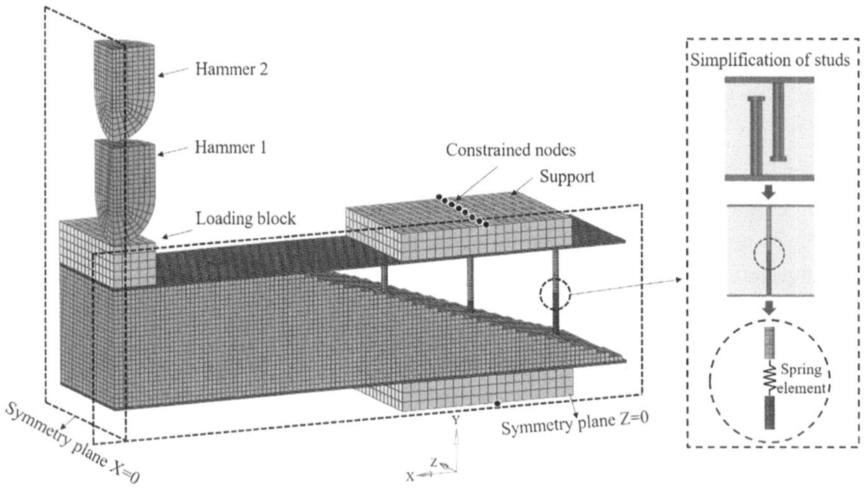

Fig. 6.15 Quarter model of composite beam under multiple impacts. (Reprinted from Zhao et al. [23], copyright 2025, with permission from Elsevier)

within a single simulation framework, as shown in Fig. 6.15. The hammers are strategically positioned at different heights with carefully calculated spacing to ensure sequential impacts without interaction, achieved by deactivating inter-hammer contact while maintaining proper gravitational acceleration and initial velocity definitions for each impactor. To optimize computational efficiency, a quarter-symmetry finite element model was developed, employing solid elements for impactors, concrete, studs, and supports, while utilizing thick-shell elements for steel plates. The modeling approach incorporates several key simplifications. Overlapped headed studs are represented by equivalent-diameter straight rods, which are connected via nonlinear spring elements whose constitutive properties are derived from experimental pull-out tests [3] and defined using *SPRING-NONLINEAR-ELASTIC parameters.

6.4.1 Calibration of Material Constitutive Model

The Continuous Surface Cap Model (CSCM), originally developed by the Federal Highway Administration for vehicle crash simulations [4], has been adapted in this study to analyze low-velocity impact responses in concrete structures. This advanced constitutive model uniquely captures three critical concrete behaviors: shear dilation, strain-softening, and strain-rate effects [5]. The model's failure surface is mathematically described through a three-invariant formulation presented in Eq. (6.4), incorporating Rubin's reduction factor (R) along with specialized functions for failure (F_f) and cap hardening (F_c). The compression meridian expressed in Eq. (6.5), characterized by parameters (α, β, λ, θ) derived from triaxial testing, plays a pivotal role in defining material strength under multiaxial stress states.

$$f(I_1, J_2, \mathrm{J}_3) = I_2 - R^2 F_\mathrm{f}^2 F_\mathrm{c} \tag{6.4}$$

where I_1, J_2, J_3 are the first invariant of stress tensor, the second and third invariant of stress deviator tensor, respectively; R is the Rubin three-invariant reduction factor [4]; F_f, F_c are the failure surface function and hardening cap surface function. The compression meridian equation of the failure surface is as follows.

$$F_\mathrm{f}(I_1) = \alpha - \lambda \exp(-\beta I_1) + \theta I_1 \tag{6.5}$$

where, α, β, λ, and θ are the parameters of the compression meridian, which can be determined by triaxial compressive tests.

The CSCM characterizes material strength through a three-dimensional failure surface defined by compressive, tensile, and shear meridians. As illustrated in Fig. 6.16a, four critical control points establish this surface: Point A ($-3f_\mathrm{tt}$, 0) represents triaxial tension; Point B ($-2f_\mathrm{bt}$, $f_\mathrm{bt}/\sqrt{3}$) corresponds to biaxial tension; Point C $\left(f_\mathrm{c}', f_\mathrm{c}'/\sqrt{3}\right)$ indicates uniaxial compression; and Point D ($\sigma_1 + 2\sigma_3$, $(\sigma_1 - \sigma_3)/\sqrt{3}$) reflects triaxial compression states. For LHDCC, the triaxial (f_tt) and biaxial (f_bt) tensile strengths approximate the uniaxial tensile strength (f_t) [6], while the uniaxial compressive strength $\left(f_\mathrm{c}'\right)$ derives from material tests. Through triaxial compression testing, three additional fitting points ($D1$, $D2$, $D3$) were established at values of $0.5 f_\mathrm{c}'$, $1.5 f_\mathrm{c}'$, and $3 f_\mathrm{c}'$ respectively. Least-squares regression of these six control points (A, B, C, D1–D3) to Eq. (6.5) yielded the compression meridian parameters (α, β, λ, θ). Figure 6.17 demonstrates LHDCC's distinct compression meridian compared to conventional concrete, revealing that standard CSCM parameters substantially overpredict LHDCC's triaxial strength without proper calibration. This comprehensive calibration process ensures accurate representation of LHDCC's unique mechanical behavior under complex stress states.

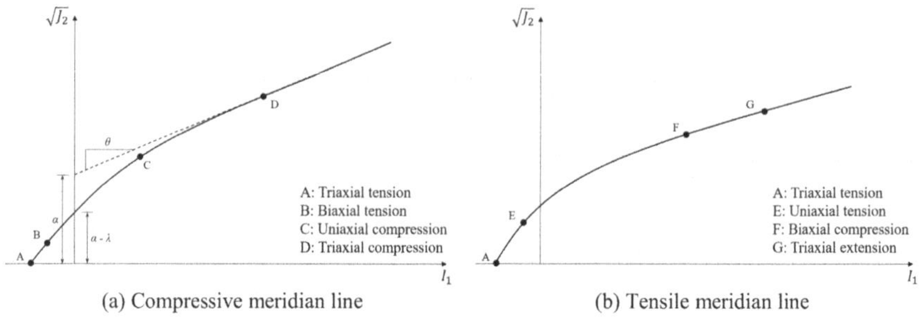

Fig. 6.16 Compressive and tensile meridian lines of the failure surface. (**a**) Compressive meridian line. (**b**) Tensile meridian line. (Reprinted from Zhao et al. [23], copyright 2025, with permission from Elsevier)

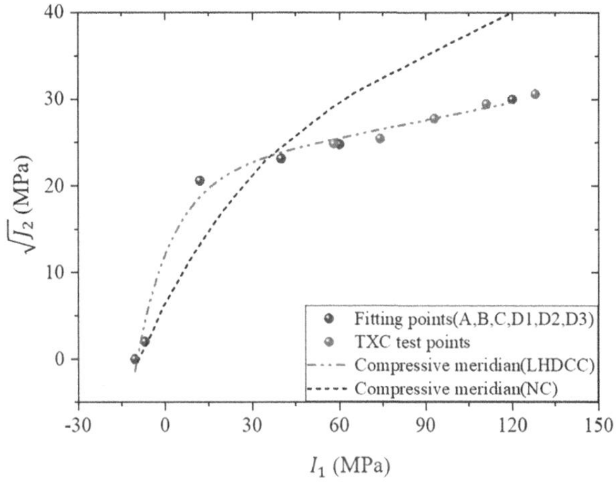

Fig. 6.17 Compression meridian fitting results. (Reprinted from Zhao et al. [23], copyright 2025, with permission from Elsevier)

The tensile meridian of the failure surface, illustrated in Fig. 6.16b, is defined through four characteristic control points: Point A ($-3\ f_{tt}$, 0) represents triaxial tension; Point E ($-f_t$, $f_t/\sqrt{3}$) corresponds to uniaxial tension; Point F $\left(f'_{bc}, f'_{bc}/\sqrt{3}\right)$ indicates biaxial compression; and Point G ($\sigma_1 + 2\sigma_3$, $(\sigma_3 - \sigma_1)/\sqrt{3}$) describes triaxial extension states. While no explicit analytical function for the tensile meridian currently exists in published literature, it can be effectively derived from the compressive meridian function (F_f) through Rubin's scaling approach (Q_2), expressed as $F_{ft} = Q_2 F_f$. This scaling relationship enables the tensile meridian (F_{ft}) to be systematically determined from its compressive counterpart while accounting for material-specific differences in strength behavior under varying stress states. The Rubin scaling factor Q_2, mathematically defined in Eq. (6.6), provides the necessary transformation between these two fundamental components of the failure surface.

$$Q_2 = \alpha_2 - \lambda_2 \exp(-\beta_2 I_1) + \theta_2 I_1 \tag{6.6}$$

The calibration process for the scaling function parameters (α_2, β_2, λ_2, θ_2) incorporates key material properties from experimental testing. The uniaxial tensile strength (f_t) is obtained directly from material characterization tests, while the biaxial compressive strength $\left(f'_{bc}\right)$ for LHDCC is established as 1.16 times the uniaxial compressive strength $\left(f'_c\right)$ based on previous research [7]. Due to the scarcity of triaxial extension data for LHDCC in existing literature, the analysis adopts the empirical relationship developed by Mills and Zimmerman [8] for conventional concrete as shown in Eq. (6.7) below:

$$\frac{\tau_0}{f_c} = \frac{0.147 + 0.550\sigma_0}{f_c} \tag{6.7}$$

where, $\tau_0 = \sqrt{2J_2/3}$; $\sigma_0 = I_1/3$. This relationship enables determination of ultimate stresses under triaxial extension conditions. Three characteristic stress states (G_1, G_2, G_3) are identified at normalized first stress invariants $\left(I_1/f'_c\right)$ of 0.5, 1.5, and 3.0

respectively. These points, combined with fundamental stress states (A, E, F), form a six-point calibration set for determining the tensile meridian parameters through least-squares optimization of Eq. (6.6). The Rubin scaling factor ($Q_2 = 0.5$ at $I_1 = 0$) [4] is implemented to maintain continuity between tensile and compressive domains, ensuring a physically consistent transition in the failure surface representation. This comprehensive calibration methodology effectively addresses data limitations while preserving the model's accuracy across different stress regimes.

Positioned between the tensile and compressive meridians, the shear meridian is established through the Rubin scaling function Q_1, as presented in Eq. (6.8) below:

$$Q_1 = \alpha_1 - \lambda_1 \exp(-\beta_1 I_1) + \theta_1 I_1 \tag{6.8}$$

The Rubin scaling factor Q_1 equals 0.5774 under zero hydrostatic pressure ($I_1 = 0$) [9], while for other stress states ($I_1 = 0.5 f'_c$, $1.5 f'_c$, $3 f'_c$), it is determined through the Willians-Warnke function given in Eq. (6.9) [4]. The shear meridian parameters (α_1, β_1, λ_1, θ_1) are subsequently derived by fitting these values to Eq. (6.8).

$$Q_1 = \frac{\sqrt{3}(1 - Q_2^2) + (2Q_2 - 1)\sqrt{3(1 - Q_2^2) + 5Q_2^2 - 4Q_2}}{3(1 - Q_2^2) + (1 - 2Q_2)^2} \tag{6.9}$$

Based on the above calibration process, the modeling parameters for LHDCC are shown in Table 6.4.

6.4.2 Contact Definition and Boundary Conditions

The numerical model employs *CONTACT-AUTOMATIC-SURFACE_TO_SURFACE definitions to characterize interfacial interactions between components. Critical contact parameters include static (FS) and dynamic (FD) friction coefficients, along with penalty stiffness coefficients for master (SFS) and slave (SFM) surfaces. Particular attention is given to the hammer-loading plate interface, where parametric studies confirm the significant influence of penalty stiffness values on peak impact force prediction [10]. Through systematic trial-and-error calibration, optimal SFS and SFM values were established for each contact pair, as documented in Table 6.5.

Table 6.4 Failure surface modeling parameters for LHDCC

Parameter	α (MPa)	β (MPa^{-1})	λ (MPa)	θ	α_1	β_1 (MPa^{-1})
Value	21.547	8.327e^{-2}	9.526	7.601e^{-2}	0.747	5.852e^{-2}
Parameter	λ_1	θ_1 (MPa^{-1})	α_2	β_2 (MPa^{-1})	λ_2	θ_2 (MPa^{-1})
Value	0.170	8.531e^{-4}	0.660	5.486e^{-2}	0.160	1.019e^{-3}

Table 6.5 Contact parameters

Contact	FS	FD	SFS	SFM
Hammer-block	0.3	0.2	0.2	0.2
Plate-concrete	0.7	0.5	0.15	0.15
Stud-concrete	0.7	0.5	0.15	0.15
Plate-support	0.3	0.2	0.2	0.2

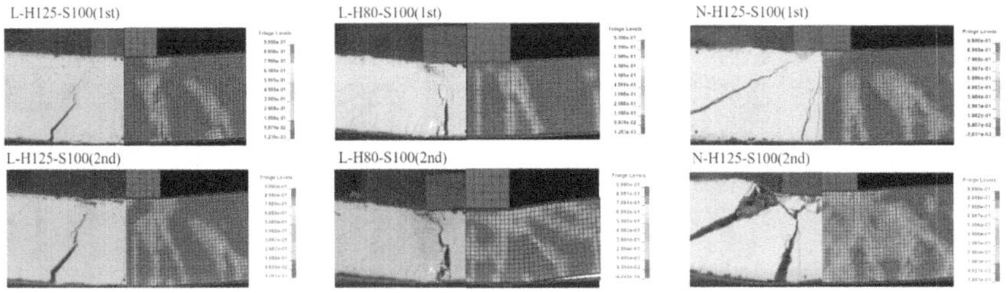

(a) L-H125-S100 (b) L-H80-S100 (c) N-H125-S100

Fig. 6.18 Comparison of FE and test results (failure mode). (**a**) L-H125-S100. (**b**) L-H80-S100. (**c**) N-H125-S100. (Reprinted from Zhao et al. [23], copyright 2025, with permission from Elsevier)

To enhance computational efficiency, symmetric boundary conditions are implemented across both XOY and YOZ planes (see Fig. 6.16). The *SET_NODE keyword identifies all nodes within the XOY plane, while *BOUNDARY_SPC_SET constrains their Z-direction displacement and X/Y-axis rotations. Support conditions are simulated by fully restraining translational degrees of freedom at the base nodes while permitting rotational freedom about the longitudinal axis, accurately representing simple support boundary conditions. This modeling approach effectively balances computational accuracy with numerical efficiency.

6.4.3 FE Results and Discussions

The numerical simulation results demonstrate strong correlation with experimental observations of the composite beam's failure mode following dual impacts, as illustrated in Fig. 6.18. The finite element model accurately predicts both the critical crack formation locations and overall failure characteristics. Quantitative analysis presented in Table 6.6 reveals close agreement between simulated and experimental values for key parameters: peak impact force (P_{max}, FE/Test ratio = 1.05 ± 0.04), average impact force (P_{ave}, 1.05 ± 0.03), maximum displacement (D_{max}, 1.00 ± 0.06), and residual displacement (D_{res}, 0.96 ± 0.12). Figure 6.19 further confirms this agreement through force-displacement curve comparisons, showing excellent matching for initial impacts. However, minor discrepancies emerge in second-impact simulations for certain specimens (e.g., L-H80-S100, N-H125-S100), attributable to asymmetric damage progression not fully captured by the symmetric modeling approach. These specimens exhibited concentrated damage in

Table 6.6 Comparison of FE results with test results

Specimen	Impact number	P_{max} (KN)			P_{ave} (KN)			D_{max} (mm)			D_{res} (mm)		
		TEST	FEM	T/F	TEST	FEM	T/F	TEST	FEM	T/F	TEST	FEM	T/F
L-H125-S100	1	415.1	441.4	1.06	236.2	243.3	1.03	15.2	16.9	1.11	11.9	11.8	0.99
	2	356.4	363.5	1.02	184.6	192.0	1.04	18.4	18.6	1.01	14.2	12.2	0.86
L-H125-S150	1	393.4	441.9	1.12	212.3	225.0	1.06	18.7	20.4	1.09	14.9	14.2	0.95
	2	343.5	364.1	1.06	163.1	176.1	1.08	24.2	23.7	0.98	21.7	19.7	0.91
L-H125-S200	1	407.8	437.8	1.07	115.8	129.7	1.12	25.4	24.9	0.98	22.7	22.2	0.98
	2	265.2	286.4	1.08	82.3	90.5	1.1	29.6	29.9	1.01	28.7	26.7	0.93
L-H80-S100	1	337.8	358.8	1.06	147.7	156.6	1.06	28.3	26.0	0.92	22.4	21.3	0.95
	2	298.5	307.5	1.03	98.3	106.2	1.08	34.2	31.1	0.91	33	29.0	0.88
L-H160-S100	1	474.5	521.3	1.10	282.6	291.1	1.03	12.9	14.2	1.10	8.9	11.2	1.26
	2	359.6	363.2	1.01	230.5	244.3	1.06	17.8	17.1	0.96	15.7	13.5	0.86
L-H125-S100(U)	1	370.8	406.6	1.10	231.6	224.7	0.97	18.2	18.4	1.01	14.4	11.2	0.78
	2	304.5	298.4	0.98	167.8	171.2	1.02	20.9	20.3	0.97	18.1	17.4	0.96
R-H125-S100	1	416.4	450.5	1.08	255.7	258.3	1.01	14.2	15.2	1.07	9	10.5	1.17
	2	368.2	375.6	1.02	189.8	201.2	1.06	15.9	16.2	1.02	10.4	9.6	0.92
U-H125-S100	1	751.6	736.3	0.98	369.6	377.0	1.02	11.2	11.3	1.01	5.4	6.4	1.19
	2	658.3	724.1	1.10	341.3	365.2	1.07	10.8	11.1	1.03	5	4.8	0.96
N-H125-S100	1	540.2	602.9	1.12	141.8	147.5	1.04	14.8	14.2	0.96	11.6	10.3	0.89
	2	456.8	470.5	1.03	115.2	121.0	1.05	27.6	25.7	0.93	29.5	28.3	0.96
L-H160-S100(D)	1	495.1	495.6	1.00	308.6	311.7	1.01	14.2	13.8	0.97	7.6	6.8	0.89
	2	414.2	430.8	1.04	277.3	285.6	1.03	13.9	14.3	1.03	7.1	6.5	0.91
Mean	–	–	–	1.05	–	–	1.05	–	–	1.00	–	–	0.96
Std. dev.	–	–	–	0.04	–	–	0.03	–	–	0.06	–	–	0.12

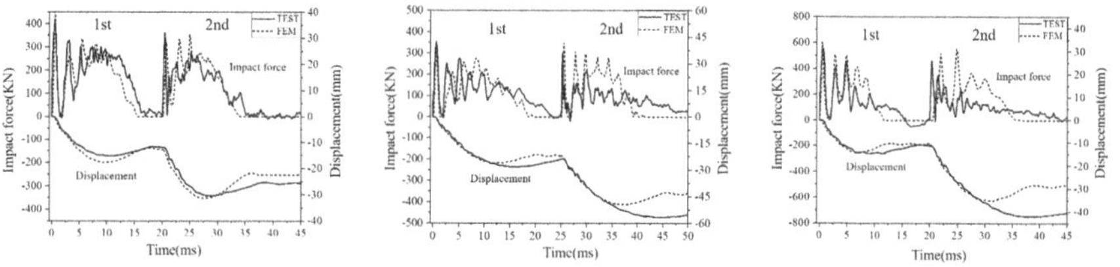

(a) L-H125-S100 (b) L-H80-S100 (c) N-H125-S100

Fig. 6.19 Comparison of FE and test results (impact force/displacement history). (**a**) L-H125-S100. (**b**) L-H80-S100. (**c**) N-H125-S100. (Reprinted from Zhao et al. [23], copyright 2025, with permission from Elsevier)

specific regions (notably the left span) during secondary impacts, creating conditions where the symmetric modeling assumptions introduce predictable variations from actual asymmetric failure patterns.

Energy dissipation analysis reveals distinct mechanisms by which the composite beams absorb impact energy. The hammer's kinetic energy transforms into internal strain energy distributed among three primary components: steel plates, shear connectors, and concrete, as quantified through *MATSUM outputs (see Table 6.7). Initial impact behavior varies significantly by failure mode. For flexural failures, steel plates dominate energy absorption (43.5% at 100 mm spacing) through significant plastic deformation, evidenced by high strain measurements. In bond-slip and shear failure modes, concrete becomes the primary energy absorber (60 + %) as steel stresses remain below dynamic yield thresholds.

Parametric studies demonstrate clear trends that increasing stud spacing from 100 mm to 200 mm reduces steel plate energy contribution by 28% (43.5–31.4%), correlating with decreased bottom plate strain measurements. Concrete thickness augmentation (80 → 160 mm) enhances concrete energy absorption by 48% (36.2–53.8%). Overlapped stud configurations show superior energy absorption compared to short stud designs. Multiple impact analysis reveals critical performance differences that single-layer specimens exhibit severe degradation (30.8–9.6% steel energy contribution). Double-layer configurations maintain consistent performance (>35% steel energy absorption across impacts). Concrete becomes increasingly dominant in subsequent impacts, absorbing >60% of energy for most specimens.

6.5 Theoretical Model

6.5.1 Governing Equation of SDOF

The maximum displacement of protective structures serves as a critical design parameter. This section develops an analytical approach to predict the dynamic response of stainless steel-LHDCC composite beams under impact loading using

Table 6.7 Energy absorption for each component in FE results

Specimen	Impact number	Concrete Energy (J)	Concrete Proportion (%)	Steel plates Energy (J)	Steel plates Proportion (%)	Studs Energy (J)	Studs Proportion (%)
L-H125-S100	1	1346	40.6	1440	43.5	525	15.9
	2	2082	61.7	1016	30.1	276	8.2
L-H125-S150	1	1725	52.4	1223	37.1	346	10.5
	2	2436	69.1	792	22.5	296	8.4
L-H125-S200	1	1812	53.9	1056	31.4	497	14.8
	2	2740	79.1	566	16.3	156	4.5
L-H80S100	1	1101	36.2	1528	50.2	414	13.6
	2	2560	78.1	488	14.9	228	7.0
L-H160-S100	1	1801	53.8	1030	30.8	518	15.5
	2	2956	81.0	352	9.6	341	9.3
L-H125-S100(U)	1	2014	61.9	996	30.6	243	7.5
	2	2812	79.6	460	13.0	262	7.4
L-H160-S100(D)	1	1380	40.8	1269	37.5	733	21.7
	2	1387	42.4	1263	38.6	621	19.0
N-H125-S100	1	1871	57.2	1028	31.4	373	11.4
	2	2116	65.8	840	26.1	261	8.1
R-H125-S100	1	1732	51.5	1053	31.3	576	17.1
	2	2600	71.3	772	21.2	273	7.5
U-H125-S100	1	1139	34.3	1673	50.3	512	15.4
	2	1016	28.3	2212	61.6	365	10.2

Note: The proportion refers to the ratio between the energy absorbed by an individual component and the total internal energy of the beam

an equivalent single degree of freedom (SDOF) model. As illustrated in Fig. 6.20a, the system assumes perfect coupling between the impact hammer and beam after initial contact, with both moving at identical velocities during the loading phase. The governing dynamic equation is expressed as:

$$m\ddot{u}(t) + c\dot{u}(t) + R[u(t)] = 0 \tag{6.10}$$

where, m is total equivalent mass (hammer + beam); c is viscous damping coefficient; R is rate-dependent structural resistance function; $u(t)$, $\dot{u}(t)$, and $\ddot{u}(t)$ are displacement, velocity, and acceleration responses.

The equivalent mass incorporates inertial contributions from both the hammer (m_h) and beam (m_b) through a mass conversion factor (μ):

$$m = m_h + m_e = m_h + \mu m_b \tag{6.11}$$

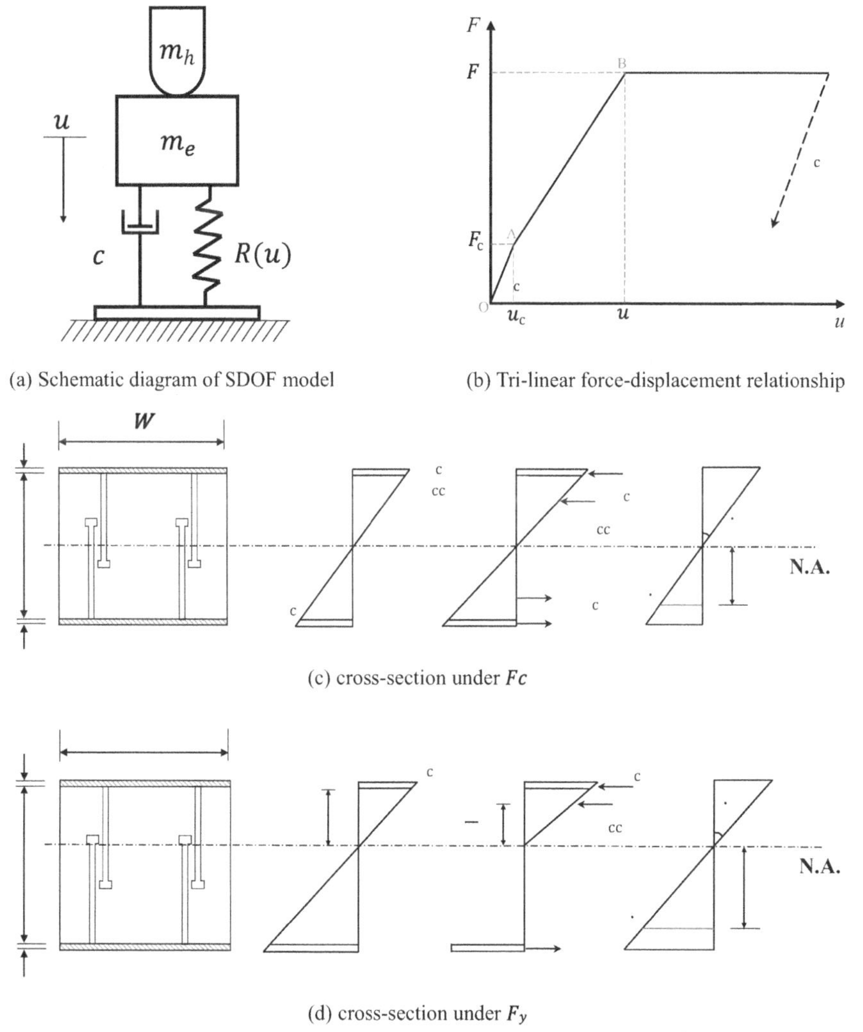

Fig. 6.20 SDOF model. (**a**) Schematic diagram of SDOF model. (**b**) Tri-linear force-displacement relationship. (**c**) cross-section under F_c. (**d**) cross-section under F_y. (Reprinted from Huang et al. [15], copyright 2025, with permission from American Society of Civil Engineers)

The conversion factor μ accounts for the beam's deformed shape characteristics and is calculated as [11]:

$$\mu = \frac{2\int_0^{l/2} \varphi^2(x)dx}{l} \tag{6.12}$$

where, $\varphi(x)$ is the shape function and calculated as:

$$\varphi(x)=\begin{cases}\frac{x}{l^3}\left(3l^2-4x^2\right) & ,u<u_y\\ \frac{2x}{l} & ,u>u_y\end{cases} \tag{6.13}$$

where, l is the clear span of the beam; u_y is the yield deflection of the beam, as discussed in the subsequent section.

6.5.2 Trilinear Model for Static Load-Displacement Behavior

The static load-displacement behavior of the composite beam is characterized by a trilinear model comprising three distinct phases (see Fig. 6.20b). The initial elastic stage (OA) terminates upon formation of flexural cracks in the concrete core, governed by three fundamental assumptions: (1) preservation of plane sections, (2) mid-height neutral axis location, and (3) negligible shear deformation effects. Figure 6.20a illustrates the strain and stress distributions at cracking load (F_c), where the critical bending moment (M_c) is derived from sectional force components: steel plate forces (N_{sc}, N_{st}) and concrete contributions (N_{cc}, N_{ct}), calculated through Eq. (6.14)–(6.17) as given below:

$$M_c=\frac{1}{2}N_{sc}(t_t+h_c)+\frac{1}{3}N_{cc}h_c+\frac{1}{2}N_{st}(t_b+h_c)+\frac{1}{3}N_{ct}h_c \tag{6.14}$$

$$N_{cc}=N_{ct}=\frac{f_t h_c W}{4} \tag{6.15}$$

$$N_{sc}=E_s\varepsilon_{sc}t_tW;N_{st}=E_s\varepsilon_{st}t_bW \tag{6.16}$$

$$\varepsilon_{sc}=\frac{f_t(h_c+t_t)}{E_ch_c};\varepsilon_{st}=\frac{f_t(h_c+t_b)}{E_ch_c} \tag{6.17}$$

The elastic stage concludes when the concrete's extreme fiber strain (ε_{ct}) reaches its ultimate tensile capacity, with steel strains (ε_{sc}, ε_{st}) determined by material properties (E_s, E_c) and geometric parameters (t_t, t_b, h_c, W). This analytical formulation establishes the baseline for subsequent nonlinear behavior, incorporating both material properties and cross-sectional geometry to predict the onset of cracking in the composite system.

The load resistance of the beam corresponding to the section bending moment at this stage is:

$$F_c=\frac{4M_c}{l} \tag{6.18}$$

The initial stiffness k_c is calculated based on the section stiffness EI_c:

$$k_c=\frac{48}{l^3}EI_c \tag{6.19}$$

$$\mathrm{EI_c} = \frac{1}{12} E_c W h_c^3 + E_s \left[\frac{W t_t^3}{12} + W t_t \left(\frac{t_t + h_c}{2} \right)^2 \right] + E_s \left[\frac{W t_b^3}{12} + W t_t \left(\frac{t_b + h_c}{2} \right)^2 \right] \tag{6.20}$$

For partially composite beams, the reduced composite action effect is accounted for by applying width reduction coefficients to the external steel plates [12]. The effective width adjustment factors for the upper (β_t) and lower (β_b) steel plates are introduced to modify the stiffness calculation, distinguishing this approach from the fully composite beam formulation where complete interaction is assumed. These coefficients quantitatively represent the degree of composite interaction between the steel plates and concrete core:

$$\beta_t = \frac{n_t K_t}{n_t K_t + 2 B t_t E_s / l} \tag{6.21}$$

$$\beta_b = \frac{n_b K_b}{n_b K_b + 2 B t_b E_s / l} \tag{6.22}$$

The parameters n_t and n_b represent the quantity of shear connectors connected to the upper and lower steel plates, respectively. The elastic stiffness values K_t and K_b, which characterize the shear-slip behavior of these connectors, are determined using established empirical relationships developed by Shim et al. [13].

$$K = \frac{1}{d(0.16 - 0.0017 f_{ck})} P_s \tag{6.23}$$

where, d is the diameter of shear connector; f_{ck} is the concrete compressive strength; P_s is the ultimate shear resistance of shear connector according to BS EN 1994-1-1 [13].

$$P_s = \min\left(0.2 f_u \pi d^2 / \gamma_v, 0.29 \alpha d^2 \sqrt{f_{ck} E_c} / \gamma_v \right) \tag{6.24}$$

where, $\alpha = 0.2(h_s/d + 1)$ when $3 \leq h_s/d \leq 4$; $\alpha = 1$ when $h_s/d > 4$; f_u, h_s, d represent the ultimate tensile strength, height, and diameter of the shear connector; γ_v represents the partial factor.

The bending stiffness of the partially composite beam section is determined by accounting for the degree of composite interaction between components as given below:

$$\mathrm{EI_c} = \frac{1}{12} E_c W h_c^3 + E_s \left[\frac{\beta_t W t_t^3}{12} + \beta_t W t_t \left(\frac{t_t + h_c}{2} \right)^2 \right] + E_s \left[\frac{\beta_b W t_b^3}{12} + \beta_b W t_b \left(\frac{t_b + h_c}{2} \right)^2 \right] \tag{6.25}$$

The yield point (point B) in the trilinear load-displacement model (see Fig. 6.20b) corresponds to either yielding of the bottom steel plate or crushing initiation in the concrete. In both scenarios, the compressed concrete region maintains a linear stress distribution, either fully intact or beginning to fail. The yield load F_y is derived from the section's yield moment M_y using Eq. (6.18), based on three fundamental assumptions described above. As illustrated in Fig. 6.20b, the neutral axis shifts upward during this stage, with its precise position determined through equilibrium of internal sectional forces. The strain and stress distributions demonstrate this transition from elastic to inelastic behavior, where the composite action significantly influences the moment capacity and neutral axis location. This analytical approach captures the critical transition in structural response while accounting for material nonlinearities in both steel and concrete components.

$$N_{sc} + N_{cc} = N_{st} \tag{6.26}$$

$$N_{sc} = \frac{\varepsilon}{x}\left(x + \frac{t_t}{2}\right)E_s t_t W; N_{st} = \frac{\varepsilon}{x}\left(h_c - x + \frac{t_b}{2}\right)E_s t_b W \tag{6.27}$$

$$N_{cc} = \frac{1}{2}E_c \varepsilon x W \tag{6.28}$$

where, x represents the distance from the neutral axis position to the top surface of concrete; ε is the strain.

Then, the section bending moment can be determined by:

$$M_y = \min\left(\begin{array}{c} f_y W t_b\left(h_c + \frac{t_b}{2} - \frac{x}{3}\right) + f_y \frac{x + \frac{t_t}{2}}{h_c - x + \frac{t_b}{2}} W t_t\left(\frac{t_t}{2} + \frac{x}{3}\right) \\ \frac{f_{ck}}{E_c} E_s W t_b \frac{h_c - x + \frac{t_b}{2}}{x + \frac{t_b}{2}}\left(h_c + \frac{t_b}{2} - \frac{x}{3}\right) + \frac{f_{ck}}{E_c} E_s W t_t\left(\frac{t_t}{2} + \frac{x}{3}\right) \end{array}\right) \tag{6.29}$$

where, f_y is the yield strength of the steel plate. Equation (6.29) derives different expressions according to the sequence of yielding of the steel plate and the crush of the concrete, respectively.

The progressive development of bending cracks creates non-uniform stiffness distribution along the beam span. To address this computational complexity, the analysis employs the minimum stiffness principle, as given in Eurocode 4 [14], for determining the beam's effective stiffness during the second stage (AB). This simplified approach accounts for the most critical section with reduced stiffness, designated as k_y, which governs the overall structural response. The methodology recognizes that crack propagation leads to localized stiffness degradation while maintaining computational efficiency by focusing on the controlling section rather than tracking variable stiffness throughout the entire span. This principle provides a conservative yet practical solution for evaluating the beam's nonlinear behavior during the post-cracking phase.

$$k_y = \frac{48}{l^3}\mathrm{EI}_y \tag{6.30}$$

$$\mathrm{EI}_y = \frac{1}{3}E_c W x^3 + \beta_t E_s\left[\frac{W t_t^3}{12} + W t_t\left(\frac{t_t + x}{2}\right)^2\right] + \beta_b E_s\left[\frac{W t_b^3}{12} + W t_b\left(h_c - x + \frac{t_b}{2}\right)^2\right] \tag{6.31}$$

where, β_t and β_b are width reduction coefficients of steel plates for partially composite beams, as shown in Eqs. (6.21)–(6.23).

When evaluating the dynamic resistance of composite beams under impact loading, proper consideration of material strain rate effects is essential. Sukontasukkul [15] demonstrated that concrete members exhibit substantially higher strain rates in one-dimensional bending compared to two-dimensional scenarios. Fujikake et al. [16] established an analytical approach relating curvature rate to loading rate, drawing parallels with static curvature-deflection relationships to determine strain rate distributions across the cross-section (see Fig. 6.20). The strain rate at any position of the cross section is:

$$\dot{\varepsilon} = z\dot{\phi} \tag{6.32}$$

$$\dot{\phi} = \frac{12}{l^2}\dot{\delta} \tag{6.33}$$

$$\dot{\delta} = \frac{m_h V}{2m} \tag{6.34}$$

where, $\dot{\varepsilon}$ is the strain rate; z represents the distance to the neutral axis; $\dot{\delta}$ represents the loading rate; $\dot{\phi}$ represents the curvature rate of the cross section; V represents the velocity of the drop hammer at the initial contact with the specimen.

The tensile and compressive strengths f_t, f_{ck} of concrete, and the yield strength and ultimate strengths f_y, f_u of steel all increase in varying degrees as the strain rate increases. The most widely used Cowper-Symonds model is used to consider the strain rate of metals [17, 18]:

$$f_y^d / f_y^s = 1 + (\dot{\varepsilon}/D)^{\frac{1}{p}} \tag{6.35}$$

where, f_y^d,f_y^s are the dynamic and static yield strengths, respectively; D and p are the strain rate related parameters, for stainless steel, which are taken as 943.6 s^{-1} and 4.626, respectively [17].

The strain rate effect of concrete is considered using the CEB-FIP code [19]:

$$f_{\mathrm{c,imp}}/f_{\mathrm{cm}} = \begin{cases} (\dot{\varepsilon_{\mathrm{c}}}/\dot{\varepsilon_{\mathrm{c0}}})^{0.014} & |\dot{\varepsilon_{\mathrm{c}}}| \leq 30s^{-1} \\ 0.012(\dot{\varepsilon_{\mathrm{c}}}/\dot{\varepsilon_{\mathrm{c0}}})^{1/3} & |\dot{\varepsilon_{\mathrm{c}}}| > 30s^{-1} \end{cases} \tag{6.36}$$

where $f_{\mathrm{c,\ imp}}$, f_{cm} represent dynamic compressive strength and static compressive strength; $\dot{\varepsilon_{\mathrm{c}}}$ is the concrete strain rate; $\dot{\varepsilon_{\mathrm{c0}}}$ is the strain rate at static loading, taken as $-30 \times 10^{-6}s^{-1}$, and the negative value indicates compression. In particular, for LHDCC and RLHDC, the following formula is used based on the previous study [20]:

$$f_{\mathrm{c,imp}}/f_{\mathrm{cm}} = 0.0106(\dot{\varepsilon_{\mathrm{c}}}/\dot{\varepsilon_{\mathrm{c0}}})^{1/3} \quad |\dot{\varepsilon_{\mathrm{c}}}| > 30s^{-1} \quad \text{(LHDCC)} \tag{6.37}$$

$$f_{\mathrm{c,imp}}/f_{\mathrm{cm}} = \beta * \lg(\dot{\varepsilon_{\mathrm{c}}}/\dot{\varepsilon_{\mathrm{c0}}}) - \gamma \quad |\dot{\varepsilon_{\mathrm{c}}}| > 30s^{-1} \quad \text{(RLHDC)} \tag{6.38}$$

where, β and γ are function of rubber volume fraction R, i.e., $\beta = 8.121R + 1.458$, $\gamma = 54.587R + 7.250$, where the constant is obtained by experiment.

$$f_{\mathrm{ct,imp}}/f_{\mathrm{ctm}} = \begin{cases} (\dot{\varepsilon}_{\mathrm{ct}}/\dot{\varepsilon}_{\mathrm{ct0}})^{0.018} & |\dot{\varepsilon}_{\mathrm{ct}}| \leq 10s^{-1} \\ 0.0062(\dot{\varepsilon}_{\mathrm{ct}}/\dot{\varepsilon}_{\mathrm{ct0}})^{1/3} & |\dot{\varepsilon}_{\mathrm{ct}}| > 10s^{-1} \end{cases} \tag{6.39}$$

where $f_{\mathrm{ct,\ imp}}$ and f_{ctm} are the dynamic tensile strength and static tensile strength of concrete; $\dot{\varepsilon}_{\mathrm{ct}}$ is the concrete strain rate; $\dot{\varepsilon}_{\mathrm{ct0}}$ is the strain rate at static loading, taken as $3 \times 10^{-6}s^{-1}$.

6.5.3 Numerical Solution of Governing Equation

The dynamic response is numerically solved using finite difference approximation, where displacement $u(i)$ at time step i is computed with time increment Δ_t. The governing relation, expressed in Eq. (6.10), is discretized via central difference schemes for velocity and acceleration (Eqs. (6.40) and (6.43)), leading to the recurrence relation, i.e., Eq. (6.43).

$$\dot{u}(i) = \frac{u(i+1) - u(i)}{\Delta t} \tag{6.40}$$

$$\ddot{u}(i) = \frac{u(i+1) - 2u(i) + u(i-1)}{\Delta t^2} \tag{6.41}$$

$$m\frac{u(i+1) - 2u(i) + u(i-1)}{\Delta t^2} + c\frac{u(i+1) - u(i)}{\Delta t} + R[u(i)] = 0 \tag{6.42}$$

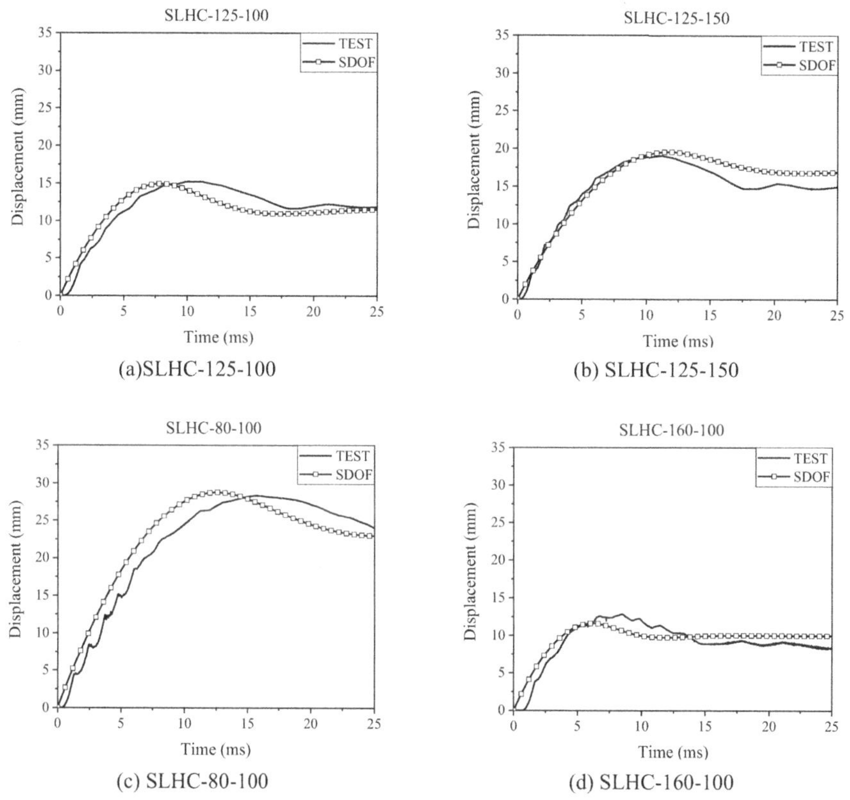

Fig. 6.21 Comparison of displacement responses between predictive and test results. (**a**) SLHC-125-100. (**b**) SLHC-125-150. (**c**) SLHC-80-100. (**d**) SLHC-160-100. (Reprinted from Huang et al. [15], copyright 2025, with permission from American Society of Civil Engineers)

$$u(i+1) = 2u(i) - u(i-1) - \frac{c\Delta t}{m}(u(i) - u(i-1)) - \frac{\Delta t^2}{m}R[u(i)] \tag{6.43}$$

Damping effects are treated differently across loading phases: neglected during initial loading ($c = 0$) following Bruhl et al. [21], while adopting 50% critical damping during rebound based on Guo and Zhao [22] findings. The piecewise formulation accounts for three behavioral regimes: pre-cracking (elastic), post-cracking (hardening), and post-yield (plastic) phases, with transition points at cracking (u_c) and yield (u_y) displacements. Comparative validation with experimental data (see Fig. 6.21) demonstrates accurate prediction of time-history response and peak displacement, though residual displacements are slightly overestimated due to the assumed constant mass coupling between hammer and beam in the SDOF idealization. The computational framework (see Fig. 6.22) provides a practical design tool for impact-resistant composite beams, balancing accuracy with computational efficiency through proper treatment of nonlinearities and phase-dependent parameters.

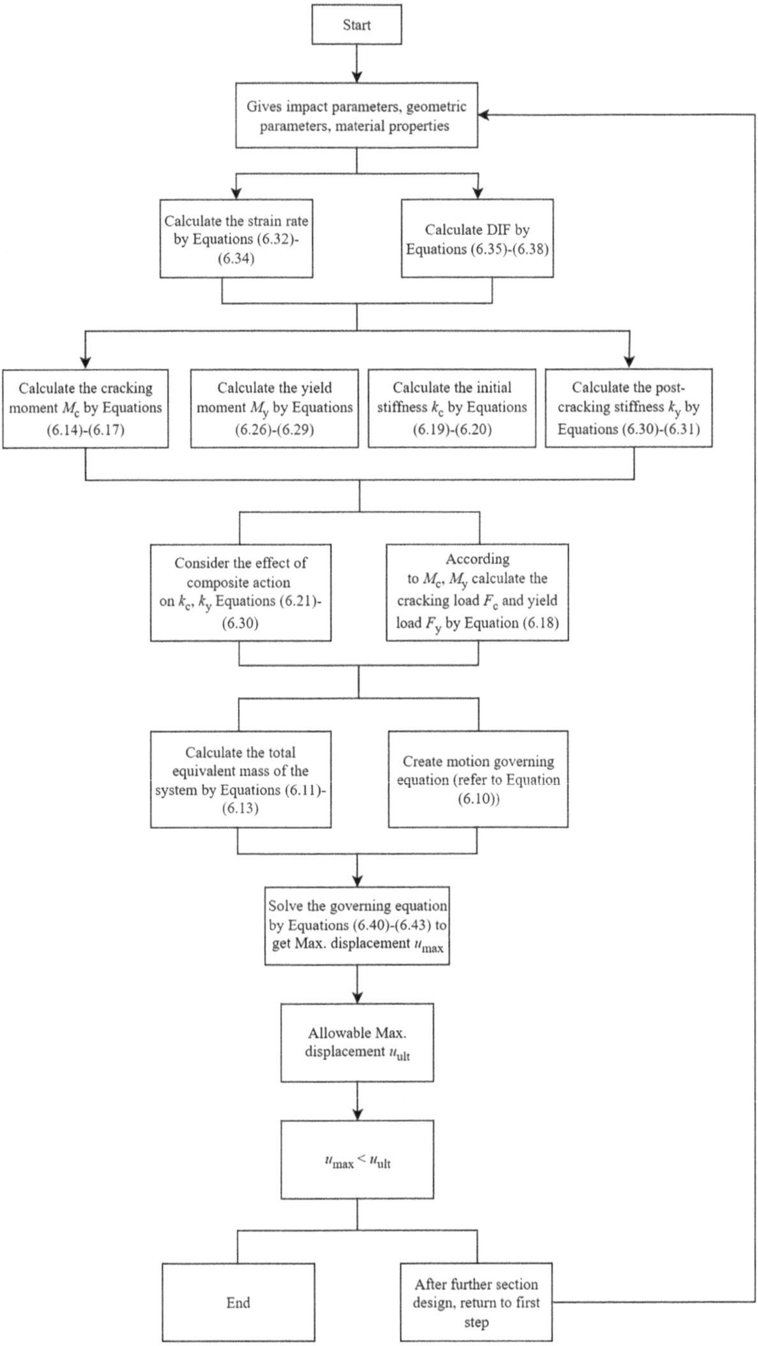

Fig. 6.22 Design flow chart. (Reprinted from Huang et al. [15], copyright 2025, with permission from American Society of Civil Engineers)

References

1. W. Zhang, Z. Huang, Z. Fu, X. Qian, Y. Zhou, L. Sui, Shear resistance behavior of partially composite steel–concrete–steel sandwich beams considering bond-slip effect, Eng. Struct. 210 (2020) 110394.
2. H. Othman, H. Marzouk, Finite-element analysis of reinforced concrete plates subjected to repeated impact loads, J. Struct. Eng. 143(9) (2017) 04017120.
3. Z. Huang, X. Zhao, W. Zhang, Flexural impact performance of stainless steel–lightweight concrete composites for marine structures, J. Struct. Eng. 148(9) (2022) 04022114.
4. Y.D. Murray, User's manual for LS-DYNA concrete material model 159: no. FHWA-HRT-05-062, Federal Highway Administration, Colorado Springs, 2007, pp. 1-74.
5. H. Hua, J. Zhao, Calibration of the continuous surface cap model for concrete, Finite Elem. Anal. Des. 97 (2015) 1-19.
6. W.F. Chen, Plasticity in reinforced concrete, J. Ross Publishing, New York, 2007, pp. 19-47.
7. Z. Huang, J.Y.R. Liew, Numerical studies of steel–concrete–steel sandwich walls with J-hook connectors subjected to axial loads, Steel Compos. Struct. 21(3) (2016) 461-477.
8. L.L. Mills, R.M. Zimmerman, Compressive strength of plain concrete under multiaxial loading conditions, ACI J. Proc. 67(10) (1970) 802-807.
9. Y.D. Murray, A.Y. Abu-Odeh, R.P. Bligh, Evaluation of LS-DYNA concrete material model 159: no. FHWA-HRT-05-063, Federal Highway Administration, Colorado Springs, 2007, pp. 5-45.
10. T.M. Pham, Y. Hao, H. Hao, Sensitivity of impact behaviour of RC beams to contact stiffness, Int. J. Impact Eng. 112 (2018) 155-164.
11. J.C. Bruhl, A.H. Varma, W.H. Johnson, Design of composite SC walls to prevent perforation from missile impact, Int. J. Impact Eng. 75 (2015) 75-87.
12. T.M. Roberts, D.N. Edwards, R. Narayanan, Testing and analysis of steel–concrete–steel sandwich beams, J. Constr. Steel Res. 38(3) (1996) 257-279.
13. C.-S Shim, P.-G. Lee, T.-Y. Yoon, Static behavior of large stud shear connectors, Eng. Struct. 26(12) (2004) 1853-1860.
14. X. Gu, X. Jin, Y. Zhou, Basic principles of concrete structures, Springer, Berlin, 2016.
15. P. Sukontasukkul, Impact behaviour of concrete under multiaxial loading, Ph.D. thesis, University of British Columbia, 2002.
16. K. Fujikake, B. Li, S. Soeun, Impact response of reinforced concrete beam and its analytical evaluation, J. Struct. Eng. 135(8) (2009) 938-950.
17. S. Jia, Q. Tan, J. Ye, Z. Zhu, Z. Jiang, Experiments on dynamic mechanical properties of austenitic stainless steel S30408 and S31608, J. Constr. Steel Res. 179 (2021) 106556.
18. W. Abramowicz, N. Jones, Dynamic axial crushing of square tubes, Int. J. Impact Eng. 2(2) (1984) 179-208.
19. CEB-FIP, Model code 2010 – first complete draft, volume 2, fib Fédération internationale du béton, Lausanne, Switzerland, 2010.
20. Z. Huang, L. Sui, F. Wang, S. Du, Y. Zhou, J. Ye, Dynamic compressive behavior of a novel ultra-lightweight cement composite incorporated with rubber powder, Compos. Struct. 244 (2020) 112300.
21. J.C. Bruhl, A.H. Varma, J.M. Kim, Static resistance function for steel-plate composite (SC) walls subject to impactive loading, Nucl. Eng. Des. 295 (2015) 843-859.
22. Q. Guo, W. Zhao, Displacement response analysis of steel–concrete composite panels subjected to impact loadings, Int. J. Impact Eng. 131 (2019) 272-281.
23. X. Zhao, Z. Huang, W. Zhang, J. Ye, Repeated impact performance of stainless steel-lightweight high ductility cement composite beams, Thin-Walled Struct. 188 (2023) 110791.

Chapter 7
Flexural and Shear Behavior of Steel–Concrete Sandwich Composites with Slotted Perforated Plate Connectors

7.1 Introduction

Steel-concrete sandwich composites represent a significant advancement in construction materials, offering an exceptional strength-to-weight ratio and inherent durability. The structural integrity of such composites, however, fundamentally relies on the effective transfer of shear forces between the concrete core and the steel faceplates. This critical role is fulfilled by shear connectors, which resist separation and enable the constituent materials to act as a unified, composite element [1–4]. Their design is therefore paramount, directly dictating the system's flexural resistance to resist bending and its shear resistance to sliding failure.

Shear connectors are vital for achieving effective composite action between steel plates and concrete in composite structures, as their type, quantity, and arrangement directly influence structural performance. Over the years, researchers have developed and evaluated various connector designs to optimize load transfer and minimize interfacial slip. Among these, overlapping single-headed studs, as proposed by Tomlinson et al. [5], utilize a staggered welding pattern to enhance load-bearing capacity by resisting slip between steel and concrete. An alternative approach, the Bi-steel system introduced by Pryer and Bowerman [6], employs welded steel rods to provide superior shear strength and prevent debonding, though it demands stringent welding quality control and imposes spacing and core thickness limitations. Another innovative solution is the J-hook connector developed by Liew et al. [7–9], which offers comparable stiffness and load resistance to stud-based systems while demonstrating higher ductility. These connectors not only resist longitudinal shear but also link upper and lower steel plates to mitigate transverse shear, reducing concrete cracking in critical regions. Further advancements include hybrid configurations combining J-hooks and studs, as studied by [10–12], which leverage the benefits of both systems while highlighting the importance of connector spacing on stiffness, strength, and impact resistance. Beyond these, alternative connector types

Z. Huang, Y. Zhou, *Innovative Lightweight High-performance Steel-concrete Sandwich Composite for Marine Structures*,
https://doi.org/10.1007/978-981-95-4681-7_7

such as angle steel [13], channel steel [14], and corrugated steel strips [15] have also been explored for their structural efficiency. Notably, perforated plate (PBL) connectors have gained prominence due to their exceptional shear resistance and stiffness, outperforming conventional stud connectors [16]. Research by Xiao et al. [17] demonstrated that steel-UHPC composite plates with PBL connectors exhibit reduced slip and higher shear resistance, leading to improved flexural performance. Similarly, Lu [18] found that PBL connectors enhance the ultimate flexural resistance and post-cracking stiffness of composite box girders by 16.2% and 25.1%, respectively, compared to stud connectors. Given their superior mechanical properties, PBL connectors represent a promising direction for developing rigid shear connectors in modern composite structures.

In steel-concrete composite structures, achieving optimal performance requires balancing two competing demands: sufficient composite action to ensure structural stiffness and strength, while maintaining adequate energy dissipation capacity which often requires reduced composite interaction. In practice, perfect full-composite action is rarely achieved due to inevitable interfacial slip under loading, making most engineering applications partial-composite structures. Researchers have extensively investigated this phenomenon through experimental, numerical, and theoretical approaches. Nie [19] observed that conventional transformed-section methods for calculating beam deformation fail to account for slip effects, leading to unsafe underestimations. In the study, the proposed stiffness reduction method provided more accurate predictions through a practical simplified formula. The critical role of connector arrangement was demonstrated by Leng et al. [20], whose shear tests revealed that sufficient stud connectors produce concrete diagonal cracking failure, while insufficient connectors lead to interfacial separation. Building on this, Lin et al. [21] developed an ultimate shear resistance model incorporating slip effects through analysis of test beams and numerical simulations, achieving good correlation with experimental results. Further extending this understanding to panel structures, Zhang et al. [22] quantified composite action degree as the ratio between total connector shear resistance and steel plate tensile resistance, developing analytical models for stiffness and strength that consider progressive failure modes under punching loads. These studies collectively advance the fundamental understanding of partial-composite behavior while providing practical design methodologies that account for true slip conditions.

Despite significant research efforts exploring various shear connectors for steel-concrete composite structures, current solutions still struggle to simultaneously achieve high bearing capacity, ductility, cost-effectiveness, and construction efficiency. This study introduces an innovative slotted perforated plate connector, developed based on PBL connector principles, specifically designed to mitigate interface slip issues in composite beams. Through comprehensive experimental investigation, the research examines the structural performance of these novel connectors under varying conditions including shear span-to-depth ratios, steel reinforcement ratios, concrete types, and connector plate configurations (thickness and spacing). The experimental program systematically evaluates how these parameters influence failure modes, ultimate load resistance, and deformation characteristics. Building on observed failure mechanisms, the study develops analytical models for predicting both flexural and shear capacities of composite beams incorporating the new connector design. These models provide engineers with practical tools for performance assessment and design optimization.

7.2 Flexural and Shear Tests of Sandwich Panels with Slotted Perforated Plate Connectors

7.2.1 Specimen and Material

This study introduces an innovative slotted perforated plate snap-in connector designed specifically for steel concrete composite beams. As illustrated in Fig. 7.1, the connector system features a unique criss-cross welding pattern that minimizes directional deformation of steel plates during fabrication. The design incorporates interlocking slots between paired perforated plates to effectively transfer transverse shear forces, while strategically placed openings facilitate the formation of concrete shear keys that prevent separation of the upper and lower steel plates. This rigid connector system demonstrates exceptional stiffness characteristics, with the interlocked steel plates creating a robust load-bearing framework that significantly enhances overall beam stiffness and resistance to interface slip. The experimental program involved testing 20 composite beam specimens, 10 each for bending and shear evaluations, to systematically investigate the structural performance under

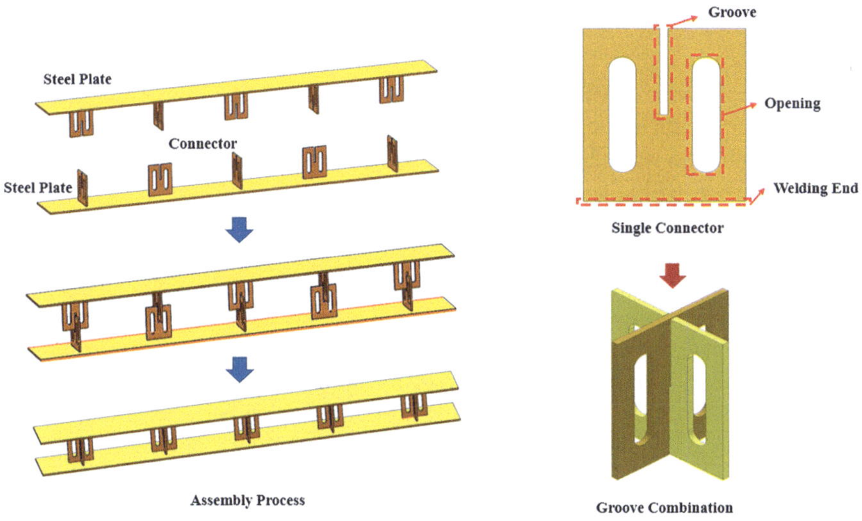

Fig. 7.1 Schematic diagram of a slotted perforated plate snap-in connector. (Reprinted from Huang et al. [23], copyright 2025, with permission from Progress in Steel Building Structures (In Chinese)

various parameters. These parameters included shear span-to-depth ratio, steel ratio, connector thickness and spacing, and concrete type, with detailed specifications provided in Table 7.1. For optimal composite action design, the shear resistance calculations employ an established PBL connector model [24] that accounts for unreinforced openings, with corresponding formulas presented in the table notes. This comprehensive approach enables thorough evaluation of the connector system's flexural and shear performance characteristics while maintaining constructability and structural efficiency.

The experimental investigation employed composite beams with standardized dimensions of 200 mm width and 150 mm concrete core thickness, incorporating both normal concrete (NC) and ultra-high performance concrete (UHPC) specimens for comparative analysis of flexural and shear behavior. Following identical curing protocols (28 days at 30 °C with 80–90% relative humidity) [25], the C45-grade NC demonstrated a cube compressive strength of 48.3 MPa with corresponding shear strength ranging from 4.6 MPa to 5.8 MPa (9.5%–12.1% of compressive strength) [26]. Comprehensive material characterization of UHPC revealed superior mechanical properties, including 142 MPa compressive strength (ASTM C39/C39M) [27], 8 MPa tensile strength, and 16.2 MPa shear strength (CECS 13-1989) [28], with detailed tensile behavior illustrated in Fig. 7.2. The steel components' material properties, rigorously tested according to GB/T 228.1-2010 [29], are systematically presented in Table 7.2, while the optimized UHPC mix formulation appears in Table 7.3. This standardized material preparation and testing protocol ensured consistent specimen quality for reliable performance evaluation.

7.2.2 Test Setup and Measurement Scheme

The test employed a structural loading system to perform both four-point bending and antisymmetric shear tests, as illustrated in Fig. 7.3. For bending tests, the shear span-to-depth ratio was adjusted by maintaining fixed loading points while varying support spacing, with a constant 100 mm distance from each support to the beam ends (denoted by 'a' as shear span length). The antisymmetric shear tests controlled this ratio through simultaneous adjustment of both loading point distance and support spacing, while maintaining fixed 100 mm offsets at the right support and left loading point. Each test protocol began with a 20 kN force-controlled preload to verify instrumentation functionality, followed by displacement-controlled loading at 0.5 mm/min until specimen failure or excessive deformation.

Comprehensive instrumentation captured load-deflection behavior through load sensors on the distribution beam and displacement transducers at critical locations (mid-span or loading points), while strategically positioned strain gauges (see Figs. 7.4 and 7.5) monitored deformation characteristics in steel plates, connectors, and concrete through a statical strain indicator. Real-time crack propagation patterns were systematically documented throughout loading to analyze failure mechanisms.

Table 7.1 Parameters and test results of specimens

Specimen type	Specimen ID	L (mm)	B (mm)	H (mm)	λ	t_s (mm)	t_c (mm)	S (mm)	Core material type	ρ (%)	η	P_u (KN)	Failure mode
Bending specimen	SCB4.5-200-6-6	1950	200	162	4.5	6	6	200	NC	7.4	1.00	213	Flexural failure
	SCB4.0-200-6-6	1800	200	162	4.0	6	6	200	NC	7.4	1.00	225	Flexural-shear failure
	SCB3.5-200-6-6	1650	200	162	3.5	6	6	200	NC	7.4	1.00	237	Flexural-shear failure
	SCB4.5-150-6-6	1950	200	162	4.5	6	6	150	NC	7.4	1.00	211	Flexural failure
	SCB4.5-250-6-6	1950	200	162	4.5	6	6	250	NC	7.4	1.00	194	Flexural-shear failure
	SCB4.5-200-4-6	1950	200	162	4.5	4	6	200	NC	4.9	1.00	119	Flexural failure + steel plate buckling
	SCB4.5-200-8-6	1950	200	162	4.5	8	6	200	NC	9.9	1.00	259	Flexural-shear failure
	SCB4.5-200-6-4	1950	200	162	4.5	6	4	200	NC	7.4	1.00	204	Flexural failure
	SCB4.5-200-6-8	1950	200	162	4.5	6	8	200	NC	7.4	1.00	204	Flexural failure
	SCB4.5-200-6-6-U	1950	200	162	4.5	6	6	200	UHPC	7.4	1.00	223	Flexural failure
Shear specimen	SCB2.5-200-6-6	950	200	162	2.5	6	6	200	NC	7.4	0.98	593	Shear failure
	SCB2.0-200-6-6	800	200	162	2.0	6	6	200	NC	7.4	0.98	507	Shear failure
	SCB3.0-200-6-6	1100	200	162	3.0	6	6	200	NC	7.4	1.00	503	Shear failure
		950	200	162	2.5	6	6	150	NC	7.4	1.00	550	

(continued)

Table 7.1 (continued)

Specimen type	Specimen ID	L (mm)	B (mm)	H (mm)	λ	t_s (mm)	t_c (mm)	S (mm)	Core material type	ρ (%)	η	P_u (KN)	Failure mode
	SCB2.5-150-6-6												With insufficient confinement shear failure
	SCB2.5-250-6-6	950	200	162	2.5	6	6	250	NC	7.4	0.98	494	Shear failure
	SCB2.5-200-4-6	950	200	162	2.5	4	6	200	NC	4.9	1.00	465	Shear failure
	SCB2.5-200-8-6	950	200	162	2.5	8	6	200	NC	9.9	0.78	597	Shear failure
	SCB2.5-200-6-4	950	200	162	2.5	6	4	200	NC	7.4	0.98	574	Shear failure
	SCB2.5-200-6-8	950	200	162	2.5	6	8	200	NC	7.4	0.98	545	Shear failure
	SCB2.5-200-6-6-U	950	200	162	2.5	6	6	200	UHPC	7.4	1.00	1266	Flexural-shear failure

Note: SCB denotes a steel–concrete composite beam. For example, SCB4.5-200-6-4 indicates that the steel-concrete composite beam specimen had a shear span-to-depth ratio of 4.5, a connector spacing of 200 mm, a connector thickness of 4 mm, and an upper and lower steel plate thickness of 6 mm each. L is the length of the specimen, B is the width of the specimen, H is the total height of the cross section, t_s is the thickness of each of the upper and lower steel plates, t_c is the thickness of the connector steel plate, NC stands for normal concrete, and P_u is the ultimate load of the specimen

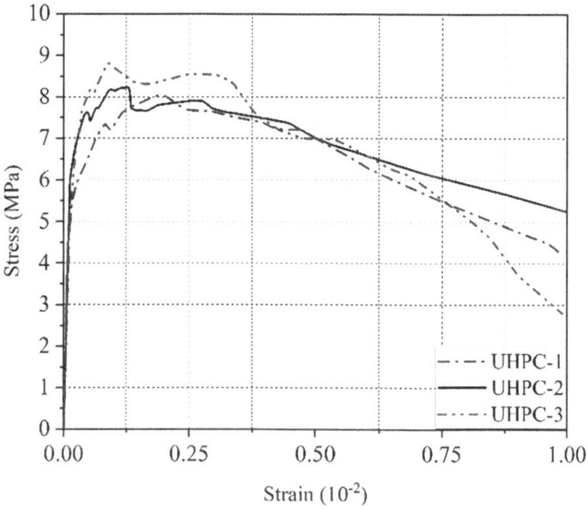

Fig. 7.2 Stress–strain curves of UHPC. (Reprinted from Huang et al. [23], copyright 2025, with permission from Progress in Steel Building Structures (In Chinese))

Table 7.2 Mechanical properties of the steel

Steel type	f_y (MPa)	f_u (MPa)	E_s (GPa)
4-mm steel plate	255	366	204
6-mm steel plate	277	407	188
8-mm steel plate	260	419	209

Note: f_y is the yield strength, f_u is the ultimate strength, E_s is the elastic modulus

Table 7.3 Mix proportions of UHPC (kg/m^3)

Mix	w/b	W	OPC	SF	MP	S	F	HWRA	SRA	Retarder
UHPC (2%)	0.14	174.3	967.6	154.8	190.9	927.5	156	33.3	5.39	1.97

Note: 2% is the steel fibre volume content, *w/b* water-to-binder ratio, *W* water (water), *OPC* ordinary Portland cement, *SF* silica fume, *MP* mineral powder, *S* sand, *F* steel fibre, *HWRA* high water reducing agent, *SRA* shrinkage reducing agent

7.3 Failure Mechanism of Sandwich Panels with Slotted Perforated Plate Connectors

7.3.1 Four-Point Bending Test

The experimental results demonstrated excellent load-bearing capacity and ductility across all ten bending specimens, with concrete crushing emerging as the predominant failure mode accompanied by extensive crack development. The innovative plate connector design effectively resisted shear failure due to its high shear strength, while the interlocking slot mechanism minimized steel-concrete interface slip (less than 2 mm in specimen SCB4.0-200-6-6, as shown in Figs. 7.6 and 7.7), thereby maximizing composite action efficiency. Analysis of specimen SCB4.0-200-6-6

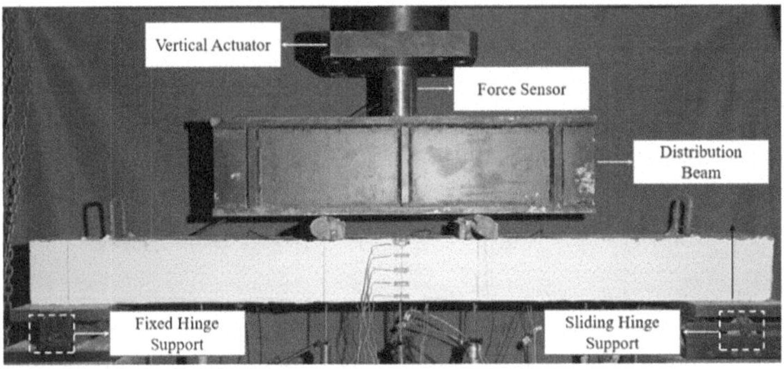

(a) Four-point loading (bending test)

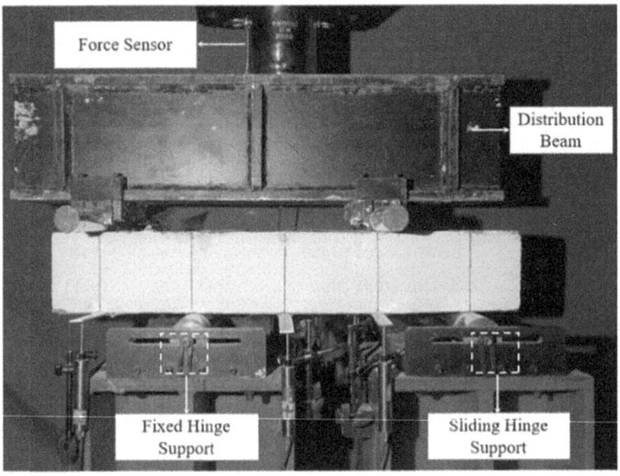

(b) Anti-symmetric loading (shear test)

Fig. 7.3 Loading setup. (**a**) Four-point loading (bending test). (**b**) Anti-symmetric loading (shear test). (Reprinted from Huang et al. [23], copyright 2025, with permission from Progress in Steel Building Structures (In Chinese))

revealed significant bending deformation with substantial section curvature prior to failure, yet maintained excellent steel-concrete bond integrity at beam ends. Strain measurements indicated the lower steel plate exceeded yield strain levels, while connector regions exhibited low strain values due to their high stiffness. Notably, critical locations near main cracks (SL2/SR2) showed substantial tensile strain, whereas regions adjacent to loading points (SL3/SR3) displayed compressive behavior. Mid-span concrete strain distribution confirmed adherence to the plane section assumption during the elastic phase, demonstrating the system's predictable deformation characteristics. The combined results validate the connector system's effectiveness in preventing slip failure while optimizing structural performance through enhanced composite action.

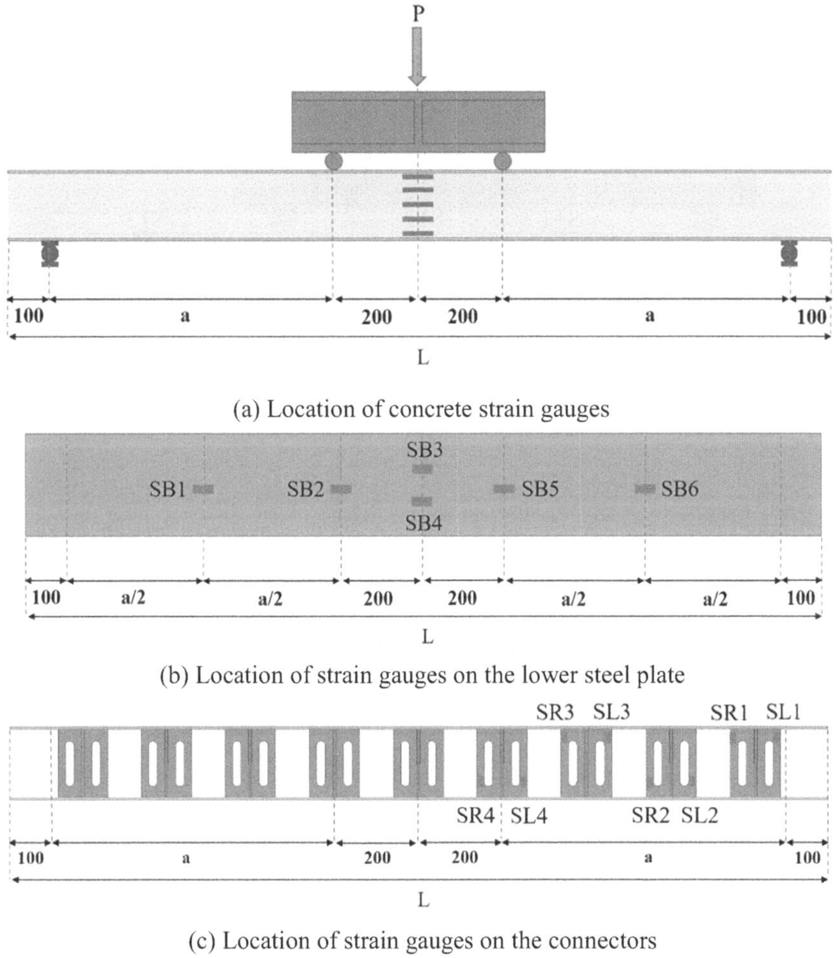

Fig. 7.4 Arrangement of strain gauges on the bending specimen (unit: mm). (**a**) Location of concrete strain gauges. (**b**) Location of strain gauges on the lower steel plate. (**c**) Location of strain gauges on the connectors. (Reprinted from Huang et al. [23], copyright 2025, with permission from Progress in Steel Building Structures (In Chinese))

The load-deflection behavior of the bending specimens, as illustrated in Figs. 7.8 and 7.9, exhibited consistent response patterns that can be characterized by four distinct behavioral phases. Initially, during the elastic stage, specimens demonstrated linear load-deflection relationships with high stiffness, showing minimal deformation and linear strain development in both steel plates and concrete prior to cracking. Upon reaching the cracking threshold, the crack development stage commenced, marked by vertical cracking in pure bending regions and initial diagonal cracking in shear spans, accompanied by gradual stiffness reduction and accelerated steel strain accumulation. The subsequent yielding stage revealed significant behavioral

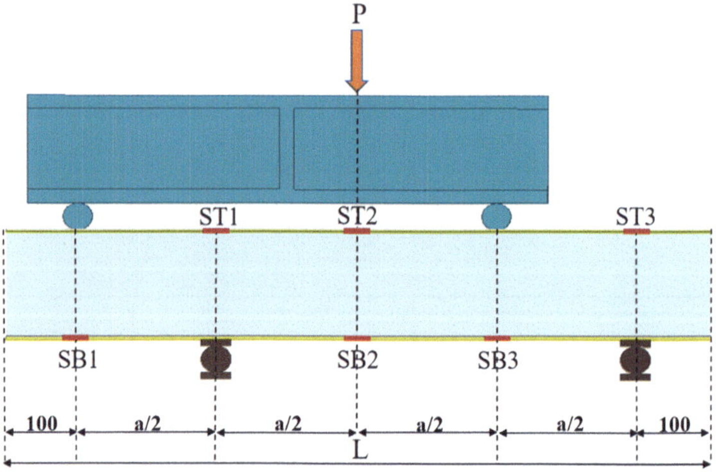

Fig. 7.5 Arrangement of strain gauges on the steel plates of the shear specimen (unit: mm). (Reprinted from Huang et al. [23], copyright 2025, with permission from Progress in Steel Building Structures (In Chinese))

Fig. 7.6 Failure mode of specimen SCB4.0-200-6-6. (Reprinted from Huang et al. [23], copyright 2025, with permission from Progress in Steel Building Structures (In Chinese))

changes as concrete tension zones became fully cracked and lower steel plates yielded, evidenced by distinct inflection points in load-deflection curves and rapid mid-span deflection growth, demonstrating the system's ductile characteristics. Final failure occurred through either compressive crushing or shear failure mechanisms, leading to abrupt load capacity reduction and test termination. This progressive failure sequence highlights the composite system's ability to maintain integrity through multiple deformation phases while developing substantial ductility prior to ultimate failure.

7.3.2 *Failure Modes*

Experimental observations revealed three distinct failure modes among the tested specimens, as illustrated in Fig. 7.10. The most prevalent was flexural-shear failure (see Fig. 7.10a) characterized by sudden diagonal cracking and concrete crushing in shear spans, particularly in specimens with thicker steel plates, wider connector spacing, or smaller shear span-to-depth ratios. These configurations enhanced

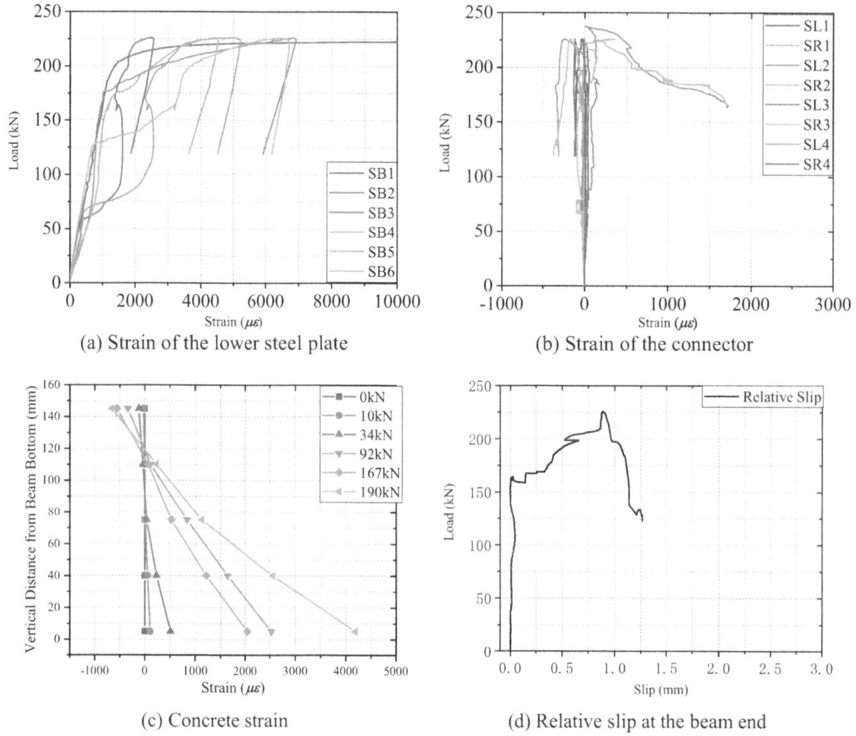

(a) Strain of the lower steel plate
(b) Strain of the connector
(c) Concrete strain
(d) Relative slip at the beam end

Fig. 7.7 Strain distribution and slippage of SCB4.0-200-6-6. (**a**) Strain of the lower steel plate. (**b**) Strain of the connector. (**c**) Concrete strain. (**d**) Relative slip at the beam end. (Reprinted from Huang et al. [23], copyright 2025, with permission from Progress in Steel Building Structures (In Chinese))

flexural resistance while reducing shear resistance, creating critical shear zones that failed after significant plastic yielding in pure bending regions. A second failure combined flexural failure with steel plate buckling (see Fig. 7.10b), occurring primarily in thin-plated specimens where yielding in tension zones transferred excessive compressive stresses to upper plates, inducing local instability. The third mode involved pure flexural failure (see Fig. 7.10c) without abrupt resistance loss, featuring extensive cracking through nearly the full section height and large deformation curvature, while maintaining compressive concrete integrity until loading termination due to excessive deflection. Each failure mechanism demonstrated the system's ability to develop substantial ductility, with failure characteristics directly influenced by plate thickness, connector spacing, and section geometry, highlighting the importance of balanced design considerations for optimal composite behaviour.

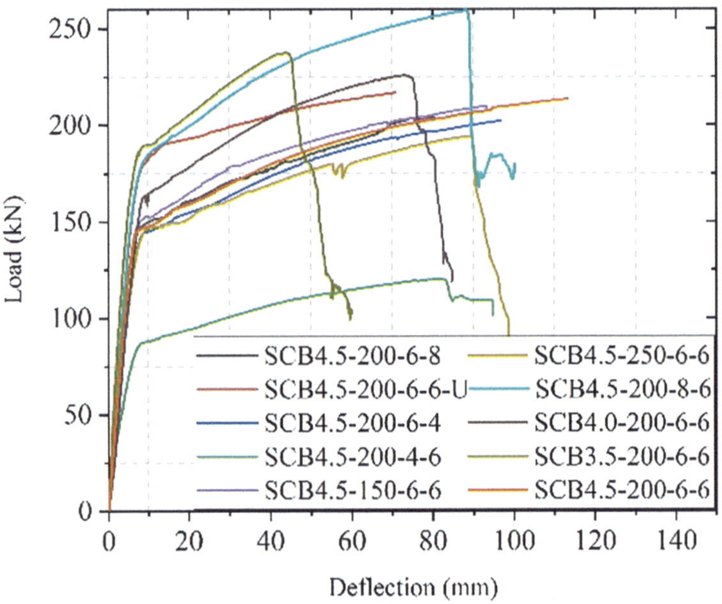

Fig. 7.8 Load–deflection curves of the bending specimens. (Reprinted from Huang et al. [23], copyright 2025, with permission from Progress in Steel Building Structures (In Chinese))

Fig. 7.9 A typical load–deflection curve. (Reprinted from Huang et al. [23], copyright 2025, with permission from Progress in Steel Building Structures (In Chinese))

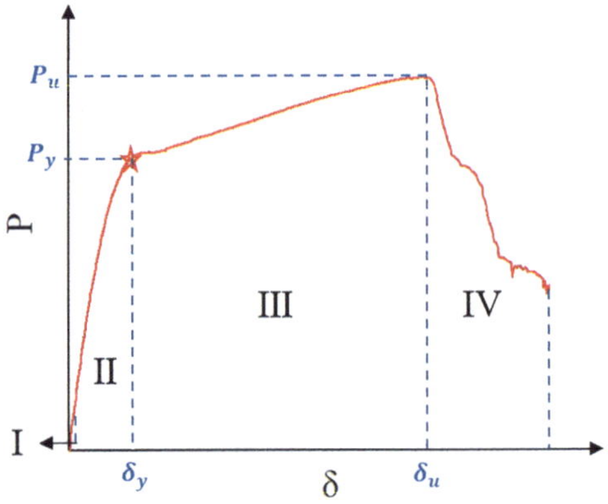

7.3.3 *Parametric Analysis*

The study examined the influence of key parameters—shear span-to-depth ratio, steel ratio, connector thickness, connector spacing, and concrete type—on the ultimate flexural resistance and deformation behavior of composite beams. Figure 7.11 presents the load-displacement curves for specimens with varying configurations. Notably, tests for specimens SCB4.5-150-6-6, SCB4.5-200-6-4/6/8, and

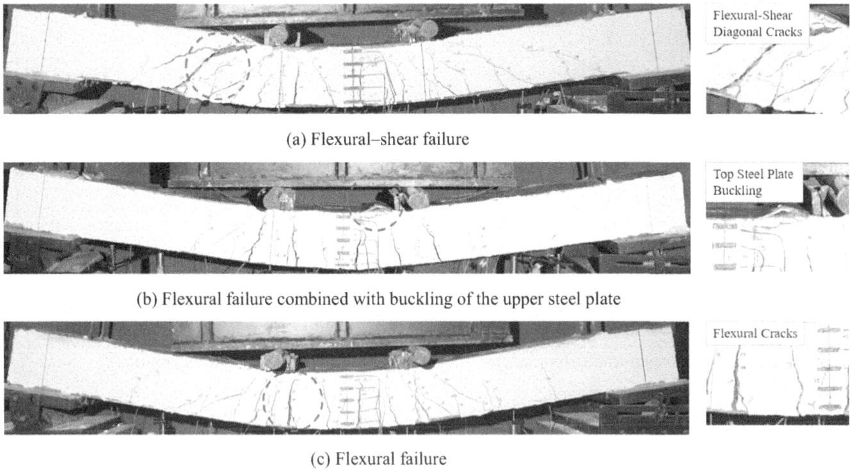

Fig. 7.10 Typical failure modes. (**a**) Flexural–shear failure. (**b**) Flexural failure combined with buckling of the upper steel plate. (**c**) Flexural failure. (Reprinted from Huang et al. [23], copyright 2025, with permission from Progress in Steel Building Structures (In Chinese))

SCB4.5-200-6-6-U were halted in later stages due to contact between the loading beam and the specimen.

1. Shear span-to-depth ratio

   Figure 7.11a compares specimens SCB3.5-200-6-6, SCB4.0-200-6-6, and SCB4.5-200-6-6, with ratios of 3.5, 4.0, and 4.5, respectively. Results indicate that as the ratio increased, yield and peak loads decreased, while ductility improved. This shift occurred because higher shear span-to-depth ratios transitioned the beam's resistance mechanism from combined flexural-shear behavior to predominantly flexural response, leading to greater deformation under bending.
2. Steel ratio

   Figure 7.11a displays curves for specimens SCB4.5-200-4-6, SCB4.5-200-6-6, and SCB4.5-200-8-6, corresponding to steel ratios of 4.9%, 7.4%, and 9.9%, respectively. Increasing the steel ratio significantly enhanced yield and peak loads. For instance, SCB4.5-200-6-6 exhibited 64% and 69% higher yield and peak loads than SCB4.5-200-6-4, while SCB4.5-200-6-8 showed increases of 108% and 117%, respectively. The relationship between steel ratio and load resistance was nearly linear.
3. Connector thickness

   Figure 7.11c illustrates the behavior of specimens SCB4.5-200-6-4, SCB4.5-200-6-6, and SCB4.5-200-6-8, with connector thicknesses of 4 mm, 6 mm, and 8 mm, respectively. The overlapping curves suggest that connector thickness had negligible effects on load resistance, stiffness, or ductility. Strain gauge data confirmed minimal strain in the connectors, indicating that even thin plates provided sufficient composite action without shear failure.

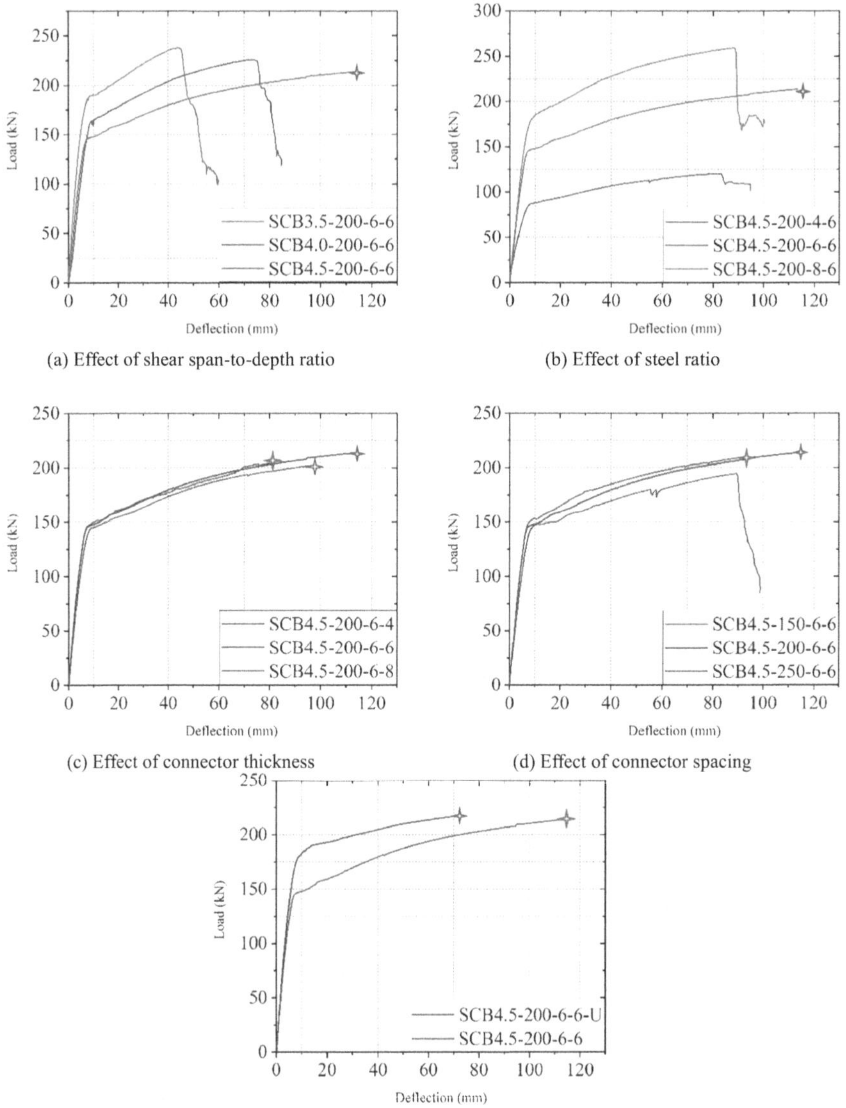

(a) Effect of shear span-to-depth ratio (b) Effect of steel ratio

(c) Effect of connector thickness (d) Effect of connector spacing

(e) Effect of concrete type

Fig. 7.11 Effects of various parameters on the flexural behaviour of the specimen. (**a**) Effect of shear span-to-depth ratio. (**b**) Effect of steel ratio. (**c**) Effect of connector thickness. (**d**) Effect of connector spacing. (**e**) Effect of concrete type. (Reprinted from Huang et al. [23], copyright 2025, with permission from Progress in Steel Building Structures (In Chinese))

4. Connector spacing

Figure 7.11d compares specimens SCB4.5-150-6-6, SCB4.5-200-6-6, and SCB4.5-250-6-6, with spacings of 150 mm, 200 mm, and 250 mm, respectively. Initial stiffness and deflection were similar, but SCB4.5-250-6-6 eventually exhibited flexural-shear failure due to insufficient shear resistance at larger spacings. The findings suggest that fewer connectors can achieve full composite action ($\eta = 1.0$), but excessive spacing may reduce ultimate resistance.

5. Concrete type

   Figure 7.11e contrasts specimens SCB4.5-200-6-6 (NC concrete) and SCB4.5-200-6-6-U (UHPC). Both failed in flexure, but UHPC increased yield resistance by 29.7% due to its superior compressive and tensile strength. Steel fibers in UHPC maintained post-cracking tensile resistance, enhancing flexural performance. Testing was stopped prematurely to avoid contact issues, so ductility comparisons were inconclusive

7.4 Four-Point Antisymmetric Shear Test

To investigate shear behavior under controlled conditions, the test beam was divided into five distinct regions: the left cantilever region, left shear span region, target shear span region, right shear span region, and right cantilever region (see Fig. 7.12). This loading configuration ensured a pure shear state in the target region by maintaining zero bending moment at its midpoint, allowing for an isolated assessment of shear performance. Strain development in the upper and lower steel plates was monitored throughout loading (see Fig. 7.13). In the NC specimen, steel strains remained relatively low without yielding (see Fig. 7.13a). Conversely, the UHPC specimen exhibited significantly higher strains, particularly at locations ST1 and SB3—near the left support and right loading point, where bending moments were greatest—exceeding yield strain levels (see Fig. 7.13b).

Load-deflection responses varied markedly between NC and UHPC specimens due to material differences (see Fig. 7.14). NC beams displayed a two-stage failure process: initial crack development followed by main diagonal crack propagation. At approximately 15% of the ultimate load (P_{su}), vertical cracks emerged on the tension side near the supports, gradually extending toward the compression side. By 50% of P_{su}, diagonal cracks formed in the target shear span region, eventually coalescing into dominant diagonal cracks at peak load. This phase was marked by intermittent load drops ("parallel downward curve shifts") as internal cracking reduced stiffness. Post-peak, crack widths expanded rapidly, accompanied by concrete spalling, steel plate bulging, and weld failures in some connectors, culminating in the failure mode depicted in Fig. 7.16.

In contrast, UHPC specimens demonstrated superior shear resistance, leveraging the material's high tensile and compressive strengths. The distinct load-deflection trends for each concrete type are further illustrated in Figs. 7.15 and 7.18, with UHPC's enhanced post-cracking performance attributed to steel fiber bridging effects. These findings underscore the critical role of concrete properties in governing shear resistance and failure mechanisms in composite beams.

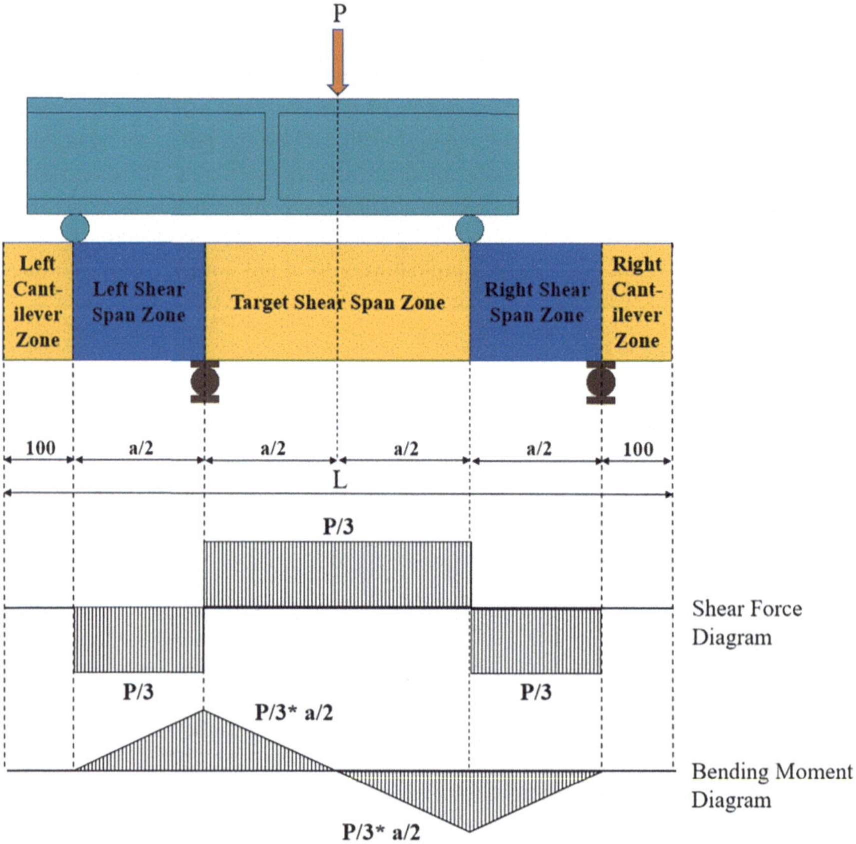

Fig. 7.12 Region division of the test beam. (Reprinted from Huang et al. [23], copyright 2025, with permission from Progress in Steel Building Structures (In Chinese))

7.4.1 Failure Modes

All NC composite beams exhibited shear failure except for specimen SCB2.5-150-6-6, which experienced a distinct failure mode due to insufficient shear connectors outside the loading point. The lack of confinement at the beam end led to severe concrete damage in the left cantilever region, resulting in a shear-slip failure characterized by significantly reduced ultimate shear resistance, as illustrated in Fig. 7.17. In contrast, UHPC specimens demonstrated a more complex load-deflection response, progressing through three distinct stages: crack development, flexural deformation, and main diagonal crack growth. Initially, the high stiffness of UHPC allowed rapid load increase until vertical cracks formed at 33% of P_{su}, primarily near the left support and right loading point. By 50%P_{su}, fine diagonal cracks with steep inclinations emerged in the target shear span region, accompanied by audible concrete cracking. As loading continued, the enhanced concrete strength

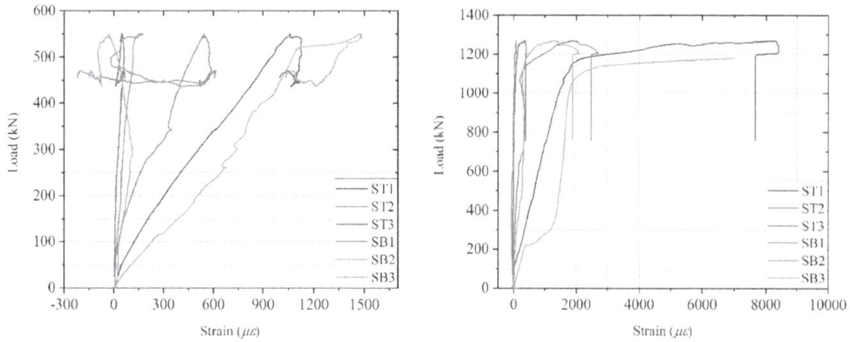

(a) SCB2.5-150-6-6 (b) SCB2.5-200-6-6-U

Fig. 7.13 Strain curves of the steel plate in the shear specimens. (**a**) SCB2.5-150-6-6. (**b**) SCB2.5-200-6-6-U. (Reprinted from Huang et al. [23], copyright 2025, with permission from Progress in Steel Building Structures (In Chinese))

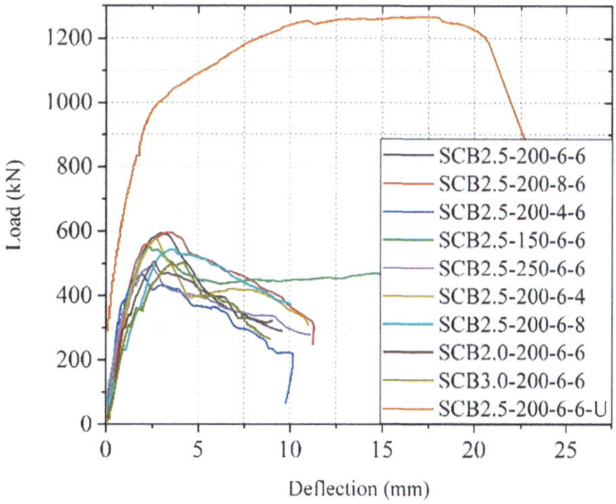

Fig. 7.14 Load–deflection curves of the shear specimens. (Reprinted from Huang et al. [23], copyright 2025, with permission from Progress in Steel Building Structures (In Chinese))

of UHPC shifted the behavior toward flexural deformation, particularly at maximum bending moment locations. Vertical flexural cracks widened to approximately 2 mm at 85% P_{su}, eventually reaching 5 mm, while diagonal cracks remained stable. However, upon reaching peak load, a sudden widening of diagonal cracks to 18 mm triggered a sharp load drop, terminating the test. Despite initial flexural characteristics, the specimen ultimately failed in shear within the target region, defining a flexural-shear failure mode (see Fig. 7.19). These observations highlight the critical influence of material properties on failure mechanisms, with UHPC's superior strength delaying shear failure through enhanced crack resistance and flexural resistance, while NC beams predominantly failed in shear due to earlier crack propagation and reduced confinement.

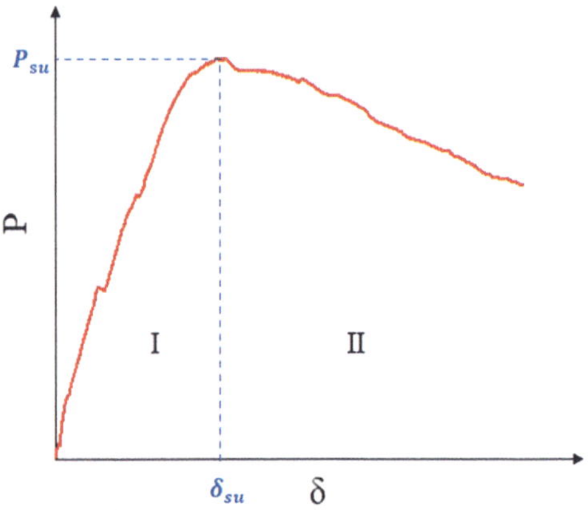

Fig. 7.15 Load–deflection curves of the NC shear specimens. (Reprinted from Huang et al. [23], copyright 2025, with permission from Progress in Steel Building Structures (In Chinese))

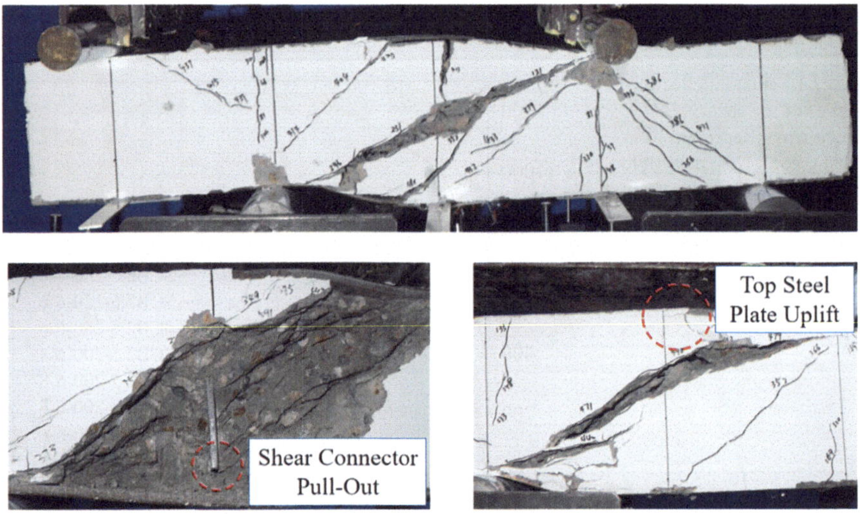

Fig. 7.16 Shear failure mode of the NC specimen. (Reprinted from Huang et al. [23], copyright 2025, with permission from Progress in Steel Building Structures (In Chinese))

7.4.2 Parametric Analysis

The study examined the influence of key parameters—shear span-to-depth ratio, steel ratio, connector thickness, connector spacing, and concrete type—on the ultimate flexural resistance and deformation performance of composite beams. Figure 7.20 presents the load-displacement curves for specimens with varying parameters. Tests for specimens SCB4.5-150-6-6, SCB4.5-200-6-4/6/8, and SCB4.5-200-6-6-U were halted in later stages due to contact between the loading beam and the specimen (indicated by asterisks). The effects of each parameter are summarized below:

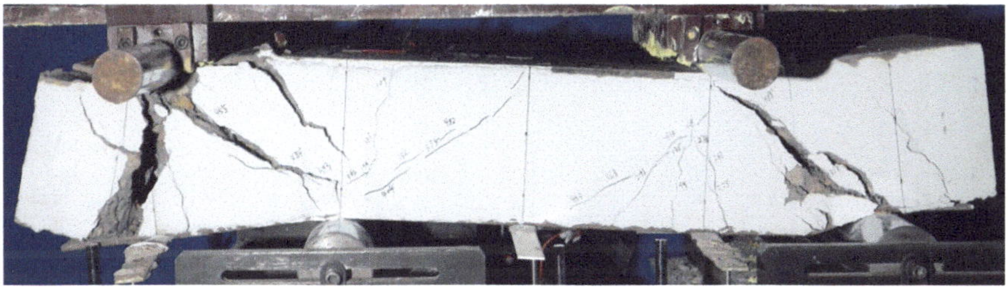

Fig. 7.17 Shear failure mode with insufficient confinement. (Reprinted from Huang et al. [23], copyright 2025, with permission from Progress in Steel Building Structures (In Chinese))

Fig. 7.18 Load–deflection curve of the UHPC shear specimen. (Reprinted from Huang et al. [23], copyright 2025, with permission from Progress in Steel Building Structures (In Chinese))

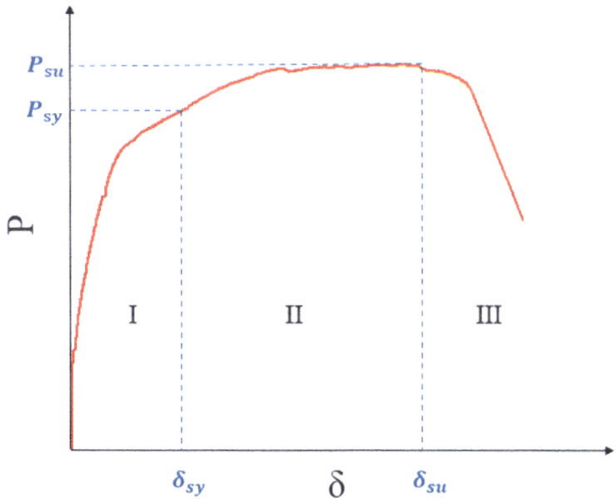

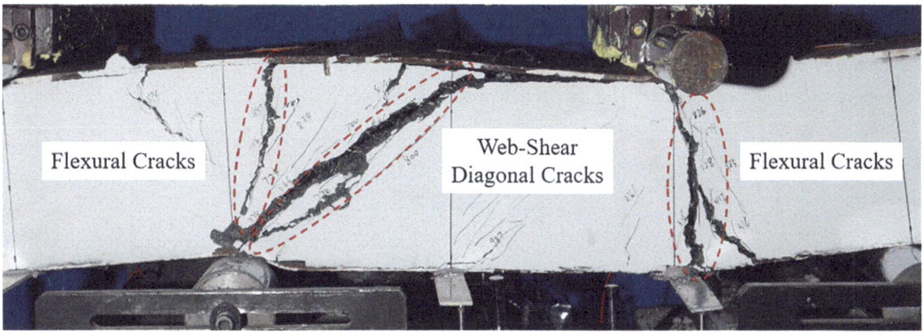

Fig. 7.19 Shear failure mode of the UHPC specimen. (Reprinted from Huang et al. [23], copyright 2025, with permission from Progress in Steel Building Structures (In Chinese))

1. Shear span-to-depth ratio

   Figure 7.20a compares the load-deflection curves of specimens SCB3.5-200-6-6, SCB4.0-200-6-6, and SCB4.5-200-6-6, with shear span-to-depth ratios of 3.5, 4.0, and 4.5, respectively. Results indicate that increasing the shear span-to-depth ratio reduced the yield and peak loads but enhanced ductility. This trend

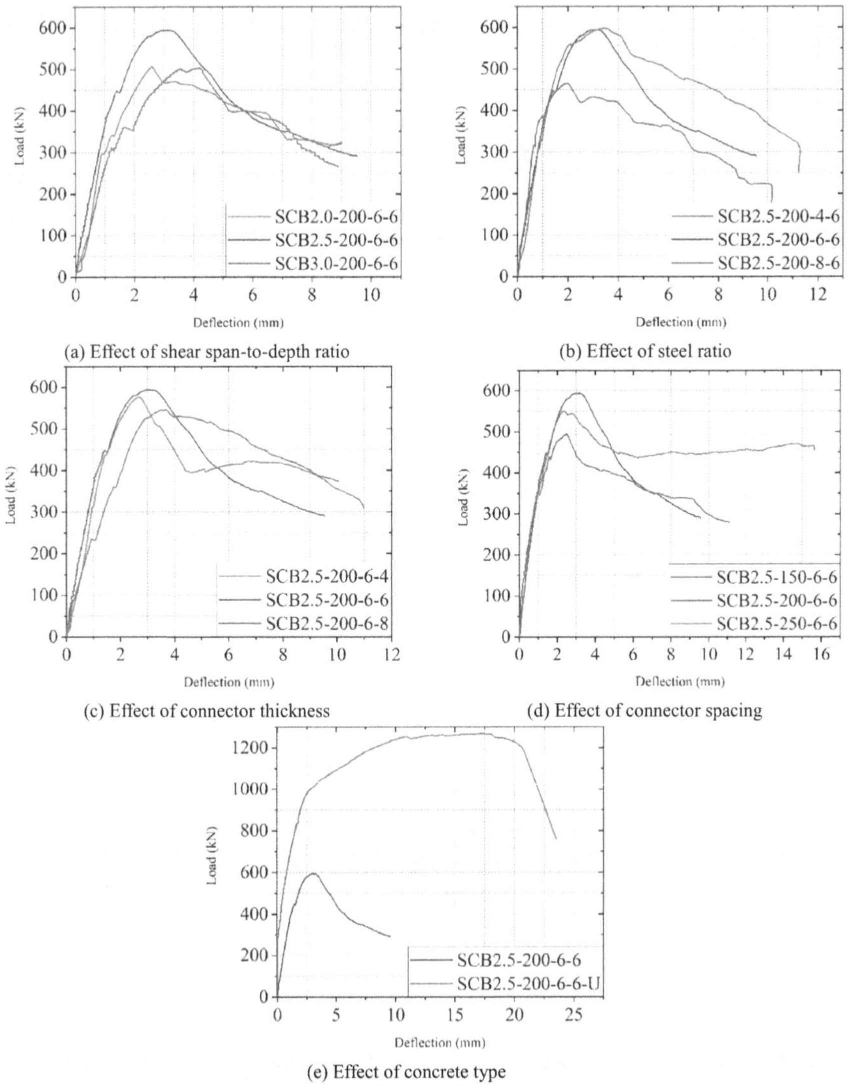

Fig. 7.20 Effects of different parameters on the shear performance of specimens. (**a**) Effect of shear span-to-depth ratio. (**b**) Effect of steel ratio. (**c**) Effect of connector thickness. (**d**) Effect of connector spacing. (**e**) Effect of concrete type. (Reprinted from Huang et al. [23], copyright 2025, with permission from Progress in Steel Building Structures (In Chinese))

occurred because, under the same bending moment, a higher shear span-to-depth ratio shifted the beam's resistance mechanism from combined flexural-shear behavior to predominantly flexural behavior, leading to greater deformation capacity.

2. Steel ratio

   Figure 7.20b illustrates the load-deflection responses of specimens SCB4.5-200-4-6, SCB4.5-200-6-6, and SCB4.5-200-8-6, corresponding to steel ratios of 4.9%, 7.4%, and 9.9% (steel plate thicknesses of 4 mm, 6 mm, and 8 mm, respectively). Both yield and peak loads increased with higher steel ratios. Specifically, SCB4.5-200-6-6 exhibited 64% and 69% higher yield and peak loads than SCB4.5-200-6-4, while SCB4.5-200-6-8 showed increases of 108% and 117%, respectively. The load-bearing capacity exhibited an approximately linear relationship with steel ratio.

3. Plate connector thickness

   Figure 7.20c displays the load-deflection curves for specimens SCB4.5-200-6-4, SCB4.5-200-6-6, and SCB4.5-200-6-8, with connector thicknesses of 4 mm, 6 mm, and 8 mm. The overlapping curves suggest that connector thickness had negligible influence on load-bearing capacity, stiffness, or ductility. Strain gauge data confirmed that connectors remained at low strain levels without yielding, indicating that even thin connectors effectively ensured composite action.

4. Plate connector spacing

   Figure 7.20d compares specimens SCB4.5-150-6-6, SCB4.5-200-6-6, and SCB4.5-250-6-6, with connector spacings of 150 mm, 200 mm, and 250 mm. Initial stiffness and deflection were similar, but SCB4.5-250-6-6 exhibited reduced shear resistance at larger spacings, leading to flexural-shear failure and a marginally lower ultimate load. The results suggest that the connectors provided sufficient longitudinal shear resistance to achieve full composite action ($\eta = 1.0$) with minimal connectors. Further reducing spacing only marginally enhanced flexural resistance.

5. Concrete type

   Figure 7.20e contrasts specimens SCB4.5-200-6-6 (NC concrete) and SCB4.5-200-6-6-U (UHPC). Both failed in flexure, but SCB4.5-200-6-6-U exhibited a 29.7% higher yield resistance due to UHPC's superior compressive and tensile strengths. Steel fibres in UHPC maintained post-cracking tensile resistance, enhancing flexural performance. Testing was intentionally stopped to prevent contact between the loading beam and specimen, so ductility comparisons were inconclusive. Failure was characterized by extensive flexural cracks near the upper steel plate, with uncrushed compression-zone concrete. At high curvature, the concrete core nearly cracked entirely, balancing tension and compression between steel plates.

7.5 Prediction of Load-Bearing Capacity of Composite Beam

7.5.1 Prediction Model of Flexural Resistance

The flexural load-displacement response of the composite beam is illustrated in Fig. 7.9. Upon reaching the yield load (P_y), the beam exhibited prolonged plastic

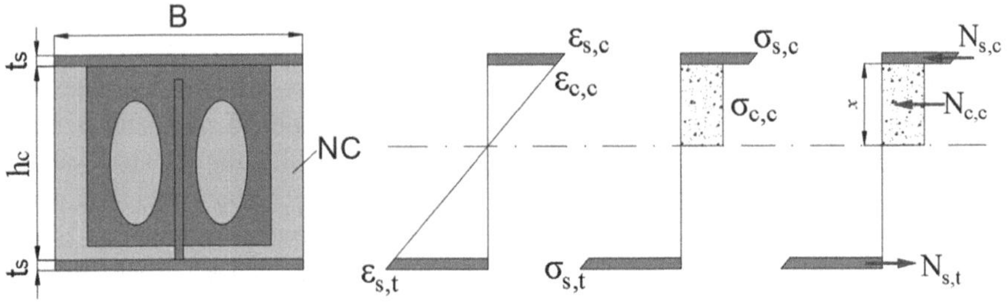

Fig. 7.21 Diagram for the flexural calculation of the cross section (NC). (Reprinted from Huang et al. [23], copyright 2025, with permission from Progress in Steel Building Structures (In Chinese))

hardening, forming a plastic hinge—a state where minor load increments resulted in significant displacement. For conservative design purposes, the flexural resistance of the composite beam was defined as its load-bearing capacity at plastic hinge formation. The corresponding strain, stress, and force equilibrium distributions across the beam's cross-section are presented in Fig. 7.21. To simplify the plastic design calculation while ensuring accuracy, the flexural resistance was evaluated based on the following assumptions:

1. Plane section assumption: Strain distribution across the section remains linear.
2. Negligible tensile contribution of NC: The tensile resistance of normal concrete (NC) is disregarded.
3. UHPC tensile contribution: Unlike NC, UHPC exhibits significant tensile strain relative to steel plate yielding; thus, its tensile contribution is considered.
4. Yielding of lower steel plate: The bottom steel plate is assumed to have fully yielded.

When a plastic hinge forms in the NC composite beam, the following equilibrium equations can be derived:

$$\begin{cases} N_{\mathrm{s,c}} + N_{\mathrm{c,c}} = N_{\mathrm{s,t}} \\ N_{\mathrm{c,c}} = f_{\mathrm{c}} B x \\ N_{\mathrm{s,c}} = f_{\mathrm{y}} (E_{\mathrm{c}} / E_{\mathrm{s}}) B t_{\mathrm{s}} \\ N_{\mathrm{s,t}} = f_{\mathrm{y}} B t_{\mathrm{s}} \end{cases} \tag{7.1}$$

The height of the concrete compression zone (x) is determined by solving the equilibrium equations,

$$x = \left(1 - \frac{E_{\mathrm{c}}}{E_{\mathrm{s}}}\right) \frac{f_{\mathrm{y}}}{f_{\mathrm{c}}} t_{\mathrm{s}} \tag{7.2}$$

where E_{c} and E_{s} represent the elastic moduli of concrete and steel plates, respectively.

The plastic bending moment of the NC composite beam is calculated as:

$$M = f_y B t_s \left[h_c + \left(1 + \frac{E_c}{E_s}\right)\frac{t_s}{2} - \left(1 - \frac{E_c}{E_s}\right)\frac{x}{2} \right] \tag{7.3}$$

The ultimate flexural resistance of the snap-in NC composite beam is then obtained as:

$$P = 2M/a \tag{7.4}$$

Unlike NC, UHPC exhibits significantly higher ultimate tensile strain. When the lower steel plate yields, the UHPC in the tension zone typically reaches its peak stress state. Consequently, the tensile contribution of UHPC must be accounted for based on its constitutive stress-strain relationship.

For analytical simplicity, the UHPC tensile behavior is idealized as a three-segment model (see Fig. 7.22), derived from the experimental stress-strain curves presented above [30]. Figure 7.23 illustrates the stress-strain distribution in the UHPC composite beam at the plastic hinge formation stage. The following equilibrium equations can be established to characterize this state:

$$\begin{cases} N_{s,c} + N_{c,c} = N_{c,t1} + N_{c,t2} + N_{c,t3} + N_{s,t} \\ N_{s,c} = f_y (E_c/E_s) B t_s \\ N_{c,c} = 0.5 f_c B (x - t_s) \\ N_{c,t1} = 0.5 f_t h_1 B \\ N_{c,t2} = f_t h_2 B \\ N_{c,t3} = 0.5 f_t h_3 B \\ N_{s,t} = f_y B t_s \end{cases} \tag{7.5}$$

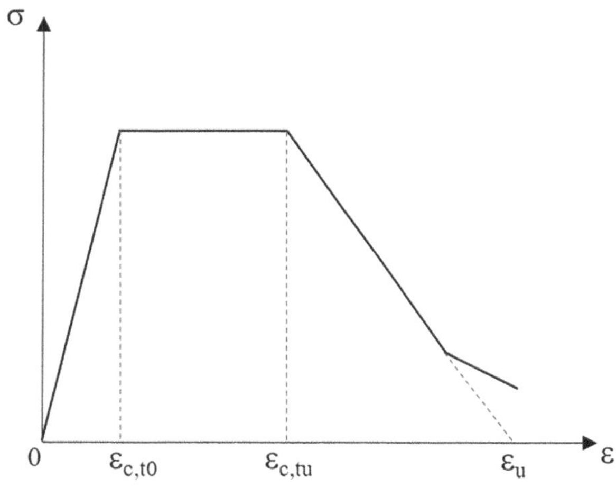

Fig. 7.22 Simplified constitutive model for UHPC in tension. (Reprinted from Huang et al. [23], copyright 2025, with permission from Progress in Steel Building Structures (In Chinese))

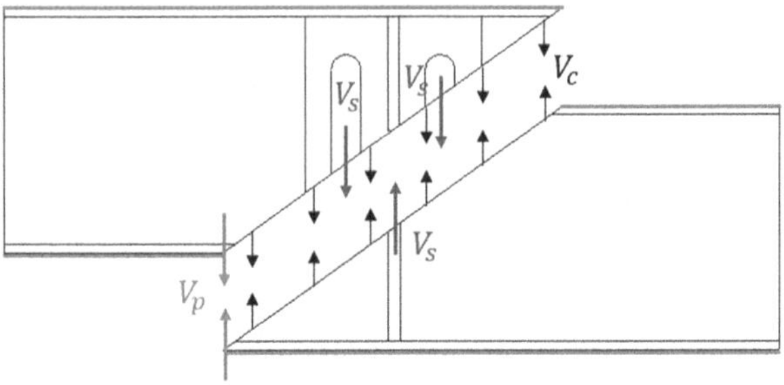

Fig. 7.23 Shear analysis of the inclined section of the steel–concrete composite beam. (Reprinted from Huang et al. [23], copyright 2025, with permission from Progress in Steel Building Structures (In Chinese))

where f_t represents the UHPC tensile strength and f_y represents the yield strength of the lower steel plate. Based on the plane section assumption, the critical section height h_1, h_2, and h_3 are determined through geometric relationships as follows:

$$\begin{cases} h_1 = \dfrac{E_c(x - t_s)\varepsilon_{c,t0}}{f_c} \\ h_2 = \dfrac{(\varepsilon_{c,tu} - \varepsilon_{c,t0})h_1}{\varepsilon_{c,t0}} \\ h_3 = \dfrac{\left(f_y - E_s\varepsilon_{c,tu}\right)h_1}{E_s\varepsilon_{c,t0}} \end{cases} \tag{7.6}$$

The equilibrium equations are solved to determine x, the height of the concrete compression zone. This solution leads to the derivation of the plastic moment resistance for the UHPC composite beam:

$$\begin{aligned} M = {} & N_{s,c}\left(x - \frac{t_s}{2}\right) + \frac{2}{3}N_{c,c}(x - t_s) + \frac{2}{3}N_{c,t1}h_1 + N_{c,t2}\left(h_1 + \frac{1}{2}h_2\right) \\ & + N_{c,t3}\left(h_1 + h_2 + \frac{1}{3}h_3\right) + N_{s,t}\left(h_c - x - \frac{t_s}{2}\right) \end{aligned} \tag{7.7}$$

The ultimate flexural resistance of the snap-in UHPC composite beam is then calculated by substituting Eq. (7.7) into the expression given in Eq. (7.4).

7.5.2 Prediction Model of Shear Resistance

The total shear resistance of composite beams with snap-in connectors comprises three primary components: (1) concrete contribution (V_c); (2) snap-in perforated plate connectors (V_s); and (3) steel plate elements (V_p). The total shear resistance is calculated as follows:

$$V_u = V_c + V_s + V_p \tag{7.8}$$

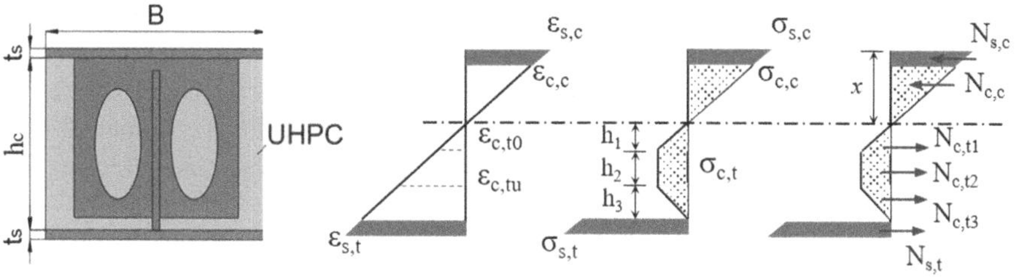

Fig. 7.24 Diagram for the flexural calculation of the section (UHPC). (Reprinted from Huang et al. [23], copyright 2025, with permission from Progress in Steel Building Structures (In Chinese))

1. Concrete shear contribution (V_c)

   Figure 7.24 illustrates the shear force distribution in the inclined section of the composite beam. The concrete shear resistance is determined using the Eurocode 2 [31] formulation:

$$V_c = \left[C_{Rd,c} k \eta_1 (100\rho_1 f_{ck})^{1/3}\right] b h_c \tag{7.9}$$

   where $C_{Rd,c}$ is taken as 0.18 for NC and 0.15 for lightweight concrete; $\eta_1 = 0.4 + 0.6\ u/2000$ represents the tensile strength reduction factor, in which u is the concrete density in kg/m^3; $k = 1 + (200/h_c)^{0.5} \leq 2$; $\rho_1 = t_s/t_c \leq 0.02$; and f_{ck} represents the compressive strength of concrete cylinder.

   Experimental parametric studies demonstrated a clear relationship between shear resistance and geometric configuration—specimens exhibited increased shear resistance with reduced shear span-to-depth ratios. This observation aligns with findings by Lin et al. [21], which established that for composite beams with shear span-to-depth ratios (λ) below 3.0, the concrete's shear contribution follows an inverse proportionality relationship (approximately $3/\lambda$). Building on this foundation, Zhang et al. [11] successfully incorporated this proportionality factor into the Eurocode 2 concrete shear resistance formulation through comprehensive composite beam testing. Their experimental validation confirmed that this modified approach yields significantly improved predictive accuracy. The enhanced concrete shear strength calculation adopts the following form:

$$V_c = \frac{3}{\lambda}\left[C_{Rd,c} k \eta_1 (100\rho_1 f_{ck})^{1/3}\right] b h_c \tag{7.10}$$

2. Connector shear contribution (V_s)

   Experimental investigations revealed that under shear loading, the interaction between slotted perforated plate connectors and surrounding concrete produces a unique mechanical response. The connectors mobilize concrete tenons that form within their openings, effectively resisting vertical separation of the steel plates. Post-test examinations consistently showed a characteristic failure pattern across all specimens: shear-induced separation of concrete tenons at the connector openings, evidenced by visible connector pull-out and distinct shear planes between the interior and exterior concrete. These observations confirm that the shear resistance contribution of the snap-in connectors stems from concrete tenon shear resistance rather than connector yielding. Based on the experimental evidence, the connector contribution to composite beam shear resistance can be quantified through the following relationship:

$$V_s = 2n\tau_c A_{s1} \tag{7.11}$$

where, n represents the number of concrete tenons in the shear span region; τ_c represents the concrete shear strength, calculated by $\tau_c = 0.12f_c$; and A_{s1} represents the area of a single opening of the connector.

3. Steel plate shear contribution (V_p)

   Experimental observations demonstrate that composite beams exhibit near-perfect composite action under loading conditions. The relative slip between steel plates and concrete remains minimal, confirming full composite behavior throughout the loading history. The steel plate contribution to ultimate shear resistance can be determined using the following relationship:

$$V_p = \frac{1}{\sqrt{3}} \frac{E_c}{E_s} f_y A_s \tag{7.12}$$

   where, $f_y A_p/\sqrt{3}$ represents the pure shear resistance of the lower steel plate, which is multiplied by the factorE_c/E_s to represent the effective shear resistance provided by the lower steel plate, and A_s represents the cross-sectional area of the steel plate.

7.5.3 Analytical Model Verification

Table 7.4 compares the test values and theoretically calculated values for the flexural resistance of the ten composite beams. The ratio of the calculated flexural resistance to the test value, P_{prop}/P_{test}, has a mean of 0.95 and a standard deviation of 0.02. Overall, the flexural resistance of the composite beams estimated by the theoretical model is very close to the test value, and the theoretical values are slightly lower than the test values, indicating that the flexural resistance estimated using the theoretical model proposed in the present study is conservative.

The test values and theoretical estimates of the shear resistance of the ten shear composite beams are compared in Table 7.5. The ratio of the calculated shear resistance to the test value has a mean of 1.05 and a standard deviation of 0.07. The theoretical model can satisfactorily predict the shear resistance of the composite beam specimen. However, the test values for SCB2.5-150-6-6 show a significant discrepancy with the calculated values. The reason for this discrepancy has been explained in the parametric analysis in Sect. 7.4.2, was the insufficient confinement

Table 7.4 Test and predictive results for flexural resistance

Specimen ID	B (mm)	t_s (mm)	h_c (mm)	a (mm)	f_y (MPa)	f_c (MPa)	E_s (GPa)	E (GPa)	Test value		Theoretical value		P_{prop}/P_{test}
									M_{test} (kNm)	P_{test} (kN)	M_{prop} (kNm)	P_{prop} (kN)	
SCB4.5-200-6-6	200	6	150	675	277	48.3	210	30	49.3	146	46.8	139	0.95
SCB4.0-200-6-6	200	6	150	600	277	48.3	210	30	48.6	162	46.8	156	0.96
SCB3.5-200-6-6	200	6	150	525	277	48.3	210	30	49.4	188	46.8	178	0.95
SCB4.5-150-6-6	200	6	150	675	277	48.3	210	30	50.6	150	46.8	139	0.93
SCB4.5-250-6-6	200	6	150	675	277	48.3	210	30	49.3	146	46.8	139	0.95
SCB4.5-200-4-6	200	4	150	675	255	48.3	210	30	29.7	88	29.6	88	1.00
SCB4.5-200-8-6	200	8	150	675	260	48.3	210	30	61.4	182	57.7	171	0.94
SCB4.5-200-6-4	200	6	150	675	277	48.3	210	30	48.6	146	46.8	139	0.95
SCB4.5-200-6-8	200	6	150	675	277	48.3	210	30	49.3	146	46.8	139	0.95
SCB4.5-200-6-6-U	200	6	150	675	277	142	210	45	62.8	188	58.4	173	0.92
Mean													0.95
Standard deviation													0.02

Table 7.5 Test and predictive results for shear resistance

	Test value	Theoretical value				
Specimen ID	$V_{u,test}$ (kN)	$V_{c,prop}$ (kN)	$V_{p,prop}$ (kN)	$V_{s,prop}$ (kN)	$V_{u,prop}$ (kN)	$V_{u,\ prop}/V_{u,\ test}$
SCB2.5-200-6-6	198	59	34	101	194	0.98
SCB2.0-200-6-6	169	73	34	67	175	1.04
SCB3.0-200-6-6	168	49	34	101	184	1.10
SCB2.5-200-6-4	191	59	34	101	194	1.02
SCB2.5-200-6-8	182	59	34	101	194	1.07
SCB2.5-200-4-6	155	59	20	101	179	1.15
SCB2.5-200-8-6	199	59	41	101	200	1.01
SCB2.5-150-6-6	183	59	32	184	275	/
SCB2.5-250-6-6	165	59	32	67	158	0.96
SCB2.5-200-6-6-U	422	89	34	358	482	1.14
Mean						1.05
Standard deviation						0.07

of the beam end of the specimen, which caused premature concrete crushing. The actual shear resistance should be higher than that measured in the test and closer to the theoretically estimated value. Therefore, this specimen was not considered when verifying the theoretical shear resistance model. The predicted shear capacities of SCB2.5-200-4-6 and SCB2.5-200-6-6-U are 15% higher than the test values, possibly due to the lack of localised strengthening at the loading points and supports during the tests (see Fig. 7.19), resulting in local cracking that had some impact on the test results.

References

1. J.Y.R. Liew, J. Yan, Z. Huang, Steel–concrete–steel sandwich composite structures – recent innovations, J. Constr. Steel Res. 130(3) (2017) 202–221.
2. J. Zhao, X. Liu, Application overview of composite steel–concrete structures in bridges, Spec. Struct. 34(2) (2017) 99–102, 116 (in Chinese).
3. J. Yan, L. Zhang, X. Lin, T. Wang, Review on mechanisms of double-skin composite protective structures, J. Nat. Disasters 29(6) (2020) 1–12 (in Chinese).
4. J. Nie, Application of steel–concrete composite structures in ocean engineering, Steel Struct. 35 (1) (2020) 20–33.
5. M.J. Tomlinson, Shell composite construction for shallow draft immersed tube tunnels, In: Immersed Tunnel Techniques, Thomas Telford, 1990, pp. 209–220.
6. J.W. Pryer, H.G. Bowerman, The development and use of British steel bi-steel, in: Proc. 8th Int. Offshore and Polar Eng. Conf., (Vol. IV), 1998.
7. J.Y.R. Liew, T.Y. Wang, K.M.A. Sohel, Separation prevention shear connectors for sandwich composite structures, US Provisional Patent Application No. 61/047,130, 2008.

8. J.Y.R. Liew, K.M.A. Sohel, Lightweight steel–concrete–steel sandwich system with J-hook connectors, Eng. Struct. 31(5) (2009) 1166–1178.
9. J.Y.R. Liew, K.M.A. Sohel, C.G. Koh, Impact tests on steel–concrete–steel sandwich beams with lightweight concrete core, Eng. Struct. 31(9) (2009) 2045–2059.
10. Z. Huang, X. Zhao, W. Zhang, Z. Fu, Y. Zhou, L. Sui, Load transfer mechanism of novel double-layer steel–LHDCC–steel sandwich panels under punching loads, Eng. Struct. 226 (2021) 111427.
11. W. Zhang, Z. Huang, Z. Fu, X. Qian, Y. Zhou, L. Sui, Shear resistance behavior of partially composite steel–concrete–steel sandwich beams considering bond-slip effect, Eng. Struct. 210 (2020) 110394.
12. W. Zhang, Z. Huang, R. Li, X. Zhao, J. Ye, Impact response of double-layer steel–RULCC–steel sandwich panels: Experimental, numerical and analytical approaches, J. Struct. Eng. 148 (10) (2022) 04022165.
13. K. Khorramian, S. Maleki, M. Shariati, A. Jalali, M.M. Tahir, Numerical analysis of tilted angle shear connectors in steel–concrete composite systems, Steel Compos. Struct. 23(1) (2017) 67–85.
14. M. Shariati, N. H. Ramlisulong, H. Sinaei, M.M. Arabnejad, and P. Shafigh, Behaviour of channel shear connectors in normal and lightweight aggregate concrete (experimental and analytical study), in: Adv. Mater. Res. 168-170 (2011) 2303–2307.
15. M. Leekitwattana, S.W. Boyd, R.A. Shenoi, Evaluation of the transverse shear stiffness of a steel bi-directional corrugated-strip-core sandwich beam, J. Constr. Steel Res. 67(2) (2011) 248–254.
16. S. He, Z. Fang, L. Zhang, Research on mechanical performance of PBL shear connectors for steel–concrete joint section of hybrid girder bridge, J. China Railway Soc. 37(10) (2015) 100–109 (in Chinese).
17. J.-L. Xiao, M. Zhou, J.-G. Nie, T.-Y. Yang, J.-S. Fan, Flexural behaviour of steel–UHPC composite slabs with perfobond rib shear connectors, Eng. Struct. 245 (2021) 112912.
18. Y. Lu, Effect of different connectors on concrete composite box-girders with corrugated steel webs, J. Harbin Inst. Technol. (New Ser.) 25(2) (2018) (in Chinese).
19. J. Nie, J. Shen, Z. Yu, A reduced rigidity method for calculating deformation of composite steel–concrete beams, China Civ. Eng. J. (6) (1995) 11–17 (in Chinese).
20. Y. Leng, X.-B. Song, Experimental study on shear performance of steel–concrete–steel sandwich beams, J. Constr. Steel Res. 120 (2016) 52–61.
21. Y. Lin, J. Yan, Y. Wang, F. Fan, C. Zou, Shear failure mechanisms of SCS sandwich beams considering bond-slip between steel plates and concrete, Eng. Struct. 181 (2019) 458–475.
22. W. Zhang, Z. Huang, J. Ye, Youtam, Numerical-based analytical model of double-layer steel–LHDCC sandwich composites under punching loads, Compos. Struct. 286 (2022) 115271.
23. Z. Huang, X. Zhao, W. Zhang, X. Yan, The flexural and shear performance of steel–concrete composite beams with perfobond slotted plate snap-in connectors, Prog. Steel Build. Struct. 25 (10) (2023) 1–15 (in Chinese).
24. C. Zhao, Y. Liu, Experimental study of shear capacity of perfobond connector, Eng. Mech. 29 (12) (2012) 349–354 (in Chinese).
25. Z. Huang, W. Zhang, S. Fan, L. Sui, J. Ye, Axial load resistance of a novel UHPFRC grouted SHS tube-sleeve connection: Experimental, numerical and theoretical approaches, J. Struct. Eng. 147(11) (2021) 04021184.
26. S. Shi, Shear strength, modulus of rigidity and Young's modulus of concrete, China Civ. Eng. J. (2) (1999) 47–52 (in Chinese).
27. ASTM C39/C39M-01, Standard test method for compressive strength of cylindrical concrete specimens, ASTM International, USA, 2017.
28. CECS 13-1989, Steel fibre reinforced concrete test method (in Chinese).
29. China National Standardization Management Committee, Metallic materials – tensile testing – part 1: Method of test at room temperature (GB/T 228.1-2010), Standards Press of China, Beijing, 2010.
30. J.-B. Yan, H. Hu, T. Wang, Flexural behaviours of steel–UHPC–steel sandwich beams with J-hook connectors, J. Constr. Steel Res. 169 (2020) 106014.
31. Eurocode 2, Design of concrete structures—Part 1-1: General rules and rules for buildings, British Standards Institution, London, 2004.

Chapter 8
Conclusion and Future Work

This book chapter has consolidated extensive experimental, numerical, and analytical research on ultra-lightweight cement composites (ULCC) and their sustainable, highly ductile and rubberized variants such as ultra lightweight high ductility cement composite (ULHDCC), LC3 based ULCC and rubberized ultra-lightweight cement composites (RULCC), which integrate low content polyethylene fibers (PE) and recycled rubber powder to enhance sustainability, ductility and impact resistance. Alongside their application in steel–concrete composite sandwich systems and advanced composite connectors, these materials form the core focus of the work. Across seven chapters, it demonstrated how novel material formulations and structural innovations can simultaneously reduce weight, enhance mechanical performance, and improve durability in demanding marine and offshore applications. The work also demonstrates how lightweight composites and optimized structural systems can absorb impact energy effectively and remain resilient under extreme offshore dynamic loading. By combining sustainable cementitious composites, structural optimization strategies, and robust modeling techniques, the work provides an integrated framework for developing next-generation resilient marine infrastructure. Section 8.1 states down the major conclusions drawn from each chapter, while Sect. 8.2 outlines directions for future research.

8.1 Conclusions

8.1.1 Key Developments and Properties of ULCC

Chapter 1 examined the development and performance of ULCC, emphasizing their potential as high performance and sustainable alternatives to conventional lightweight concretes. The investigations revealed that the unique incorporation of fly ash cenospheres (FACs), silica fume, and PE fibers allows ULCC to combine low

Z. Huang, Y. Zhou, *Innovative Lightweight High-performance Steel-concrete Sandwich Composite for Marine Structures*,
https://doi.org/10.1007/978-981-95-4681-7_8

density with superior mechanical and durability characteristics. Depending on the mix design, ULCC achieves densities between 1072 and 1691 kg/m^3 and compressive strengths ranging from 18.6 MPa to 87.3 MPa, demonstrating specific strength well above those of conventional lightweight concretes. This strength-to-weight advantage is attributed to the rigid shell structure of FACs, their hollow microstructure, and the enhanced interfacial transition zone (ITZ) formed through pozzolanic reactions with silica fume.

The chapter also highlighted the flexibility of ULCC mix design. Variations in binder type, water-to-binder ratios, and FAC content were shown to strongly influence compressive strength, elastic modulus, and ductility. Optimized FAC/binder ratios of 0.36–0.45 were particularly effective in maintaining high strength while reducing weight. Fiber reinforcement, especially with PE fibers, further enhanced tensile strain capacity and post-peak toughness, enabling strain-hardening behavior and mitigating brittle failure modes. ULCC achieved comparable ductility while reducing fiber content significantly, thus improving cost-effectiveness compared to conventional engineered cementitious composites (ECC).

LC3-based binders reduced embodied carbon by 42% and improved chloride resistance, while PVA fiber reinforcement enhanced ductility, toughness, and fire resistance, withstanding temperature up to 1000 °C without spalling due to the creation of vapor-release channels upon fiber melting. These findings confirm ULCC's suitability for demanding structural applications where lightweight, durability, and resilience to extreme conditions are critical.

Lastly, the chapter highlighted ULCC's sustainability and long-term potential. Life cycle assessment revealed superior performance in transportation and operational phases of prefabricated modular construction, with overall sustainability indices 25% higher than conventional ordinary Portland cement (OPC) concretes. The experimental and analytical findings presented in the chapter demonstrate that ULCC is not only structurally efficient but also environmentally advantageous. In crux, ULCC demonstrate that lightweight, high-performance, and sustainable cementitious systems can be engineered for structural use.

8.1.2 Dynamic Performance and Modeling

Chapter 2 investigated the dynamic response of RULCC reinforced with low volume PE fibers through split Hopkinson pressure bar (SHPB) testing and advanced numerical modeling. The need for impact-resistant lightweight concretes was addressed by incorporating recycled rubber powder from waste tires into the ULCC matrix, thereby enhancing ductility and energy absorption. With a density of approximately 1450 kg/m^3 and compressive strengths reaching 52.2 MPa, RULCC demonstrated a remarkable balance between reduced weight and high performance. Tensile strain capacities of 3–5% were achieved with only 0.7% PE fiber dosage. With the same fiber dosage and 10% rubber replacement, the material exhibited a compressive strength of 24.9 MPa and showed improved ductility with multiple microcracking. Compared to conventional ECC that typically require 2% polymer fibers, RULCC achieved similar strain-hardening effects with a 65% reduction in fiber content, offering both mechanical efficiency and economic feasibility.

Static compressive tests showed that increasing rubber content led to reductions in strength and elastic modulus due to lower stiffness of rubber particles and weaker interfacial bonding with the cement matrix. However, SHPB testing revealed that small volumes of rubber powder (<10%) enhanced impact resistance and improved energy dissipation capacity. Stress-strain responses confirmed that rubberized mixes absorbed significantly greater energy through crack propagation and fiber bridging, while PE fibers-maintained specimen integrity under high strain-rate loading. RULCC also showed pronounced strain-rate sensitivity, with dynamic increase factors (DIFs) notably higher than those of normal-weight concrete (NWC), lightweight aggregate concrete (LWAC), and high-strength concrete (HSC).

A modified DIF analytical model incorporating both strain rate and rubber replacement ratio was proposed, outperforming established formulations in the literature. Furthermore, a modified Holmquist-Johnson-Cook (H-J-C) constitutive law was developed and successfully calibrated against SHPB test results, which led to accurate finite element (FE) simulations of RULCC behavior under dynamic loads. Similarly, a data-driven meta-modeling framework was also introduced using machine learning (ML) algorithm. Genetic programming (GP), a type of evolutionary algorithm, was applied to derive an explicit relationship for predicting dynamic impact strength (DIS) by taking into account the effects of fiber ratio, rubber ratio, and impact strain rate. The proposed relationship provides a practical and computationally efficient tool for design purposes. This chapter establishes RULCC as a viable material for impact-resistant infrastructure and offers validated modeling strategies for impact strength.

8.1.3 Load Transfer Mechanism and Design of Sandwich Members

Chapter 3 investigated the quasi-static behavior of SCS sandwich walls, beams, and curved shells infilled with ULCC and subjected to patch loads, using experimental testing, numerical simulations, and analytical studies. The study examined sandwich walls under axial and combined loads. The introduction of J-hook connectors effectively restrained outward face plate buckling and confined the concrete core, enhancing axial resistance and ductility. Walls with sufficient connector density achieved stable load-shortening behavior, while those with sparse connectors experience premature buckling, brittle post-peak responses, and reduced ductility.

For sandwich beams, three distinct failure modes were identified: flexural failure, shear-compression, and shear-tension. Flexural failure initiated with yielding of the tensile steel face plate and progressed with ductile unloading accompanied by wide core cracks. Shear-compression failures were governed by the crushing of diagonal concrete struts following the development of inclined cracks, while shear-tension failures were dominated by inclined cracking and pull-out of shear connectors. The ultimate resistance of the beams was shown to depend strongly on the steel contribution ratio, plate slenderness, and connector efficiency.

Curved sandwich shells exhibited a characteristic two-peak load response under patch loading. The first peak corresponded to punching shear of the concrete core, while the second peak was associated with tensile membrane action of the outer face plate at large deflections. Failure was ultimately governed by punching fracture of the steel plates. Both peaks confirmed compliance with ISO 19906 ice load requirements, highlighting the suitability of the proposed system for offshore Arctic applications. Curvature introduced a beneficial arch effect, which significantly improved the contact pressure capacity compared to flat panels. Experimental results showed that curved SCS panels exhibit a wide range of failure modes, which includes: flexural failure, shear compression due to diagonal strut crushing, shear tension, and plastic hinge mechanisms. The governing failure modes were strongly controlled by geometric parameters, particularly the rise-to-span ratio (r/L) and span-to-core thickness (L/h_c). At low r/L values, failure resembled that of flat sandwich beams, with failure primarily governed by flexural and transverse shear modes. However, at higher r/L values, failure shifted toward shear compression or shear tension mechanisms. In extreme cases, a three-hinged mechanism was also observed. Based on this study, the optimal ratio of r/L is proposed to be approximately 0.17, where shear resistance reached its peak before declining at higher values. Similarly, L/h_c ratios between 8 and 20 were shown to provide balance between shear and compression performance, preventing premature core crushing/ instability-driven failures.

The validated FE models accurately reproduced experimental observations for walls, beams, and shells. Crack propagation, damage evolution in the concrete, and local buckling of the face plates were captured accurately, making FE a reliable tool for extended parametric analyses. Analytical models were developed to predict the compressive resistance and compression-bending interaction resistance of sandwich walls. For pure compression case, the proposed formulations demonstrated reasonable accuracy in predicting the ultimate compressive resistance of SCS sandwich walls, with average test-to-prediction ratios of 1.09 and 1.08, accompanied by low coefficients of variance. For the compression-bending case, the proposed analytical model offered conservative yet reliable predictions, with minor deviations in cases involving premature flexural-shear failure, when compared with Eurocode 4 and validated against test/FE data. Furthermore, analytical expressions were also derived for the shear resistance of the sandwich beam and the punching resistance of sandwich shell. For beam, the predictions showed good agreement with test results. For sandwich shells, Eurocode 2 provisions were found to significantly underestimate the shear resistance of curved SCS sandwich panels since the code is primarily formulated for reinforced concrete beams, neglecting the critical contribution of steel face plates to shear resistance in sandwich systems. Overall, this chapter establishes that SCS sandwich composites provide a lightweight yet resilient solution for resisting patch loads and ice pressures, making them as a promising candidate for Arctic offshore structures.

8.1.4 Punching Shear Mechanism and Design of Sandwich Panels

Chapter 4 experimentally, numerically, and analytically examined the structural performance of single- and double-layer steel-lightweight high- ductility cement composite (LHDCC)-steel (SCS) sandwich panels under concentrated punching loads. The punching cone model was shown to effectively represent the load transfer mechanism of double-layer SCS panels. Before punching shear failure of concrete, the concrete core significantly contributed to stiffness and load resistance. After concrete failure, the steel plates' membrane action dominated the structural performance. Partial composite panels exhibited global deformation due to bond-slip, whereas full composite panels showed localized indentation under the loading patch.

It was found that the stiffness response at both elastic (k_e) and plastic (k_p) stages was sensitive to shear span, steel plate thickness, and loading patch size. k_e was found to be additionally influenced by concrete core height and shear connector spacing, while k_p was governed by steel plate yield strength. Similarly, load resistance parameters such as the first peak resistance (P_1) and residual resistance at large displacement ($P_{\delta\,=\,75\ \mathrm{mm}}$) were found to be governed by a combination of core thickness, plate thickness, connector spacing, and loading patch size.

Failure modes were shown to depend on the shear span-to-depth ratio (λ) and the degree of composite action (η). Partial composite panels ($\eta < 1$) displayed bond-slip and ductile global responses, while fully composite panels ($\eta \geq 1.0$) demonstrated localized punching with higher stiffness and minimal slip. The double-layer configurations significantly improved structural resilience compared to single-layer designs. The addition of a middle steel plate enhanced ultimate resistance, redistributed load paths after top plate fracture, and maintained residual capacity, ensuring ductility and damage tolerance even after severe local failure.

FE simulations effectively reproduced the load-displacement curves and failure patterns observed in tests, providing strong validation of the experimental findings. Similarly, the proposed analytical stiffness models and an idealized four-stage load-displacement framework accurately captured elastic stiffness, plastic stiffness, peak resistance, and post-peak behavior. These formulations, which accounted for bond-slip in partial composites, showed strong agreement with experimental observations. This chapter demonstrated that combining high-ductility cementitious cores with double-layer SCS configurations yields panels with superior punching resistance, residual strength, and ductility. It provides a strong foundation for developing resilient and damage tolerant structures when exposed to concentrated or localized loading.

8.1.5 Impact Resistance and Residual Performance of Sandwich Panels

Chapter 5 integrates findings of the impact response, residual performance, and design optimization of double-layer SCS sandwich panels through experimental testing, numerical, and analytical modeling. The double-layer configuration exhibited clear advantages over single-layer counterparts, particularly in resisting localized impact forces. The intermediate steel plate redistributed impact loads across a broader area, reducing penetration depth by approximately 11% and mitigating the severity of the core damage. Post-impact testing confirmed that double-layer panels showed higher residual capacity and ductility. Furthermore, the damage was localized to the impact zone rather than extending across the full span.

Material composition was also shown to be critical. Rubberized concrete cores incorporating 5% rubber powder by volume achieved optimal balance between strength and ductility, improving crack resistance and minimizing damage to the cement matrix. The rubberized core further enhanced the energy dissipation behavior of material and structure, limiting the deformation of sandwich panels from dynamic loading. These results demonstrated that RULCC and rubberized ultra-lightweight high-ductility cement composites (RLHDCC) can be considered as promising core materials for impact-resistant SCS panels.

The dominant failure mode was punching shear when the specimens were subjected to multiple impacts. The concrete cores compacted within the impact zone, dissipating energy effectively through localized crushing and contributing to the overall impact resistance of the panels. Synergistic yielding of steel plates and controlled cracking of concrete dissipated over 95% of input energy, highlighting the efficiency of the steel-concrete interaction. Composite action is governed by connector spacing and proved to be a key design parameter. Partially composite panels with larger connector spacing exhibited approximately 17% higher impact resistance than fully composite designs, as increased flexibility enabled greater ductility and reduced vertical cracking. However, this improvement in impact resistance came at the cost of slightly reduced residual flexural capacity.

Numerical modeling, particularly restart-based FE simulations, accurately reproduced both impact and post-impact behavior including crack propagation and load redistribution. Analytical modeling demonstrated that single-degree-of-freedom (SDOF) models could predict global response but underestimated localized effects. On the other hand, two-degree-of-freedom (TDOF) models more reliably captured displacement histories and residual strength, providing improved predictive capability for impact design scenarios.

Design optimization studies revealed that increasing steel plate thickness was the most effective strategy for enhancing residual capacity, while connector diameter exerted a strong influence on post-impact performance. Rubberized concrete cores, controlled composite action, and double-layer configurations collectively emerged as the optimal design pathway for protective structures subjected to repeated impacts. In summary, this chapter provides a robust foundation for developing structures that are durable, possess increased residual capacity, and remain damage-tolerant in harsh marine environments.

8.1.6 Flexural and Impact Behavior of Stainless Steel–LHDCC Sandwich

Chapter 6 presented a comprehensive investigation of the flexural and post-impact performance of stainless steel-LHDCC composite beams using experimental testing, numerical simulations, and analytical modeling, to evaluate their suitability for marine structures. Primarily, three distinct failure modes were identified. Partially composite beams ($\eta < 1$) primarily exhibited flexural cracks and interfacial bond-slip, while fully composite beams ($\eta = 1$) developed pure flexural failures characterized by concrete cracking and steel yielding. Beams constructed with low shear-strength concrete cores showed shear failure governed by diagonal cracking, highlighting the influence of material properties on global response. Double-layer configurations enhanced impact resistance by limiting crack propagation through the intermediate steel plate. Rubberized LHDCC further reduced residual deformation by 60% compared to conventional cores, which is attributed to the energy absorption capacity of rubber particles. Short-headed studs were effective in dissipating impact energy, though they reduced load capacity. Overlapped studs enhanced strength, but decreased flexibility.

Numerical simulations successfully simulated dynamic responses and failure modes. Parametric studies further revealed that higher impact momentum shortens the response period, while increased energy elevates peak displacements. Furthermore, a trilinear load-displacement model was developed to incorporate strain-rate effects and degree of composite action, providing an accurate analytical representation of beam behavior. The derived SDOF model predicted displacement-time histories and peak displacements with good agreement, though it slightly overestimated residual displacements. A design flowchart was also proposed to enhance practical applicability.

The beams demonstrated performance degradation under repeated impacts through reduced average impact force, increased peak displacement, reduced global stiffness, and diminished energy absorption. Core thickness was identified as the most critical parameter affecting these responses, while the number of layers mainly influenced maximum displacement. Concrete type played a decisive role in impact resistance. Normal concrete (NC) specimens suffered severe degradation under repeated impacts due to low tensile strength and brittle fracture. Ultra-high-performance concrete (UHPC) cores maintained superior integrity across multiple impacts but offered lower specific energy absorption due to their higher density. LHDCC and its rubberized variant (RLHDCC) achieved the most balanced performance, combining reduced structural mass with enhanced impact resistance, higher specific energy absorption, and improved tolerance to damage under repeated impacts.

The reliability of the FE framework was ensured by the calibration of the Continuous Surface Cap Model (CSCM) for LHDCC based on extensive uniaxial and triaxial testing. Based on the analysis of energy dissipation mechanisms, it was noted that steel plates absorbed most energy during flexural failures through plastic deformation, while concrete cores contributed significantly via crack propagation in shear-type failures. Connector spacing was also shown to affect performance, with wider spacing reducing steel plate energy absorption by up to 35%. These findings provide both a technical foundation and practical design tools for impact-resistant marine infrastructure.

8.1.7 Development and Performance Evaluation of Slotted Perforated Plate Connectors in Composite Beams

Chapter 7 developed and evaluated a novel slotted perforated plate snap-in connector designed to enhance the performance of steel-concrete composite beams. An extensive experimental program was carried out to investigate the flexural and shear behavior of beams incorporating the proposed connector. The emphasis was on understanding the governing failure mechanisms, the role of connector geometry, and the contribution of advanced cementitious cores. Furthermore, analytical models were proposed to predict flexural and shear resistance. The models were validated against test results to establish a mechanics-based framework for design.

The experimental results showed that the snap-in connectors effectively mobilize composite action via two complementary mechanisms. First, the slots in the steel plates provide mechanical interlocking that enhances transverse shear transfer. Second, concrete tenons formed within the perforations prevent vertical separation between the steel plates. This dual mechanism minimized interfacial slip and enabled near-perfect composite action, ensuring full utilization of both steel and concrete components. Consequently, the beams achieved high stiffness, strong resistance to interfacial failure, and consistent load transfer between materials.

Under pure bending conditions, the composite beam demonstrated exceptional structural performance, combining high load-bearing capacity with remarkable ductility. Experimental results revealed a characteristic flexural failure involving sequential yielding of the tension steel plate followed by compressive crushing of the concrete flange. Parametric studies identified three primary factors governing flexural behavior: (1) steel reinforcement ratio, (2) concrete compressive strength, and (3) degree of composite action. Notably, in fully composite configurations, variations in connector thickness (4–8 mm range) and spacing (150–250 mm) showed negligible influence on ultimate moment resistance, confirming that the slotted perforated plate connectors effectively achieve complete section utilization.

Shear performance was significantly enhanced in specimens with UHPC cores. Compared to NC beams, UHPC-core composite beams achieved 100% or more increase in shear resistance and nearly 5 times improved deformation capacity. The presence of steel fibers in UHPC enhanced post-cracking toughness and generated stronger concrete tenons at connector openings, which further improved resistance to shear-induced separation.

Analytical models were developed to predict flexural and shear resistance. The formulations incorporate the elastic-plastic stress distribution of composite sections and the dual contribution of concrete tenons and connector mechanisms in shear transfer. Model predictions showed excellent agreement with test data, with correlation in load-displacement behavior and ultimate resistance. The validation demonstrates that the proposed framework can provide a reliable basis for design, making the proposed connector a promising innovation for next-generation steel-concrete composite structures.

8.2 Future Research Directions

Despite the significant advancements in ultra-lightweight cement composites (ULCC, ULHDCC, RULCC) and associated steel-concrete composite sandwich systems, several critical research gaps remain. Future efforts should prioritize long-term durability under coupled environmental-mechanical loads (e.g., cyclic marine exposure, i.e., fatigue), large-scale validation via field deployments and pilot offshore applications, and integration with renewable energy infrastructure (e.g., floating solar or offshore wind turbine structures) to enable multifunctional marine structures. Additionally, standardized design guidelines must be developed to facilitate code inclusion and ensure structural safety. Advancements in automated manufacturing—such as modular prefabrication and 3D printing—are essential to reduce production costs and enable scalable implementation. Further applied research should address limitations like low elastic modulus through innovative structural designs (e.g., sandwich beams/slabs for impact resistance) and mitigate brittleness via fiber-reinforced polymer (FRP) confinement, while also expanding applications in areas such as sprayed ULCC reinforcement and 3D-printed lightweight components.

The manufacturer's authorised representative in the EU is Springer Nature Customer Service Centre GmbH, Europaplatz 3, 69115 Heidelberg, Germany. If you have any concerns regarding our products, please contact ProductSafety@springernature.com

Printed and bound by CPI Group (UK) Ltd, Croydon, CR0 4YY
07/10/2026
02229357-0008